Drug Discovery from Herbs
Approaches and Applications

About the Centre

The Centre for Science and Technology of the Non-Aligned and Other Developing Countries (NAM S&Centre) is an inter-governmental organisation with a membership of 47 countries spread over Asia, Africa, Middle East and Latin America. Besides this, 12 S&T agencies and academic/research institutions of Bolivia, Botswana, Brazil, India, Nigeria and Turkey are the members of the S&T-Industry Network of the Centre. The Centre was set up in 1989 to promote South-South cooperation through mutually beneficial partnerships among scientists and technologists and scientific organisations in developing countries. It implements a variety of programmes including international workshops, meetings, roundtables, training courses and collaborative projects and brings out scientific publications, including a quarterly Newsletter. It is also implementing 5 Fellowship schemes, namely, NAM S&T Centre Research Fellowship, Joint NAM S&T Centre – ICCBS Karachi Fellowship, Joint CSIR/CFTRI (Diamond Jubilee)–NAM S&T Centre Fellowship, Joint NAM S&T Centre – ZMT Bremen Fellowship and Research Training Fellowship for Developing Country Scientists (RTF-DCS) in Indian institutions. These activities provide, among others, the opportunity for scientist-to-scientist contact and interaction, training and expert assistance, familiarising the scientific community on the latest developments and techniques in the subject areas, and identification of technologies for transfer between member countries. The Centre has so far brought out 69 publications and has organised 99 international workshops and training programmes.

For further details, please visit www.namstct.org or write to the Director, NAM S&T Centre, Core 6A, 2nd Floor, India Habitat Centre, Lodhi Road, New Delhi-110003, India (Phone: +91-11-24645134/24644974; Fax: +91-11-24644973; E-mail: namstcentre@gmail.com; namstct@bol.net.in).

Drug Discovery from Herbs
Approaches and Applications

— *Editors* —

Suresh Bhojraj

Tijen Talas-Ogras

Shamiem Adam

Subba Rao V. Madhunapantula

CENTRE FOR SCIENCE & TECHNOLOGY OF THE
NON-ALIGNED AND OTHER DEVELOPING COUNTRIES
(NAM S&T CENTRE)

2017
DAYA PUBLISHING HOUSE®
A Division of
ASTRAL INTERNATIONAL PVT. LTD.
New Delhi – 110 002

ISBN: 978-93-86071-46-0 (International Edition)

Centre for Science and Technology of the Non-Aligned and Other Developing Countries (NAM S&T Centre)
Core-6A, 2nd Floor, India Habitat Centre, Lodhi Road,
New Delhi-110 003 (India)
Phone: +91-11-24644974, 24645134, Fax: +91-11-24644973
E-mail: namstct@gmail.com
Website: www.namstct.org

Published by : Daya Publishing House®
A Division of
Astral International Pvt. Ltd.
– ISO 9001:2008 Certified Company –
4760-61/23, Ansari Road, Darya Ganj
New Delhi-110 002
Ph. 011-43549197, 23278134
E-mail: info@astralint.com
Website: www.astralint.com

AMITY INSTITUTE FOR HERBAL AND BIOTECH PRODUCTS DEVELOPMENT

– An Institution of Ritnand Balved Education Foundation –
Thiruvananthapuram

Prof. (Dr.) P. PUSHPANGADAN,

Ph.D., FAS, FBRS, FES, FIAT, FNRS, FNSE, FNESA, FNAASc, FNASc, FRSC (UK)
(*Padma Shri Awardee*, *UN Equator Initiative Eaureate and Borlaug Awardee*)
(*Former Director, NBRI & CIMAP, Lucknow & JNTBGRI & RGCB, Trivandrum*)
Hon. Director General & Senior Vice President, RBEF

Foreword

Herbal medicine refers to using any plant part from seeds, berries, roots, leaves, and bark to flowers for medicinal purposes. Plants in India have been used for medicinal purposes long before recorded history. India- a mega-diversity centre harboring a multitude of medicinal plant species each presumably studded with as yet unknown genetic and chemical variations of economic importance. Chinese and Egyptian papyrus writings describe medicinal uses for plants as early as 3,000 BC. In India, Ayurveda evolved during the Vedic period (2000-1000BCE). Ancient So, Medicinal plants are highly esteemed all over the world as a rich source of therapeutic agents for the prevention of diseases and ailments. Now-a-days there is a revival of interest with herbal-based medicine due to the increasing realization of the health hazards associated with the indiscriminate use of modem allopathic medicines and the shift can be termed as 'Return to Nature'. The demand for medicinal plants is steadily increasing in both developing and developed countries.

The World Health Organisation estimated that 80 per cent of people worldwide rely on herbal medicines for some part of their primary health care. Also, modern pharmacopoeia still contains at least 25% drugs derived from plants while many others are synthetic analogues built on prototype compounds isolated from plants.. India and China are the largest users of herbal medicines.

While herbal medicine can potentially contribute to the advancement of healthcare, many major challenges must be overcome prior to the successful integration of herbal remedies into mainstream medicines. There is a great concern about the safety and efficacy of herbal use. The quality assurance of these products is very important. The use of herbal products of poor quality can result in the adverse effects on the health. The main quality concerns could be contaminations with microorganisms etc or misidentification of the plant species. Thus, identification of the herbal drug is an extremely important control step for safety and efficacy reasons. It has been stated by the World Health Organization (WHO) that the most critical assessment of herbal medicine is safety evaluation. Another important associated issue is the ignorance about the adverse effects of herbal drugs on health due to overdosing or contaminated formulations to the inherent toxicity of the herbs of choice. All these issues need to be resolved on the basis of scientific advancements and adopting associated appropriate safety regulations.

I am delighted to learn that the Centre for Science and Technology of the NON-Aligned and Other Developing Countries (NAM S&T Centre) jointly with the JSS University, Mysore, India, organised a joint International Training Workshop on "Herbal Medicine: Drug Discovery from Herbs: Approaches, Innovations and Applications" during 30th March–3rd April 2015 in Mysore (Karnataka) and Ooty (Tamil Nadu), India to discuss advancements in the herbal drug sector and related issues. The present book is an outcome of these deliberations and comprises 25 scientific and technical papers from experts and professionals from 17 countries.

I compliment the NAM S&T Centre for bringing out this valuable publication on such an important topic to showcase the research efforts in herbal drugs area of various developing countries and their beneficial rice towards human health.

P. Pushpangadan

No.3, Ravi Nagar, Ambalamukku
Peroorkada P.O
Thiruvananthapuram -695 005
Kerala, India

Tel: (0471) 2432138, 3269772, 2432148
Mob: 0-98950 66816
E-mail: palpuprakulam@yahoo.co.in
www.drpalpupushpangadan.com

Preface

Use of herbal preparations for treating various diseases, also known as "Herbal Medicine" is not new, but it has gained attention recently due to various scientific observations ascribing health beneficial effects such as treatment of cancers, prevention of obesity and diabetes to different herbal preparations. More over, it is also a general belief that herbal preparations are not toxic, hence can be used in place of other medicines that might cause toxicity. Further, many pharmaceutical industries and start-up companies have also started marketing herbal preparations globally, thereby further promoting the use of these herbal drugs for not only disease treatment but also to use as age-defying agents and beautifying creams. However, still many issues pertaining to method of preparation, consistency in the composition and quality from batch-to-batch, efficacy of prepared samples in clinical trials, and implementation of universally applicable quality standards needs to be addressed in detail. In addition many initiatives have to be taken to promote the integration of herbal drug research with the scientific advancements in other areas of related fields. For example, uses of nano-science and technology advancements to more effectively deliver herbal drugs for improving the efficacy and reducing toxicity (if any) and drug dosage. Therefore, much better regulatory standards need to be implemented to further strengthen the use of herbal preparations as medicines.

Asian countries have been known for their rich source of a wide variety of herbs and herbal drug practices. For example many ayurvedic formulations have been prepared and tested for efficacy against various communicable and non-communicable diseases in India, Pakistan and Sri Lanka. Similar efforts have also been taken in other developing countries including Afghanistan, Botswana, Cambodia, Egypt, Indonesia, Iraq, Malaysia, Myanmar, Nigeria, Rwanda, South Africa, Turkey, Vietnam and Zimbabwe. More over, the governments of Asian, African and European countries have also initiated several funding mechanisms and student exchange schemes to promote herbal drug training and research. As a part of this initiative Center for Science and Technology of Non-Aligned and Other Developing Countries (NAM S&T center) consisting of 47 membership countries has been involved in organizing workshops and training programs in different fields of science. Very recently an International Training Workshop on Herbal Medicine: Drug Discovery from Herbs – Approaches, Innovations and Applications was conducted by NAM S&T center in JSS University, Mysuru, Karnataka between 30th March 2015 to 3rd April 2015. Experts, professionals, research scholars, post-graduate students and administrators from 24 NAM S&T countries have participated in this workshop and presented their research results, views and thoughts.

This book is not only a compilation of research and review articles of high-significance presented in the above-mentioned workshop but also contain articles submitted to the NAM S&T center for consideration of publication. All these articles have been peer-reviewed by experts in the respective subject fields and thoroughly scrutinized for the quality and accuracy of the results. These articles have been categorized in to 4 major sections based on the research that has been explored. The first category on Anti-Diabetic Drugs comprises of 7 articles explaining the current needs for developing a platform for screening anti-diabetic drugs, the role of *in vitro* and *in vivo* studies for identifying and assessing the potential of naturally occurring diabetes inhibiting agents. Other articles in this section covered the anti-glycation effects of herbal extracts in vitro as well as in preclinical animal models. One study also determined the efficacy of black seed extracts on various biochemical changes in Type-2 diabetes individuals.

Anti-microbial effects of herbal preparations such as inhibition of bacterial and fungal infections, transmission of viral vectors have been tested world wide and active ingredients characterized. But, many research groups are still exploring this area of research to identify much more potent, multi-functional anti-microbial agents to prevent and effectively inhibit the microorganisms' growth. The second section of this book comprising five papers focuses on the (a) potential of herbal preparations for inhibiting various opportunistic infections that usually occur in HIV infected individuals; (b) ability of plant extracts for preventing and treating the bacterial and fungal infections; and (c) preclinical toxicity and efficacy evaluation of herbal extracts for regulating anti-microbial infections in rats and mice.

Determination of toxicity and health beneficial effects of herbal preparations is one of the prerequisites to fulfill to consider a preparation for assessing it in the clinical trials. Therefore, in the third section of this book various studies determining the herbal drugs' toxicity and efficacy for preventing fertility and pregnancy related complications have been presented. In addition, sedative effects of essential oils from *Heracleum afghanicum* Kitamura seeds and the mechanism of action of *Zingiber officinale* Roscoe Var. *Rubrum* (Halia Bara) active ingredients for inhibiting psoriasis have been presented in this section. Final section of this book focuses on the research activities on traditional medicines in Egypt, Cambodia, Vietnam, and

East Mediterranean high-mountains. A separate article on how to identify herbal drugs was also discussed in this section.

In summary, this book is an excellent compilation of various research activities assessing the efficacy of herbal medicines for preventing and treating different communicable and non-communicable diseases.

Suresh Bhojraj

Tijen Talas-Ogras

Shamiem Adam

Subba Rao V. Madhunapantula

Introduction

Plants are being widely used for medicinal purposes ever since the beginning of the humankind. Traditional medicine is extensively practiced in all major world communities and modern medicine identifies 'herbalism' or 'herbology' as a form of alternative medicine. Thus even though the development and mass production of chemically synthesised drugs have revolutionised healthcare in most parts of the world, large sections of the population in developing countries still depend on traditional practitioners and herbal medicines for their primary care. Medicinal plants have been and are still the major sources of medicine worldwide. The available statistics shows that in Africa ~90 per cent, in India ~70 per cent and in China ~40 per cent of the population depends on traditional medicines to help meet their healthcare needs, and more than 90 per cent of general hospitals in China have units for traditional medicine. The annual worldwide market for the ethno-botanical products is estimated to be about US$60 billion.

Herbal medicines have so far not been strictly based on evidence gathered using the scientific methods, but lately these have gained a lot of attention and a greater scientific interest in the medicinal use of plants has arisen. Although herbal supplements may be considered safe, some are known to be toxic at high doses and others may have potential side effect after prolonged use. The general public is largely unaware that adverse health effects can be associated with the use of herbal supplements resulting from overdosing. There are the matters of great concern. The increasing use of traditional medicines, the general lack of research, the growing unease by stakeholders vis-à-vis the demands for patenting rights, evidence of safety, efficacy, good quality traditional medicines and a range of other ethical issues coupled with the need for integration and maximisation of their potential as a source of healthcare are some of the pressing challenges that must be tackled for acceptable use of traditional and alternative medicines in modern therapeutics.

The biggest challenge in this field is for identification of the constituents and standardisation of the same.

Extensive research is also needed to meet the challenges of identifying the active compounds in the plants, and there should be research-based evidence on whether whole herbs or extracted compounds are better. Smart screening methods and metabolic engineering offer exciting technologies for new natural product drug discovery. Advances in rapid genetic sequencing, coupled with manipulation of biosynthetic pathways, may provide a vast resource for the future discovery of pharmaceutical agents.

In order to discuss the above issues, the Centre for Science and Technology of the Non-Aligned and Other Developing Countries (NAM S&T Centre), jointly with the JSS University, Mysore, India, organised an International Training Workshop on 'Herbal Medicine: Drug Discovery from Herbs: Approaches, Innovations and Applications' during 30th March – 3rd April 2015 in Mysore (Karnataka) and Ooty (Tamil Nadu), India. The workshop facilitated the capacity building and exchange of information and expertise among the developing countries.

There were 31 experts, professionals, researchers and administrators from 24 countries - Afghanistan, Botswana, Cambodia, Cameroon, Egypt, Indonesia, Iran, Iraq, Malaysia, Mongolia, Myanmar, Nepal, Nigeria, Oman, Pakistan, Qatar, Rwanda, South Africa, Slovenia, Tanzania, Turkey, Uganda, Vietnam and Zimbabwe in the training workshop and 83 from the host country India, including specials guests, speakers and co-chairs.

The present book edited by Prof. Suresh Bhojraj, Dr.Tijen Talas-Ogras, Dr. Shamiem Adam and Dr. Subba Rao V. Madhunapantula is a follow up of the above scientific event of the NAM S&T Centre and includes 25 scientific and technical papers from 17 countries presented at the above workshop as well as a few papers contributed by other experts on the subject. The contents of the book are divided into four sections, respectively, (i) Anti-Diabetic Drugs; (ii) Anti-Microbial Agents from plants; (iii) Beneficial Effects on Health and Toxicity Studies of Herbal drugs; and (iv) Herbal Drug Research in Developing Countries. I am grateful to Dr. Palpu Pushpangadan, Director General, Amity Institute for Herbal and Biotech Product Development, for writing a Foreword for this book.

I greatly appreciate the efforts put in by the four co-editors by contributing valuable suggestions to incorporate appropriate modifications in the manuscripts submitted by various authors and in the overall value addition of the publication. I am also indebted to the entire team of the NAM S&T Centre, particularly to Dr. (Mrs.) Kavita Mehra, Mr. M. Bandyopadhyay, Ms.Vaneet Kaur, Ms. Harsha Doriya and Mr. Pankaj Buttan in compiling the manuscripts, liaising with the authors, cover page designing, proof reading, formatting and taking other actions in giving a shape to this volume.

I hope that this book will provide valuable information and resource material to researchers and professionals who are dealing with herbal drugs research on its various dimensions.

Prof. Dr. Arun P. Kulshreshtha
Director General,
NAM S&T Centre

Contents

II. Anti-Microbial Agents from Plants

III. Beneficial Effects on Health and Toxicity Studies of Herbal Drugs

I. Anti-Diabetic Drugs

Chapter 1

Anti-Hyperglycemic Agents from Natural Sources: *In vitro* and *In vivo* Studies

Mona H. Hetta

Pharmacognosy Department,
Fayoum University, 63514, Egypt
E-mail: monahetta@gmail.com; mhm07@fayoum.edu.eg

ABSTRACT

This work highlighted the importance of natural sources as anti-diabetic *in vivo* and *in vitro*. Selected plants growing in Egypt and belonging to different families were tested *in vitro* on carbohydrates hydrolyzing enzyme activities (α-amylase, α–glucosidase and β-galactosidase). Results showed that when compared to Acarbose standard, the extracts of *Eruca sativa* Mill. leaves (Rocket) exhibited inhibitory effect on the enzymes as dose dependent with the concentration of inhibitors. *Arachis hypogaea* L. (Peanuts) pericarp showed higher percentage of inhibition of α–amylase, α–glucosidase and β-galactosidase with two different extracts but lower than the standard Resveratrol and Acarbose. The root extract of *Conyza discoridis* exerted the most significant decrease in serum glucose and the most significant increase in serum insulin levels. *Ficus platypoda* leaves was more active when tested orally in diabetic rats than *Ficus lyrata* species in decreasing the blood glucose level at 200 mg/kg/day. *Citrus medica* L. (Etrog) exerted a significant reduction in blood glucose level in diabetic rats when compared to Gliclazide standard.

Keywords: *Eruca sativa, Arachis hypogaea, Conyza discoridis, Ficus species, Citrus medica,* Anti-diabetic, *In vitro, In vivo.*

1. INTRODUCTION

The number of people suffering from diabetes mellitus (DM) in the world reached 347 million (Danaei *et al.*, 2011). This number is expected to increase to 592 million in 2035. (http://www.idf.org/worlddiabetesday/current-campaign). Estimated 1.5 million deaths in 2012 were caused directly by diabetes. The latter may become the seventh leading cause of death in the world by the year 2030. It is a leading cause of kidney failure, amputation and blindness (http://www.who. int/mediacentre/factsheets/fs312/en/). More than 80 per cent of diabetes deaths occur in middle and low -income countries. The prevalence of DM reaches 20 per cent in the urban areas of Egypt (Bos and Agyemang, 2013). The Northern Africa (MENA) and the Middle East have the highest prevalence of diabetes as a world region. Egypt is considered as number nine among the top 10 ranking (MENA) countries. This disease is a life threatening often with increasing incidence in rural populations throughout the world.

There is a trend nowadays towards the use of herbal medicine and get back to nature with the confidence that they are cheaper and safer. The use of medicinal herbs as a remedy started early since ancient times where they found in Tombs of Ancient Egyptian. Flora of Egypt contains about 2000 species of plants, beside the successfully acclimatized plants.

Eruca sativa Mill. (Synonym: Water cress, Rocket, Rucola, Taramira), is belonging to family Cruciferae or Brassicaceae. It is important as a vegetable and spice. Investigation of the leaves aqueous extract revealed the presence of flavonoidal components (Michael *et al.*, 2011). The leaves contain unique structure of glucosinolates. It was reported that also that the seed oil ameliorated hyperglycemia and oxidative stress.

The peanuts (*Arachis hypogaea* L.), is a dehiscent legume, collected under soil. It is used as a source of seed oil or its edible shelled nuts as food. The peanut pericarp, as by- products of the peanut industry, still does not have any significant value or use (Sobolev and Cole 2003). The antioxidant activity of peanut hulls methanolic extracts was previously reported (Lee *et al.*, 2006). The hypoglycemic and hypolipidemic activity of the aqueous extract of *A. hypogaea* seeds were also reported as a significant hypolipidemic which decrease the total cholesterol, serum triglycerides, HDL and LDL-cholesterol in normal and diabetic rats (Moreno *et al.*, 2006).

Conyza dioscoridis (L.) Desf., Family Asteraceae, is a highly branched shrub that grows wild and is characterized by being glandular and hairy. The plant is widely distributed in the Middle East and surrounding African countries. In Egypt, it occurs mainly in Nile region, Eastern and Western Deserts, Oases of the Mediterranean coastal strip and Sinai Peninsula (Boulos, 2002). This plant is known in folk medicine as a popular remedy to relieve rheumatic pains (Boulos and E1-Hadidi, 1989), in treatment of epilepsy in children, colic, ulcer, carminative and cold (Ibn El Bitar, 1890). It is called "mosquito tree" due to its insect repellent effect (Shaltout, and Slima, 2007). It is reported that *C. dioscoridis* extract exhibited a significant anti-diarrheal activity (Atta, 2004), and effective as diuretic (Atta,

2010). It has antimicrobial activities due to its volatile constituents (El-Hamouly and Ibrahim, 2003). The extract of the aerial parts showed anti-inflammatory activity (Awaad *et al.*, 2011).

Genus Ficus which belongs to Family Moraceae containing around 2000 species of woody plants, trees, erect shrubs, and climbers. Ficus is known as figs or fig trees. They are grown for their ornament and fragrance. They are native to the tropics with a few species in the semi-warm temperate zone. Ficus species are widely used in folk medicine in Egypt as anti-diabetic (Chopra, *et al.*, 1950). *Ficus platypoda* (Miq) A.Cunn, (desert fig or rock fig) is endemic to central, northern Australia, and Indonesia (Wikipedia, 2012). *Ficus lyrata* Warb species is also known as *Ficus pandurata* Sander or Chinese fig, with banjo-shaped leaves and is indigenous to tropical, central, and West Africa.

Citrus medica L. (Etrog, Citron, itranj in arabic) is native to Southeast Asia. It has been cultivated since ancient times. Etrog is called in antiquity the Persian or Median Apple. It was considered as the symbol of resistance (Ronit Treatman, 2010). Diamante–Citron variety is reported as hypoglycemic and insulin secretagogue bioactive (Peng *et al.*, 2009).

2. MATERIALS AND METHODS

Plant Material

The leaves of *Eruca sativa* Mill. (Figure 1.1a) were collected from Beni-Suef Gogernorate, in September 2012. It was kindly authenticated by a specialist in Botany Department - Faculty of Sciences, Beni-Suef University, Egypt. A voucher specimen no. BUPD 33 is desposited in the Department of Pharmacognosy, Faculty of Pharmacy, Beni- Suef University, Egypt.

Fruits of *Arachis hypogaea* L. (Peanuts, Figure 1.1b), were collected in 2010 from El-Behera Gogernorate. It was kindly authenticated by Mrs. Therese Labib, Senior Botanist in El-Orman Garden, Egypt. A voucher sample no. BSP25 is desposited the department of Pharmacognosy, Faculty of Pharmacy, Beni-Suef University, Egypt. The seeds were manually separated from fruits. The pericarp was air-dried in shade then powdered and stored in a glass container.

Conyza dioscoridis was collected from Egypt (El-Fayoum Gogernorate) in 2009. The plant (Figure 1.1c) was identified by Prof. Dr. Mounir M. Abd Elghani, Faculty of Sciences, Department of Botany, Cairo University, Egypt. A voucher specimen no. C1-2012, is deposited in the department of Pharmacognosy, Faculty of Pharmacy, Beni-Suef University, Egypt.

The two species of Ficus: *F. platypoda* (Miq) A. Cunn. and *F. lyrata* Warb. (Figure 1.1d and 1.1e respectively) used in this study were collected in March 2009 from Giza Zoo, Cairo, Egypt. The plants were kindly identified by Dr. Mohamed Gibali, Senior Botanist. Voucher specimens of both species were deposited in the department of Pharmacognosy, Faculty of Pharmacy, Beni-Suef University (2009BUPD18 and 2009BUPD19 respectively). The leaves and stems of both species were air-dried, powdered and then stored for further chemical and biological studies.

Figure 1.1a: *Eruca sativa* Mill. Leaves.

Figure 1.1b: *Arachis hypogaea* L. Pericarp.

Figure 1.1c: *Conyza discoridis* (L.) Desf. Roots.

Figure 1.1d: *Ficus platypoda.*

Figure 1.1e: *Ficus lyrata.*

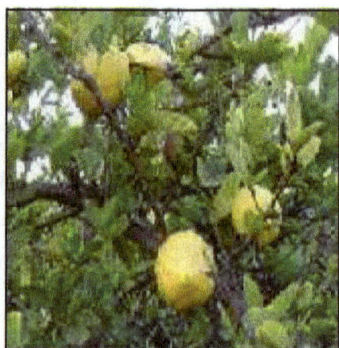

Figure 1.1f: *Citrus medica* Leaves.

The leaves of *C. medica* L. (Etrog, Figure 1.1f) were collected from El Qualyobia Governorate, Egypt, in April, 2009. The plant was identified by Mrs. Therese Labib, Senior Botanist, El Orman Garden, Egypt. A voucher specimen (BUPD20) is deposited in Pharmacognosy Department, Beni-Suef University, Egypt.

Preparation of Extracts

An amount of 250g of fresh leaves of *Eruca sativa* were extracted till exhaustion with ethanol on cold by maceration. The solvents were distilled off from the combined extracts and the residue left was stored in a dark container, in a refrigerator to be used in the biological study. Another amount of fresh leaves (250g) were extracted on cold with distilled water (1 L) using a sonicator. The dried residue after lyophillization was also stored as the previous amount till the biological study.

The powdered Peanuts pericarp (100 g) was extracted on cold by maceration with 70 per cent MeOH in a dark container till exhaustion. Combined extracts were filtered and concentrated under reduced pressure and the residue left was shacked with H_2O and successively fractionated with solvents of increasing polarity, using a separating funnel, starting with *n*-hexane, EtOAC then MeOH. Each extractive was concentrated and the residue left was weighed and kept in an amber glass container and then stored in a refrigerator until used for the phytochemical and biological studies.

Conyza discoridis powdered organs (leaves, flowers and roots, 500g each) were, separately extracted on cold with ethanol 70 per cent, by maceration, till exhaustion.

The air-dried powdered leaves of both species of Ficus (500 g each) were extracted with 80 per cent ethanol till exhaustion. The solvent was evaporated under reduced pressure and the residues obtained were kept for biological study.

Fresh leaves (2 kg) of Etrog were air dried and powdered (500g). The latter was defatted with petroleum ether and tested for their phytochemical constituents. The defatted powder (250g) was extracted, on cold, by maceration, with 70 per cent MeOH, till exhaustion. The collected extracts were concentrated under reduced pressure.

Chemicals, Drugs and Kits

All the chemicals used were of analytical grade (Merck, Sigma and Aldrich), the kits from (Sigma Chemical Company (USA), Biosystems (Spain) and Biodiagnostic (Egypt), the carbohydrate metabolizing enzymes (purified enzymes); α–amylase, α-glucosidase and β-galactosidase (EC3.2.1.1, EC3.2.1.20 and EC3.2.1.23 respectively) were obtained from Sigma Chemical Company (NY), USA. Streptozotocin (STZ) and Alloxan, used for induction of Diabetes, were purchased from Sigma- Aldrich chemical company, USA, Gliclazide (Diamicron®), Servier Laboratories, France, was obtained from local pharmacies and served as a standard. Ascorbic acid standard was purchased from Merck, Germany Co., Egypt. Glucose kit used for the enzymatic determination of glucose was purchased from (BIOLABO SA, France) and (Biodiagnostic, Egypt). Rat insulin ELISA kit (H type) Biovendor, Czech Republic, was used for determination of the concentration of insulin in samples.

Carbohydrates Hydrolyzing Enzymes Inhibition Assay

α-Amylase inhibitory activity was determined according to the method described by Ali *et al.* (2006) by measuring the absorbance at 540 nm. The inhibitory activity of α-galactosidase was measured by the method of Sanchez and Hardisson

(1980) and the resulting color of *O*-nitrophenolate ions was measured at 420 nm. A standard curve was performed using different concentrations of O-nitrophenol. α-glucosidase inhibitory activity was determined according to glucose-oxidase method of Kim *et al.* (2005) and the reducing activity was estimated by measuring the release of p-nitrophenol from p-nitrophenyl α-D-glucopyranoside at 405 nm. A standard curve was carried out using different concentrations of p-nitrophenol.

In vivo Study

Experimental Animals

The animals were obtained from a breeding colony at Faculty of Veterinary Medicine, Beni-Suef University and National Research Center, Egypt. Albino mice (20 – 25 g) of both sexes were used for Determination of LD50 and acute toxicity study. Adult male albino rats (120 - 160g) were used for the antihyperglycemic activity. The animals were kept under the same hygienic conditions, fed with well balanced normal diet.

Anti-hyperglycemic Activity

For Conyza discoridis

Thirty six male adult albino rats were divided into six groups (6 rats for each group). The first group served as a negative control, receiving normal saline orally. The rats in groups no. 2-6 were fasted for 12 hours before the induction of diabetes with freshly prepared solution of STZ (50 mg/kg body weight (b.wt) by an intraperitoneal injection. STZ was dissolved in 0.05 M sodium citrate buffer, pH 4.5 (Montilla *et al.*, 2004). Blood samples were collected after 72 hours of induction of diabetes for measuring blood glucose and insulin level. Group 2 served as STZ-diabetic control, and other groups 3, 4 and 5 were treated daily by oral route, with total alcohol extracts of leaf, flower and root respectively daily for two weeks by doses of 50mg/kg b.wt each using an intragastric tube. Group 6 was treated with a dose of 10 mg/kg b.wt. of standard Gliclazide. At the end of the experiment blood samples were collected for determination of the biochemical parameters.

For Ficus Species

The rats were treated with a single intraperitoneal injection of Alloxan monohydrate in a dose of 150 mg/kg body weight to be rendered diabetic and then anesthetized by ether. The blood samples were collected for glucose level determination from the retro-orbital venous plexus. The rats were divided into seven groups (6 rats each) and treated as follows: Group 1: negative control; Group 2: as positive control diabetic non-treated rats; Group 3 and 4: orally treated with 80 per cent ethanolic extracts of *F. platypoda* at two doses (200 and 400 mg/kg body weight), respectively diabetic rats,; Group 5 and 6: orally treated with 80 per cent ethanol extracts of *F. lyrata* at two doses (200 and 400 mg/kg body weight), respectively diabetic rats. All previous doses represent 1/10 and 1/5 of the maximum soluble concentration. Group 7: diabetic rats were treated with a single oral dose (20 mg/kg body weight) of a standard anti-diabetic drug (Gliclazide). After 28 days of treatment, blood samples were collected and biochemical parameters were measured.

For *Citrus medica*

Five groups (7 rats each) and all rats, except group 1 (normal non-treated rats, negative control), were rendered diabetic after a single intraperitoneal injection of Alloxan monohydrate (dose of 135 mg/kg b.wt). Group 2 served as a positive control, non-treated diabetic. Groups 3 and 4 were orally treated for one month with 70 per cent MeOH extract of the plant at doses of 200 and 400 mg/kg b.wt., respectively and group 5 was treated with Gliclazide (Diamicron®) reference drug (20 mg/kg b.wt). Glucose levels were estimated in serum after enzymatic glucose oxidase-method.

Statistical Analysis

Analysis of the data was performed by one way Analysis of Variance (ANOVA) and subsequent analysis was performed using Tukey test. p-values <0.05 were selected to indicate statistical significance between the groups. Statistical analysis of results, was carried out using named SPSS statistics 17.0, release (Aug. 23, 2008), Chicago, USA as analytical software.

Ethics

All animal procedures were performed after approval from the Ethics Committee of Beni-Suef University and the Research Ethics Committee of National Research Centre, Cairo, Egypt and in accordance with the recommendations of the proper care and use of laboratory animals.

3. RESULTS AND DISCUSSION

The suppression of glucose occurs by different mechanisms in investigation of natural herbs. One of these approaches is to decrease the post–prandial hyperglycemia by retarding the absorption of glucose through the inhibition of the carbohydrate hydrolyzing enzymes α- amylase, α-glucosidase and β–galactosidase in the digestive tract. Inhibition of these enzymes delay carbohydrate digestion and prolong overall carbohydrate digestion time, causing a reduction in the rate of glucose absorption and consequently blunting the post prandial plasma glucose rise (Rhabasa and Chiasson; 2004). The fact that α–glucosidase and α-amylase showed different inhibition kinetics seemed to be due to structural differences related to the origin of the enzymes (Kim *et al.*, 2005).

Percentage of inhibition of the ethanol extract of *Eruca sativa* Mill. showed more prominent effect in carbohydrate inhibitory activity at different concentrations when compared to the standard Acarbosethan the water extract (Table 1.1). A dose–response relationship was found in the carbohydrate inhibitory activities as the activity increased on increasing of the concentration of extracts increased for each individual one (Hetta *et al.*, 2014c).

The powdered pericarp of *Arachis hypogea* L. yielded 27 per cent of total 70 per cent MeOH and 19 per cent, 28 per cent and 27.5 per cent for successive extractives (*n*-hexane, EtOAC and MeOH respectively). Polar extracts (EtOAC and MeOH)

Table 1.1: Carbohydrates Metabolizing Enzymes Inhibition Percentage of *Eruca sativa* Mill. Leaves

Concentration	α-amylase			α-glucosidase			β-galactosidase		
	Acarbose	Water Extract	Ethanol Extract	Acarbose	Water Extract	Ethanol Extract	Acarbose	Water Extract	Ethanol Extract
10 µg/ml	33.11±1.23[a]	18.26±3.97 [a]	32.00±1.21[a]	45.10±3.12[a]	18.47±4.56[a]	28.05±1.63 [a]	65.20±10.12[a]	17.24±6.39[a]	51.35±5.24[a]
50 µg/ml	39.13±3.22[b]	24.55±4.03 [b]	36.35±2.69[b]	78.30±8.10[b]	25.49±1.76[b]	37.80±2.55 [b]	76.90±12.11[b]	32.35±8.72 [b]	56.62±6.53[b]
100 µg/ml	48.36±4.10[c]	32.56±2.07 [c]	55.22±4.92[c]	85.30±10.80[c]	33.55±3.08[c]	44.21±2.87 [c]	80.33±13.80[b]	33.53±11.38[b]	64.49±11.82[c]
500 µg/ml	56.56±1.69[d]	37.14±0.89 [d]	69.38±8.22[d]	90.20±5.98[d]	39.01±1.78[d]	49.19±1.59 [d]	88.23±5.98[c]	42.13±2.44 [c]	67.82±3.94 [c]
1000 µg/ml	72.34±2.67[e]	44.59±1.55 [e]	78.40±7.29 [e]	98.54±11.55[c]	55.11±2.70[e]	62.5±2.24 [e]	90.44±10.90[c]	40.08±5.11 [c]	66.86±3.79 [c]
LSD 5 per cent	5.88	5.18	5.80	5.90	5.93	5.80	6.1	5.93	5.80

showed appreciable carbohydrate inhibitory activities (Table 1.2). Resveratrol standard was the most active (5.18 per cent, 5.94 per cent and 13.26 per cent) followed by EtOAc extract (4.32 per cent, 5.93 per cent and 13.7 per cent) in inhibition of α-amylase, α-glucosidase and β-galactosidase respectively when compared to the standard Acarbose (5.88 per cent, 5.9 per cent and 13 per cent respectively). However MeOH extract showed the lowest inhibitory activity (3.9 per cent, 4.9 per cent and 14.1 per cent respectively). The activity increased as the concentration of the tested extract increased in each case as a dose–response relationship. (Table 1.2). We can deduce a linear relationship with a significant increase in reducing activity with the increase in concentrations of the extracts (Hetta *et al.*, 2014b).

Table 1.2: Carbohydrates Metabolizing Enzymes Inhibition per cent of EtOAc and MeOH Extracts of Peanuts Pericarp Compared to Resveratrol Standard

Concentrations	Acorbase	EtOAc Extract	MeOH Extract	Reseveratrol
α-amylase inhibition per cent				
10 µg/mL	32.21±1.23[a]	17.59±0.60[e]	16.48±0.98[e]	28.20±1.97[d]
50 µg/mL	35.13±2.44[c]	28.70±1.04[a]	24.78±2.24[b]	34.10±3.03[c]
100 µg/mL	47.36±3.10[g]	34.29±3.11[d]	28.34±1.89[f]	45.16±2.12[g]
500 µg/mL	52.56±1.69[c]	45.12±3.33[a]	39.65±2.80[b]	52.23±2.29[c]
1000 µg/mL	71.34±2.67[d]	58.10±2.14[a]	50.50±1.10[b]	67.97±3.86[c]
LSD 5 per cent	5.88	4.32	3.9	5.18
α-glucosidase inhibition per cent				
10 µg/mL	29.95±2.00[b]	25.15±3.00[c]	23.00±3.06[c]	28.95±2.63[b]
50 µg/mL	43.23±3.08[c]	34.81±23.16[b]	33.99±2.70[b]	37.90±2.15[a]
100 µg/mL	52.33±4.07[d]	45.20±2.39[d]	41.00±2.00[a]	46.21±2.87[c]
500 µg/mL	69.12±3.10[c]	48.00±2.00[d]	42.61±1.79[a]	55.79±1.50[b]
1000 µg/mL	85.89±2.28[c]	64.39±1.90[d]	59.91±2.90[a]	69.70±2.28[b]
LSD 5 per cent	5.90	5.93	4.90	5. 94
β-galactosidase inhibition per cent				
10 µg/mL	45.85±2.90[a]	23.12±5.31[b]	17.90±4.32[c]	34.35±3.20[d]
50 µg/mL	55.90±4.70[a]	34.78±4.90[b]	35.35±9.00[b]	39.62±8.90[c]
100 µg/mL	59.00±6.10[a]	39.59±8.35[b]	36.80±9.38[c]	53.49±8.80[d]
500 µg/mL	78.56±5.78[a]	62.23±2.99[b]	54.70±4.41[c]	69.78±4.90[d]
1000 µg/mL	83.66±6.51[a]	65.10±5.07[b]	56.00±5.01[c]	72.86±5.01[d]
LSD 5 per cent	13.00	13.7	14.1	13.26

For *Conyza discoridis* powdered organs (leaves, flowers and roots, 500g each) were, separately extracted with ethanol 70 per cent and yielded 4.97, 3.66 and 3.11 per cent, respectively). A significant decrease in serum glucose levels for roots and leaves were showed (106.45±8.25 and 109.36±9.83 µmol/ml, respectively) when compared to Gliclazide standard (109.28±8.00 μmol/ml). The root extract was more significant than the standard itself. A significant increase in insulin levels for roots and leaves (6.53±0.83 and 6.17±0.68 ng/ml, respectively) were noticed when compared to the standard (6.26±0.66 μg/ml), Table 1.3.

The ethanol 70 per cent extract of of *C. dioscoridis* roots significantly decreased blood glucose level of STZ diabetic rats (Tables 1.3), (El Zalabani *et al.*, 2012) and their effects were comparable to Diamicron® (oral anti-diabetic reference drug). The attributed antihyperglycemic activity of this organ is due probably to a synergistic effect between its sterols, triterpenes and phenolic compounds components (Saha *et al.*, 2011, Prajapati *et al.*, 2008, Shahwar *et al.*, 2012).

Table 1.3: Effect of Administration of 50 mg/kg of Ethanolic 70 per cent Extracts of Three Organs (Leaves, Flower and Roots) of *C. dioscoridis* on Serum Glucose and Insulin Levels of Streptozotocin (STZ) – Induced Diabetic Male Rats

Serum Insulin Level $\bar{x}$±SE (ng/ml)	Serum Glucose Level $\bar{x}$±SE(nmol/ml)	Treated Groups
7.05±0.76	94.80±7.60	gp1: Normal control
2.51±0.43*	333.77±28.00 *	gp2: STZ-diabetic control
6.17±0.68 [a]	109.36±9.83 [a]	gp3: STZ-alcoholic extract of leaf
2.55±0.57*	284.81±39.82*	gp4: STZ-alcoholic extract of flower
6.53±0.83 [a]	106.45±8.25 [a]	gp5: STZ-alcoholic extract of root
6.26±0.66 [a]	109.28±8.00 [a]	gp6: STZ-Gliclazide (10 mg/kg)

Data was expressed as mean±SE (n=6 rats per group).

Values are statistically significant at P <0.05.

Statistical analysis was carried out using one way analysis of variance (ANOVA) followed by Tukey-Kramer multiple comparisons test.

*: Significantly different from the normal control group at p <0.05.

a: Significantly different from the diabetic control group at p < 0.05.

For *Ficus* species, results showed that 80 per cent ethanol extract of leaves of *F. platypoda* induce a significant reduction in blood glucose level (107.9±5.817, 64.11±4.358) at the tested doses 200, 400 mg/kg/day, respectively, when compared to *F. lyrata* (127.2±4.359, 127.7±6.889) and the standard Gliclazide (110.8±7.240) after 1 month of treatment (Table 1.4); (El-Kashoury *et al.*, 2013).

The tested extract of Etrog succeeded to completely restore normal glucose level and it significantly prevented further elevation of blood sugar. The powdered Etrog (500g) were extracted with 70 per cent MeOH and concentrated under reduced pressure to give a semisolid mass (36 g). At doses of 200 and 400mg/kg.b.wt of 70 per cent MeOH extract of the defatted powder of *Citrus medica* L. leaves, results showed that there was a significant and a dose dependant decrease in serum level of glucose using 200mg (105.2±8.35) and 400mg (87.4±6.30) treated doses when compared to Gliclazide (110.8±7.24) and the diabetic positive control group (172.3±82.09) mg/dl, (Table 1.5). At 400 mg/kg normalization of the elevated serum triglycerides after

1 month of administration was observed (Hetta *et al.*, 2014a).

Table 1.4: Effect of the 80 per cent Ethanol Extracts of *Ficus platypoda* (Miq) A. Cunn. and *Ficus lyrata* Warb Leaves on Blood Glucose Level in Rats

Groups	Dose/Route of Administration	Glucose (mg/dl)		
		Basal	After 72 h	1 Month of Treatment
Normal control	Saline	93.09±4.626	109.0±3.710	112.5±3.620[@]
Diabetic control	Saline	98.70±5.636	246.4±10.56[*]	172.3±2.089[*†]
F. platypoda	200 mg/kg/day, p.o	99.54±6.416	256.1±10.12[*]	107.9±5.817[@]
	400 mg/kg/day, p.o.	99.53±4.979	261.8±15.63[*]	64.11±4.358[*@†]
F. lyrata	200 mg/kg/day, p.o	103.8±8.262	252.1±14.35[*]	127.2±4.359[@]
	400 mg/kg/day, p.o.	104.8±9.699	242.3±10.13[*]	127.7±6.889[@]
Gliclazide	20 mg/kg/day, p.o.	101.8±4.233	237.8±8.486[*]	110.8±7.240[@]

Values represent the mean±S.E. of seven rats for each group.

[*] $p < 0.05$: Statistically significant from normal control.

[@] $p < 0.05$: Statistically significant from diabetic control.

[†] $p < 0.05$: Statistically significant from Gliclazide.

The 70 per cent MeOH extract of the defatted leaves of *C. medica* L. proved to be rich in flavonoids as flavanones and flavanols which were reported as significant biological antioxidant in cell cultures (Bhuiyan *et al.*, 2009) and offer a protection in the early stage of diabetes (Akiyama *et al.*, 2010). They also normalize the blood glucose altering the glucose regulatory enzymes (Akiyama *et al.*, 2010). They proved to decrease in glucose levels and improve glycolytic and gluconeogenetic enzymes in tissues (Kamalakkanna and Prince; 2006). Flavonoids could also ameliorate hyperglycemia by protein tyrosine phosphatase 1B (PTP1B) expression in liver (Lu *et al.*, 2010).

Table 1.5: Effect of 70 per cent MeOH extract of *Citrus medica* L. (Etrog) Leaves on Blood Glucose Level in Rats

Groups	Dose/Route of Administration	Glucose (mg/dl)		
		Basal	After 72 h	After 1 Month
Control	saline	93.1±4.62	109.0±3.70 [bc]	112.5±3.62 [b]
Diabetic control	saline	98.7±5.63	246.4±10.56 [a]	172.3±2.09 [ac]
70 per cent MeOH	200 mg/kg/day	106.8±5.87	247.0±12.78 [a]	105.2±8.35 [bc]
extract	400 mg/kg/day	109.3±5.04	240.5±7.83 [a]	87.4±6.30 [abc]
Diamicron ®	20 mg/kg/day	101.8±4.23	237.8±8.49 [a]	110.8±7.24 [b]

Values represent the mean±S.E. of seven rats for each group.

a $p < 0.05$: Statistically significant from normal control (Dunnett's test).

b $p < 0.05$: Statistically significant from diabetic control (Dunnett's test).

c $p < 0.05$: Statistically significant from diamicron® (Dunnett's test).

4. CONCLUSIONS

The tested plants of different have the potential to be used as new natural products for the management of diabetes.

5. ACKNOWLEDGEMENTS

I would like to thank all the co-authors participated in the different research papers accomplished on the tested plants.

6. REFERENCES

1. Atta, A., 2004. Antidiarrhoeal activity of some Egyptian medicinal plant extracts. *Journal of Ethnopharmacology*, 92: 303-309.

2. Atta, A., 2010. Evaluation of the diuretic effect of *Conyza dioscoridis* and *Alhagi maurorum*. *International Journal of Pharmacy and Pharmaceutical Sciences*, 2: 162-165.

3. Awaad, A.S., El-meligy, R.M., Qenawy, S.A., Atta, A.H, Soliman, G.A., 2011. Anti-inflammatory, antinociceptive and antipyretic effects of some desert plants. Journal of Saudi Chemical Society. 15(4): 367–373.

4. Bos, M., Agyemang, C., 2013. Prevalence and complications of diabetes mellitus in Northern Africa, a systematic review. *BMC Public Health*, 13: 387

5. Boulos, L., 2002. Flora of Egypt. Volume 3: Verbenaceae - Compositae. Al-Hadara Publishing: Cairo; 189.

6. Boulos, L., E1-Hadidi, M., 1989. The Weed Flora of Egypt. The American Univ. Cairo Press: Cairo; 361.

7. Chopra, R.N., Chubra, I.C., Handa, R.L., Kapin, L.D., 1950. Indigenous Drugs of India. Calcutta: U.N. Dhur and Sons Private Ltd.

8. Danaei, G., Finucane, M.M., Lu, Y., Singh, G.M., Cowan, M.J., Paciorek, C.J. *et al.*, 2011. National, regional, and global trends in fasting plasma glucose and diabetes prevalence since 1980: systematic analysis of health examination surveys and epidemiological studies with 370 country-years and 2.7 million participants. *Lancet*, 378(9785): 31–40.

9. El Zalabani, S., Hetta, M., Ismail, A., 2012. Genetic Profiling, Chemical Characterization and Biological Evaluation of Two *Conyza* Species Growing in Egypt. *Journal of Applied Pharmaceutical Science (JAPS)*, 2 (11): 54 – 61.

10. El-Hamouly, M.M.A., Ibrahim, M.T., 2003. GC/MS analysis of the volatile constituents of individual organs of *Conyza dioscoridis* (L.) Desf. growing in Egypt. *Alex. J. of Pharmaceutical Sci.*, 17(1): 75-80.

11. El-Kashoury, E.A., Hetta, M.H., Yassin, N. Z., Hassan, H.M., El Awdan, S.A., Afifi, N.I., 2013. Comparative DNA Profiling, Phytochemical Investigation and Biological Evaluation of Two *Ficus* Species Growing in Egypt. *Pharmacognosy Research*, 5(4): 45-53.

12. Hetta, M.H., Aly, H.F., Ali., N.W., 2014b. Estimation of Resveratrol content in Peanuts pericarp and its relation with the *in vitro* inhibitory activity on carbohydrate metabolizing enzymes. *Die Pharmazie, 69* (2): 92-95.

13. Hetta, M.H., El-Alfy, T.S., Yassin, N.Z., Abdel-Rahman, R.F., Kadry, E.M., 2014a. Phytochemical and Antihyperglycemic Studies on *Citrus medica* L. Leaves (Etrog) Growing in Egypt. *International Journal of Pharmacognosy and Phytochemical Research (IJPPR)*, 5(4); 271-277.

14. http://www.who.int/mediacentre/factsheets/fs312/en

15. Ibn El Bitar, 1890. Mofradat Al Adwiah wal Aghzia. Boulac Press: Egypt.

16. International Diabetes Federation (IDF.). http://www.idf.org/worlddiabetesday/current-campaign

17. Kim, Y.M., Jeong, Y.K.,Wang, M.H., Lee, W.Y., Rhee, H.I., 2005. Inhibitory effect of Pine extract on alpha-glucosidase activity and postprandial hyperglycemia. *Nutrition, 21*: 756–761.

18. Lee, S.C., Jeong, S.M., Kim, S.Y., Park, H.R., Nam, K.C., Ahn, D.U., 2006. Effect of far-infrared radiation and heat treatment on the antioxidant activity of water extracts from peanut hulls. *Food Chem*, 94: 489–493.

19. Michael, H.N., Shafik, R.E., Rasmy, G.E., 2011. Studies on the chemical constituents of fresh leaf of Eruca sativa extract and its biological activity as anticancer agent *in vitro*. *Journal of Med Plants Res*, 5(7): 1184-1191.

20. Hetta, M.H., Aly, H,F., Arafa, A., 2014c Inhibitory effect of *Eruca sativa* Mill. on carbohydrate metabolizing enzymes *in vitro*. *International Journal of Pharmaceutical Sciences Review and Research (Int. J. Pharm. Sci. Rev. Res)*, 26(2); 205-208.

21. Montilla, P., Barcos, M., Muñoz, M.C., Muñoz-Castañeda, J.R., Bujalance, I., Túnez, I.,.2004. Protective effect of Montilla-Moriles appellation red wine on oxidative stress induced by streptozotocin in the rat. *J. Nutr Biochem.*, 15(11): 688-93.

22. Moreno, D.A., Ilic, N., Poulev, A., Raskin, I., 2006. Effects of *Arachis hypogaea* nutshell extract on lipid metabolic enzymes and obesity parameters. *Life Sci*, 78: 2797–2803.

23. Peng, C.H., Ker, Y.B., Weng, C.F., Peng, C.C., Huang, C.N., Lin, L.Y., Peng, R.Y.,2009. Insulin Secretagogue Bioactivity of Finger Citron Fruit (*Citrus medica* L. var. Sarcodactylis Hort., Rutaceae). *Journal of Agricultural and Food Chemistry*, 57(19): 8812 – 9.

24. Prajapati, D.D., Patel, N.M., Savadi, R.V., Akki, K.S., Mruthunjaya, K., 2008. lleviation of alloxan-induced diabetes and its complications in rats by *Actinodaphne hookeri* leaf extract. *Bangladesh J Pharmacol.*, 3: 102-106.

25. Rhabasa-Lhoret, R., Chiasson, J.L., 2004. Alpha glucosidase inhibitors. In Defronzo RA, Ferrannini E, Keen H, Zimmet P (Eds.) *International Textbook of Diabetes mellitus* (Vol. 1) (3rd ed), JohnWiley and Sons Ltd., UK, pp. 901–914.

26. Ronit Treatman, Thu Aug 26, 2010, The Etrog: The Father of All Lemons. http://blog.pjvoice.com/diary/44/the-etrog-the-fatherof- all-lemons

27. Saha, P., Mazumder, U.K., Haldar, P.K., Sen, S.K., Naskar, S., 2011. Antihyperglycemic activity of *lagenaria siceraria* aerial parts on streptozotocin induced diabetes in rats. *Diabetologia Croatica*, 40 (2): 49 - 60.

28. Sánchez, J., Hardisson, C., 1980. Glucose inhibition of galactose-induced synthesis of β-galactosidase in *Streptomyces violaceus. Arch. Microbiol.*, 125: 111–114

29. Shahwar, D., Raza, M.A., Saeed, A.M., Riasat, F.I. Chattha, M. Javaid, S. Ullah, S. Ullah, 2012. Antioxidant potential of the extracts of *Putranjiva roxburghii, Conyza bonariensis, Woodfordia fruiticosa* and *Senecio chrysanthemoids. African Journal of Biotechnology*, 11(18): 288-4295.

30. Shaltout, K.H., Slima, D.F., 2007. The biology of Egyptian woody perennials. *Ass. Univ. Bull. Environ.Res.*, 10(1): 85-103.

31. Sobolev, V.S., Cole, R.J., 2003. Note on utilisation of peanut seed testa. *J Sci Food Agricult*, 84: 105–111.

32. Wikipedia. Wikimedia Foundation, Inc. Available from: http://en.wikipedia.org/wiki/Ficus_platypoda [Last accessed on 2012 May 11].

Chapter 2

Anti-diabetes Compounds from *Aspergillus terreus*

Rizna Triana Dewi

Research Center for Chemistry,
Indonesian Institute of Sciences
Kawasan PUSPIPTEK, Serpong-Indonesia, 15431
E-mail: rtriana_dewi@yahoo.com

ABSTRACT

Three strains of *Aspergillus terreus* fungi isolated from an Indonesian soil sample and one mutan strain of *A. terreus* were cultured in liquid media yielded eight secondary metabolites (1-8). The natural products as well as three synthetic butyrolactone I derivatives (3a-3c) were assayed for α-glucosidase inhibitory and scavenging free radical DPPH activities. Compounds 3, 3a, 4, 7, and 8 were actives against yeast α-glucosidase and could scavenger DPPH free radical. Compound 5, 3b, and 3c were displayed strong inhibitory activity on α-glucosidase but no activity on DPPH free radical. Compound 1, 2, and 7 only showed medium activity on DPPH free radicals. The results of this study revealed that *A. terreus* could be considered as a potential source of natural antioxidant and anti-diabetic drug candidates.

Keywords: Aspergillus terreus, α-glucosidase inhibitor, Antioxidant, Diabetes mellitus.

1. INTRODUCTION

Diabetes mellitus (DM) is now becoming a common metabolic disorder resulting from the inability of our body's response to high blood glucose levels. There were approximately 382 million adult aged 20-79 years with diagnosed DM in 2013 (Ramachandran *et al.*, 2013). Type II diabetes is estimated to account for 90-95 per cent of all diabetes and is characterized by insulin resistance, a relative insulin deficiency, and hyperglycemia. There is considerable evidence that hyperglycemia results from the generation of reactive oxygen species (ROS) production, ultimately leading to increased oxidative stress in a variety of tissues such as nepropathy, retinopathy, neuropathy, and macro-and micro vascular damage (Evans *et al.*, 2002). Therefore, the controlling postprandial hyperglycemia is critical in the early treatment of diabetes mellitus and in reducing chronic vascular complications (Shihabudeen *et al.*, 2011). One of the therapeutic approaches used to decrease postprandial hyperglycemia is the retardation of glucose absorption by inhibiting carbohydrate hydrolyzing enzymes, such as α-glucosidase, in the digestive organs and consequently blunting the postprandial plasma glucose rise (Kim *et al.*, 2005). Furthermore, the combination of α-glucosidase inhibitors and antioxidants was recently shown to be effective on treating diabetes mellitus and preventing its development (Shibano *et al.*, 2008; Takahashi and Miyazawa, 2012).

Glycosidases are well known targets in the design and development of anti-diabetic, antiviral, antibacterial, and anticancer agents (Du *et al.*, 2006; Kim *et al.*, 2008). α-Glucosidase inhibitors competitively bind to the carbohydrate-binding region of α-glucosidase enzymes, and thereby compete with oligosaccharides to prevent their cleavage to absorbable monosaccharides (Baron AD, 1998; Cheng and Josse, 2004). In type II diabetes, delaying glucose absorption after meal by inhibtion of α-glucosidase is known to be beneficial in therapy.

There are reports of α-glucosidase inhibitors such as acarbose and voglibose from microorganism and nojirimycin and 1-deoxynojirimycin from plants (Kim *et al.*, 2005). Acarbose, a pseudo tetrasacharide isolated from the fermentation broth of *Actinoplanes* spp. SE-50, has been utilized as medicine for treatment of type 2 diabetes mellitus (Zhu *et al.*, 2008). Continuous used of acarbose may cause side effects such as diarrhea, abdominal gaseous, liver toxicity and adverse gastrointestinal symptoms, thereby raising the symptoms and risk factor of heart disease (Kim *et al.*, 2008), therefore the search of new sources of α-glucosidase inhibitors from natural sources still required. There are many reports on screening of α-glucosidase inhibitors from plant source; however microorganism source might be more promising for scale up production of the desired compounds. Interest in the isolation of α-glucosidase inhibitors from certain microorganisms has increased due to fast growing characteristic of microorganisms. For example, validamycin A was isolated from *Streptomyces hygroscopicus* var. *limoneus*, broth of *Bacillus subtilis* B2 also possessed strong α-glucosidase activity (Zhu *et al.*, 2008), the new N-containing maltooligosaccharide GIB-638 was isolated from a culture filtrate of *Streptomyces fradiae* PWH638 (Meng and Zhou, 2012) and Aspergillusol A was isolated from marine-derived fungus *Aspergillus aculeatus* (Ingavat *et at.*, 2009). However, there have been relative few studies on α-glucosidase inhibitors and antioxidants from species of *Aspergillus* sp.

The genus *Aspergillus* represents a diverse group of fungi that are among the most abundant fungi in the world (Krijgsheld *et al.*, 2013). Aspergillus spp, have been widely applied as a starter in the koji industries and are ussually inoculated into the solid-culture of steamed rice, barley soybean, and the mixture of wheat flour and soyflour. In the traditional fermented food *Aspergillus* spp. Possess the functional of saccharification and can be used to obtain special colour and flavours

(Yen *et al.*, 2003). In our previous study, etyl acetate extract of rice koji of *A. terreus* RCC1 (KEE) was found to exhibit significant yeast α-glucosidase and suppresed postprandial hyperglicemia in mice (Dewi *et al.*, 2007). The active compound of KEE was isolated and identified as sulochrin (Dewi *et al.*, 2009 prosiding). However, α-glucosidase and antioxidant compounds from the submerged culture fermented of *A. terreus* have not been investigated. The objected of this study was to isolate and identify the active compounds in the ethyl acetate extract of three strains of *A. terreus* which isolated from Indonesian soil (*A. terreus* LS01, MC751, and LS07), and *A. terreus* RCC1 of a mutant developed from ATCC 20542.

2. MATERIALS AND METHODS

Fungal Material

The fungi *A. terreus* cultures used in this study were:

☆ The isolates of *A. terreus* LS01 and LS07 were isolated from the Teluk Kodek, Pemenang area, West Nusa Tenggara Province, Indonesia in April 2009. It was identified as *Aspergillus terreus* Thom., according to its morphological characteristics and 28 SrDNA sequence.

☆ The isolates of *A. terreus* MC751 was isolated from a leaf litter in Lawu Mountain, Central Java province, Indonesia, in July 2006. This fungus was identified as *A. terreus* on the basis of the sequence data of ITS rDNA.

☆ The isolates of *A. terreus* RCC1, a mutant developed from ATCC 20542, was obtained from research Center for Chemistry, Indonesian Institute of Sciences, Indonesia.

All the voucher specimens have been deposited in the culture collection of Microbial Collection (LIPIMC) Microbiology Division Research Center for Biology, Indonesian Institute of Sciences.

Fermentation, Extraction, and Isolation

A. terreus LS01

The liquid culture experiments were conducted in a 250 mL Erlenmeyer -flask containing 50 mL of PMP medium (2.4 per cent potato dextrose broth, 1 per cent malt extract, and 0.1 per cent peptone). The flasks were incubated at 25°C with shaking at 100 rpm for seven days. EtOAc (3x1 L) was added into the culture broth (3 L) and extracted for 20 min by vigorously shaking. The EtOAc extract (800 mg) obtained from the liquid fermentation of *A. terrus* LS01 was subjected to column chromatography on Silica Gel 60 using a stepwise gradient from *n*-hexane: EtOAc yielded compound **1** (terreic acid) and **2** (teremutin) (Dewi *et al.*, 2012).

Compound 1. (1S,6R)-4-hydroxy-3-methyl-7-oxabicyclo[4.1.0]hept-3-ene-2,5-dione (*terreic acid*). Pale yellow needles, m.p. 126-127°C. UV spectra (MeOH) λ_{max} (log å) 213 (4.03) and 314 (3.88). $[\alpha]^{28.6}_{D}$ -34 (c, 0.046 in MeOH). HRFABMS: $[M+H]^+$ m/z 155.0358, calcd: 155.0342 for $C_7H_7O_4$. Data ^1H (500 MHz; $CDCl_3$, δ-values, *J* in Hz): 1.93 (s,3H); 3.86 (d,1H, J=3.9); 3.89 (d,1H, J=3.35); 6.87 (s,1H) and ^{13}C NMR (100 MHz; $CDCl_3$ δ-values): 8.85, 51.63, 53.84, 120.45, 151.92, 187.55, 190.79.

Compound 2. (1S, 2S, 6S)-2,5-dihydroxy-4-methyl-7-oxabicyclo[4.1.0]hept-4-en-3-one [(±)-*terremutin*]. Colorless solid, m.p 164-166°C. UV spectra (MeOH) λ_{max} (log å) 272 (4.05). $[\alpha]^{28.6}_{D}$ -283 (c, 0.16 in MeOH). HRFABMS: [M+H]⁺ m/z 157.0498, calcd: 157.0498 for $C_7H_9O_4$. Data ¹H (500 MHz; Acetone-d_6 δ-values, J in Hz): 1.65 (s,3H); 3.34 (d,1H, J=2.55); 3.64 (dd,1H, J=1.3); 4.68 and ¹³C NMR (100 MHz; Acetone-d_6, δ-values) 7.50, 52.35, 55.28, 66.17, 108.87, 168, 191.

A. *terreus* MC751

Five discs (5mm) of fungal mycelia were used to inoculate 150 mL of seed medium in an Erlenmeyer flask (500 mL) in CzY medium (Czpek-dox medium with addition of 0.5 per cent yeast extract). Each culture was maintained at 25°C under static conditions for 15 days. Fifteen-liter cultures were separated from mycelia by filtration. The filtrate was extracted with EtOAc (5x1 L), and the extract was dried under reduced pressure to obtain a brown solid (2.18 g). The filtrate extract was fractioned by column chromatography on silica gel using a stepwise gradient from 100 per cent n-hexane, to 100 per cent EtOAc, to 50 per cent EtOAc in MeOH to obtain compound **3** (butyrolactone I, 500 mg) and compound **4** (butyrolactone II, 60 mg) (Dewi *et al.*, 2014a).

Compound **3**, Butyrolactone I, as a yellowish gum. UV spectra (MeOH) λ_{max} 307 (log ε 4.3). $[\alpha]^{22}_{D}$ .5 +68.333 (c, 0.3 in MeOH) HRFABMS: [M+H]⁺ m/z 425.1607, calcd: 425.1601 for $C_{24}H_{25}O_7$], indicated 13 degrees of unsaturation. ¹H (Acetone-d6) δ 1.56 (3H, s), 1.64 (3H, s), 3.10 (2H, d, J=6.8), 3.43 (2H, d, J=14.8), 3.76 (3H, s), 5.51 (1H, br, J=7.3), 6.49 (1H, dd, J=8.0), 6.52 (1H, d, J=8.0), 6.53 (1H, d, J=3.4), 6.95 (2H, d, J=9.0), 7.61 (2H, d, J=9.0). ¹³C NMR (Acetone-d6) δ 171.0 (C-5), 168.7 (C-1), 158.9 (C-4'), 154.8 (C-4''), 139.1 (C-2), 132.5 (C-9''), 132.4 (C-2''), 130.2 (C-6' and C-3'), 129.6 (C-6''), 122.9 (C-1'), 128.3 (C-3''), 124.9 (C-1''), 123.4 (C-8''), 128.0 (C-3), 116.7 (C-5' and C-2'), 115.1 (C-5''), 86.0 (C-4), 53.8 (OCH₃), 39.3 (C-6), 28.6 (C-7''), 26.0 (C-10''), 17.8 (C-11'').

Compound **4**, Butyrolactone II. Colorless gum, UV (MeOH) λ_{max} (λ) nm (log ε): 307.5 (4.17). $[\alpha]^{28}_{D}$ .7 +4.78 (c=0.45, Acetone) FABMS: [M+H]⁺ m/z 357 for $C_{19}H_{17}O_7$. ¹H NMR (Acetone-d6) δ 3.40 (2H, d, J=14.8, H-6), 3.76 (3H, s, 5-OCH₃), 6.59 (2H, d, J=8.0, H-3'',5''), 6.69 (2H, d, J=8.0, H-2',6'), 6.98 (2H, d, J=8, H-3',5'), 7.66 (2H, d, J=8.0, H-2'',6''). ¹³C NMR (Acetone-d6) δ 170.9 (C-5), 168.8 (C-1), 158.9 (C-4'), 157.4 (C-4''), 139.3 (C-2), 132.3 (C-2'and C-6') 130.1 (C-2'' and C-6''), 128.0 (C-1''), 124.9 (C-1'), 122.8 (C-3), 115.5 (C-3' and C-5'), 116.7 (C-3'' and C-5''), 85.97 (C-4), 53.8 (OCH₃), 39.2 (C-6).

A. *terreus* LS07

Two discs (8mm) of fungal mycelia were used to inoculate 150 mL of CzY and incubated at 25°C on shaking condition (60 rpm) for seven days. After harvesting, fermentation broth (10 l) was extracted with EtOAc (10x2 L), followed by concentration in vacuo to afford 2.7 g, as oily brown gummy. The EtOAc (2.7 g) was fractionated on silica gel column and eluted by n-hexane: CHCl₃ gradient to give compound **5** (linoleic acid, 30 mg), compound **6** (ergosterol, 20 mg), compound **3** (700 mg) and **4** (100 mg) (Dewi *et al.*, 2014b).

Compound **5**, oleic acid: colorless oil, ^{1}H-NMR (CDCl$_3$ 500 MHz) δ: 5.33 (2H, m, H-9,10), 2.34 (2H, t, H-2), 2.01 (4H, m, H-8, 11), 1.62 (2H, m, H-3), 1,25-1.30 (22H, m, H-4-8, 11-17), 0.88 (3H, t, H-18), MS m/z (per cent)= 282 ([M], 264 ([M-H$_2$O]), 97 (64), 83 72), 69 (80), 55 (84), 41 (13.5); deduced for C$_{18}$H$_{34}$O$_2$.

Compound **6**, ergosterol: colorless crystal, mp 166-168°C, EIMS m/z (per cent) = 396 ([M], 87), 378 ([M-H$_2$O]$^+$, 12), 363 ([M-(H$_2$O+CH$_3$)]$^+$, 100), 271 (25), 253 (52), 211(33).^{1}H-NMR (CDCl$_3$, 500 MHz) δ:5.57 (dm, 1H, H-6), 5.38 (dm, 1H, H-7), 5.17 (m, 2H, H-22,23), 3.62 (m, 1H, H-3), 2.46 (dm, 1H, H-5), 2.35 (m, 2H, H-20, 24), 2.09-1.93 (m, 3H), 1.92-1.89 (m, 4H), 1.88-1.55 (m, 4H), 1.50-1.40 (m, 3H), 1.38-1.16 (m, 3H), 1.02 (d, J = 7.2, 3H, CH$_3$-21), 0.93 (s, 3H, CH$_3$-19), 0.91(d, J = 7.2, 3H, CH$_3$-28), 0.82 (d, J = 6.8, 3H, CH$_3$-27), 0.80 (d, J = 6.8, 3H, CH$_3$-26), 0.61 (s, 3H, CH$_3$-18).

A. *terreus* RCC1

A. terreus was maintained at MY agar for seven days at 25°C. The agar was cut into small plugs and inoculated into 500 ml Erlenmeyer flasks containing 150 ml of PMP broth (potato 2.5 per cent, malt extract 1 per cent, peptone 0.1 per cent, and glucose 2 per cent). The fungus was incubated at shaking condition (60 rpm) at 25°C. Fifteen days old fermentation broth (5 l) was extracted with ethyl acetate (5 x 1 L) and concentrated under reduced pressure to give crude extract (8 g). The gum fraction (6 g) was subjected to column chromatography on silica gel (*n*-hexane-EtOAc, step gradient elution from (hexane: ethyl acetate; 0:100 to 100: 0) to obtain compound **3** (980 mg), compound **7** (isoaspulvinone E, 20 mg), compound **8** (aspulvinone E, 15 mg) (Dewi *et al.*, 2015).

Compound **7**, Isoaspulvinone E: Yellow crystalline, mp >250°C, $[\alpha]_D^{25}$. **6** -5.14 (c=0.35, MeOH), UV (MeOH) λ$_{max}$ (λ) nm (log ε): 314.5 (4.63); HRFABMS [M+H]$^+$: m/z 297.0748 (calcd. for C$_{17}$H$_{13}$O$_5$, 297.0759). ^{1}H-NMR (500 MHz, acetone-*d6*) δ: 8.35 (2H, d, J=9.0, H-7a, 7b), 8.23 (2H, d, J=9.0, H-12a, 12b), 6.82 (2H, d, J= 8.4, H-13a, 13b), 6.68 (2H, d, J=8.4, H-8a, 8b), 6.02 (1H, s); ^{13}C-NMR (125 MHz, acetone-*d6*) δ: 178.8 (Cq-4), 173.9 (Cq-2), 157.0 (Cq-14), 153. 5 (Cq-9), 149.2 (Cq-5), 132.8 (CH-12a, 12b), 128.4 (Cq-6, 11), 126.2 (CH-7a, 7b), 116.1 (CH-13a, 13b), 114.8 (CH-8a, 8b), 109.1 (Cq-3), 99.2 (CH-10).

Compound **8**, Aspulvinone E: yellow powder solid, mp >250°C, +4.00 (c=0.2, MeOH), UV (MeOH) λ$_{max}$ (λ) nm (log ε): 324.5 (4.61); HRFABMS[M+H]$^+$: m/z 297.0762 (calcd. for C$_{17}$H$_{13}$O$_5$, 297.0759).^{1}H-NMR (500 MHz, acetone-*d6*) δ: 8.16 (2H, d, J=9.0, H-7a, 7b), 8.09 (1H, d, J=9.0, H-12a), 7.59 (1H, d, J=9.0, H-12b), 6.82 (2H, d, J= 8.4, H-8a, 8b), 6.75 (1H, d, J=9.0, H-13b), 6.71 (2H, d, J=8.4, H-8a, 8b), 6.18 (1H, s); ^{13}C-NMR (125 MHz, acetone-*d6*) δ: 171.5 (Cq-2), 156.6 (Cq-4), 155.6 (Cq-9), 158.1 (Cq-14), 149.3 (Cq-5), 132.5 (CH-12a), 132.9 (CH-12b), 128.4 (CH-7a, 7b), 127.1 (Cq-6, 11), 116.3 (CH-8a, 8b), 115.5 (CH-13a), 115.4(CH-13b), 110.8 (CH10), 104.9 (Cq-3).

Preparation of Butyrolactone I Derivatives

Derivatization of butyrolactone I (**3**) was conducted by cyclization and acetylation. Cyclization of **3** was carried out according to the method of Parvapkar *et al.* (2009). Compound **3** (85 mg) was dissolved in MeOH (10 mL) containing conc. HCl (0.2 mL). The mixture was stirred at room temperature for 2 h until complete conversion of compound **3** as indicated by TLC. The solvent was removed under vacuum and the resulting residue was extracted with $CHCl_3$. The $CHCl_3$ soluble fraction was separated by preparative TLC elution with *n*-hexane: EtOAC (2:3), to yield compound **3a** (60 mg, 70 per cent). The butyrolactone I acetate was obtained treating 212 mg of compound **3** with acetic anhydride in pyridine. After the usual work-up, compounds **3b** (90 mg, 32 per cent) and **3c** (60 mg, 42 per cent) were obtained.

Compound **3a**: Yellowish solid crystal, mp: 93-95°C, UV (MeOH) λ_{max} (λ) nm (log ε): 307.5 (4.92); 226.5 (4.85); $[\alpha]_D^{28} \cdot 7$ +88.73 (c=0.58, $CHCl_3$) FABMS: [M+H]+ m/z 425 for $C_{24}H_{25}O_7$. ^{1}H NMR (Acetone-*d6*) δ 1.16 (6H, s, H10″,11″), 1.63 (2H, br, J=6.5, H8″), 2.50 (2H, m, H7″), 3.36 (2H, J=15, H6), 3.68 (3H, s, 5-OMe), 6.38 (1H, s, H2″), 6.50 (1H, s, H5″), 6.56 (1H, d, J=3.4, H6″), 6.86 (2H, d, J=9.0, H3′, 5′), 7.65 (2H, d, J=9.0, H2′, 6′). ^{13}C NMR (Acetone-*d6*) δ 173.5 (C-5), 172.5 (C-1), 157.9 (C-4′), 154.3(C-4″), 144.9 (C-2), 75.0 (C-9″), 126.4 (C-2″), 129.6 (C-6′ and C-2′), 125.5 (C-6″), 133.0 (C-1′), 117.5 (C-3″), 130.6 (C-1″), 33.8 (C-8″), 121.2 (C-3), 116.6 (C-5′ and C-3′), 123.7 (C-5″), 86.8 (C-4), 53.8 (OCH_3), 40.0 (C-6), 23.3 (C-7″), 27.4 (C-10″), 27.6 (C-11″).

Compound **3b**: Colourless gum: UV ($CHCl_3$) λ_{max} (λ) nm (log ε): 281 (3.74). $[\alpha]_D^{20}$ +7.154 (c=1.3, $CHCl_3$).FABMS: [M+H]+ m/z 551 for $C_{30}H_{31}O_{10}$.

Compound **3c**: Colourless gum, UV ($CHCl_3$) λ_{max} (λ) nm (log ε):294 (3.84). $[\alpha]_D^{20}$ +39.957 (c=1.15, $CHCl_3$). FABMS: [M+H]+ m/z 509, for $C_{28}H_{29}O_9$.

Chemical Analysis

The structure of compounds isolated by column chromatography of the extract were identified by different spectroscopic techniques ^{1}H, ^{13}C NMR and DEPT spectra were recorded on a JEOL 500 with TMS as an internal standard. The chemical shift values (δ) are given in parts per million (ppm), and coupling constant (J) in Hz. The mass spectra of the compounds were obtained using high resolution FAB-MS.Melting points were recorded in a Micro melting point apparatus and are uncorrected. Optical rotation values were measured with a Jasco P-2100 polarimeter. The UV-Vis absorption spectra of the active compounds in MeOH were recorded on a Hitachi U-1600 spectrophotometer.

Biological Activity

α-Glucosidase Inhibition Assay

The inhibitory activity for α-glucosidase was assessed as reported by Kim (2005) with minor modifications. The reaction mixture contained 250 µL of 3mM *p*-NPG and 495 µL of 100mM phosphate buffer (pH 7.0) added to a tube containing 5 µL of sample dissolved in DMSO at various concentrations (5 to 100 µg/mL). The reaction mixture was pre-incubated for 5 min at 37°C, the reaction was started by adding 250 µl of α-glucosidase (0.065units/mL) and the incubation was continued for 15 min. The reaction was stopped by adding 1mL of 0.2 M Na_2CO_3. The inhibitory effect on α-glucosidase activity was determined by measuring the amount of *p*-nitrophenol released at λ 400 nm. Individual blanks for test samples were prepared to correct background absorbance where the enzyme was replaced with 250 µL of phosphate buffer. All the tests were run in triplicate.

The percent inhibition of α-glucosidase and α-amylase was assessed using the following formula: per cent Inhibition = [1- (As/A0)] x 100, where A0 was the absorbance of the control reaction and As was the absorbance in the presence of the sample. The IC_{50} values were calculated from the mean inhibitory values by applying logarithmic a regression analysis.

Determination of the Inhibition Pattern on α-glucosidase

To evaluate the type of inhibition against yeast α-glucosidase, increasingly higher concentrations of *p*-NPG were used as a substrate in the absence and presence of butyrolactone I. The inhibition type was determined from Lineweaver-Burk plots.

DPPH Free Radical-Scavenging Assay

The free radical-scavenging activities of samples were measured by using DPPH according to Yen and Chen, 1995. Each sample in MeOH (4 mL) with a concentration of 10-200 µg/mL was mixed with 1 mL of 1 mM DPPH solution in MeOH. The mixture was shaken vigorously and left to stand for 30 min in the dark, and then the absorbance was measured at λ 517 nm against a blank. The ability to scavenge the DPPH radical was calculated using the following equation: DPPH-scavenging effect (per cent) = [1-(As/A0)×100], where A0 was the absorbance of the control reaction and As was the absorbance in the presence of the sample. The percentage of scavenging activity was subsequently plotted against the sample concentration. The half maximal inhibitory concentration (IC_{50}) was calculated from the graph of percent antioxidative activity against sample concentration. The assays were carried out in triplicate and the results expressed as mean values±standard deviations. Quercetin was used as a reference compound.

3. RESULTS AND DISCUSSION

Isolation of Active Compounds

Bioassay-guided fractionation of EtOAc extracts of four strains of the terrestrials *Apergillus terreus* through chromatographic methods give to ten isolated compounds including three derivatives of butyrolactone I (Figure 2.1). Sstructures of the isolated compounds were deduced by intensive studies of their 1D and 2D NMR, MS data and comparison with related structure.

α-Glucosidase Inhibitory Activity of Isolated Compounds

In the present study, all isolated compounds were tested for α-glucosidase inhibitory activity. Quercetin was used as a positive control based on the reports

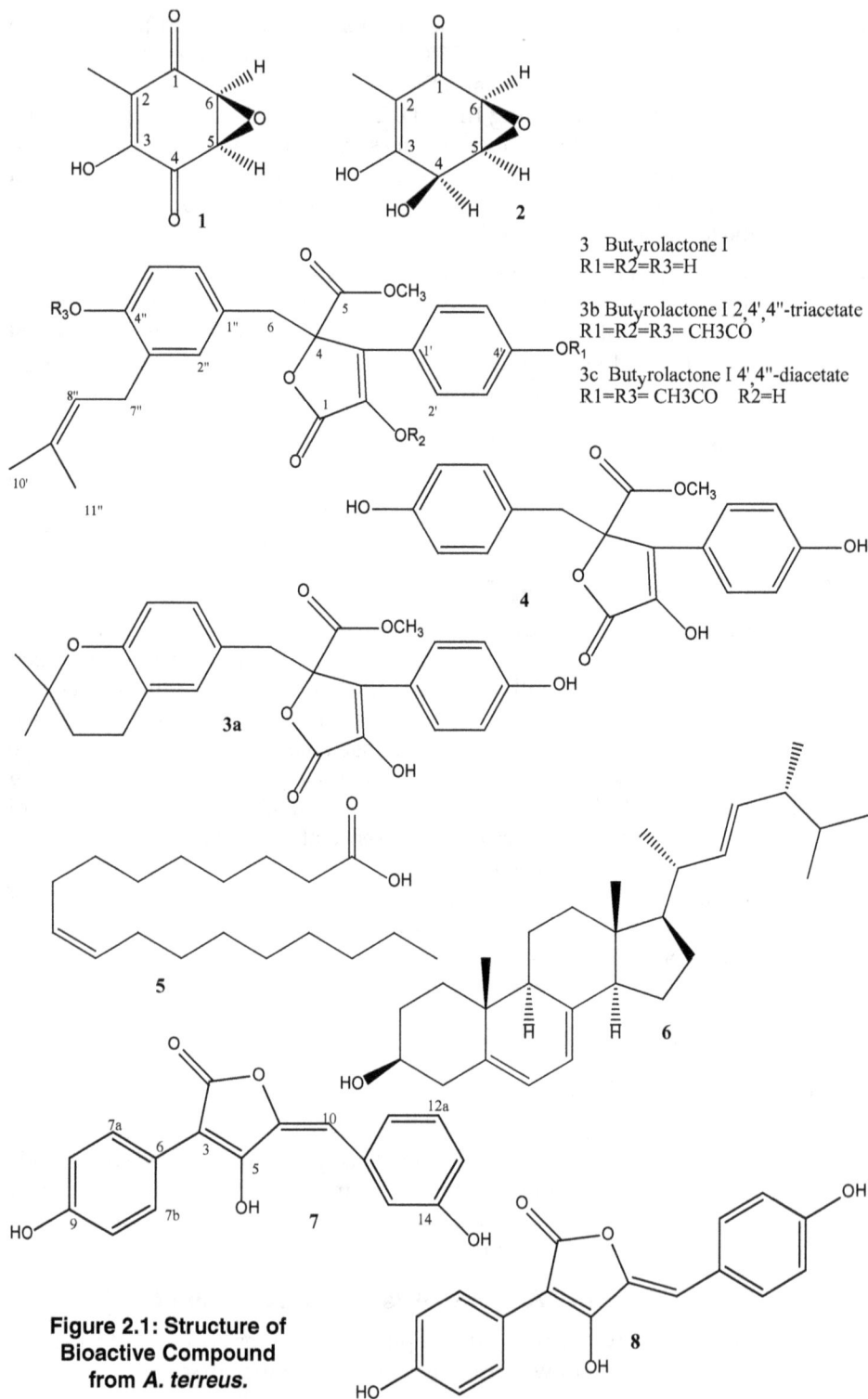

3 Butyrolactone I
R1=R2=R3=H

3b Butyrolactone I 2,4',4"-triacetate
R1=R2=R3= CH3CO

3c Butyrolactone I 4',4"-diacetate
R1=R3= CH3CO   R2=H

Figure 2.1: Structure of Bioactive Compound from *A. terreus*.

that it is a phenolic compound with a stronger inhibitory effect on α-glucosidase from *S. cereviceae* than acarbose (Tadera *et al.*, 2006).

The α-glucosidase inhibitory activities of isolated compounds of *A. terreus* are tabulated in Table 2.1. The result showed that compound **3** was a potent inhibitor of the α-glucosidase with an IC_{50} of 52.17µM. In contrast, compound **4**, which lack a prenyl side chain, exhibited less inhibitory activity. Converting the prenyl side chain to a dihydropyran ring in compound **3a** caused a significant decrease in the inhibitory activity. Hence, it was assumed that the prenyl side chain of compound **3** contributed to the inhibitory effect. However, the substitution of any hydroxyl group with an acetyl group in butyrolactone I led to a dramatic reduction in inhibitory activity in compound **3b**, a finding consistent with Gao *et al.*, 2004 which reported that the removal of hydroxyls in flavonoids decreased α-glucosidase inhibitory activity.

**Table 2.1: α-Glucasidase Inhibitory and Antioxidant Activity of
Active Compounds from *A. terreus* MC751**

Compound	IC_{50} (µM)[a]	
	Inhibition of α-glucosidase	Antioxidant
1	n.d	115.0±4.04
2	n.d	114.0±2.19
3	52.17±5.68	51.39±3.68
3a	175.18±5.95	47.55±3.08
3b	>300	n.d
3c	84.18±8.98	n.d
4	96.01±3.70	17.64±6.41
5	8.54±0.61	n.d
6	n.d	378.79±3.42
7	8.92±3.51	167.82±5.29
8	2.70±2.73	114.86±3.27

Quercetin 14.6±3.72 39.63±5.21

n.d not detected

[a]IC_{50} value are shown as mean±S.D. from three independent experiment

Compound **3c**, which retained one OH-bond as an alphahydroxy-lactone, showed significantly higher activity against α-glucosidase than compound **3b**. Based on the α-glucosidase inhibitory activities of compounds **3, 4**, SAR inference could be made. Compounds **3** and **3c** showed stronger activity than the others, wich suggested that the inhibitory effect of these butyrolactones was influenced by both the prenyl side chain and alpha hydroxy-lactone group. The influence of the prenyl side chain in butyrolactones was reported previously (Parvatkar *et al.*, 2009; Cazar *et al.*, 2005), however no SAR study of the α-glucosidase inhibitory activity of butyrolactone I derivatives has been reported before.

Oleic acid (**5**) showed excellent inhibition on yeast α-glucosidase with IC_{50} value of 8.54 μM which is lower than that of quercetin (IC_{50} value of 14.6 μM), however no activity on DPPH radicals.To our best knowledge, this is the first report on the inhibitory activity against α-glucosidase of oleic acid which isolated from *A. terreus*. However, other unsaturated fatty acid were reported have potential activity toward α-glucosidase such as 7(Z)-octadecanoic acid and 7(Z),10(Z)-octadecanoic acid with IC50 were 1.81 and 2.86 μM, respectively. These compounds were purified from the body wall of *Stichopus japonicas* (Nguyen *et al.*, 2011) and 10-hydroxy-8(E)-octadecanoic acid, an intermediate of bioconversion of oleic acid (Hsia *et al.*, 1984).

In addition, we evaluated the inhibitory activities of saturated and unsaturated fatty acid to clarify whether inhibitory activity of compound **5** due to double bond in fatty acid or not. The inhibitory activity of oleic acid (**5**), linoleic acid, and linolenic acid exhibited high inhibition at 10 μg, with 91.92±0.85 per cent, 82.84±1.51 per cent, and 76.26±2.41 per cent, respectively. On the other hand, stearic acid (C18:0) exhibited poor inhibitory activity (8±2.01 per cent at 10μg/mL) (Figure 2.2).

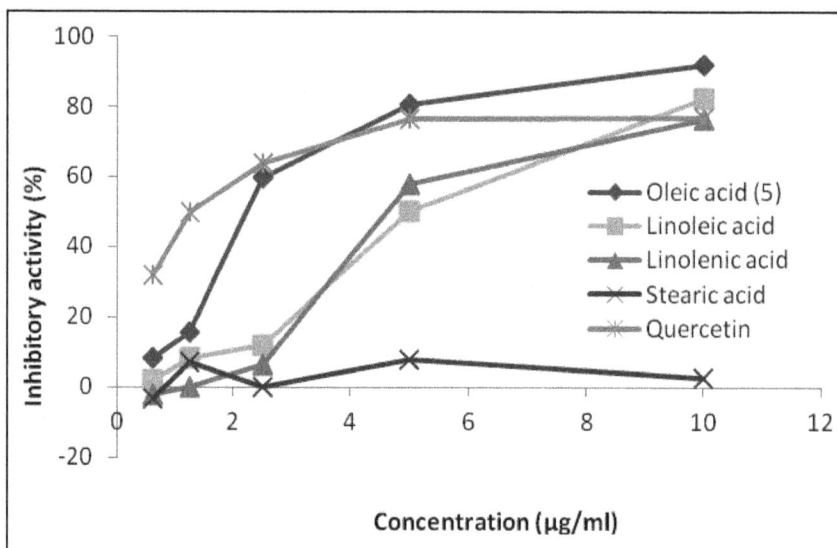

Figure 2.2: α-Glucosidase Inhibitory Activity of Unsaturated and Saturated Fatty Acid.

The inhibitory activity against α-glucosidase of the unsaturated fatty acid were ranked as follows; oleic acid > linoleic acid ≥ linolenic acid, while stearic acid, saturated fatty acid, did not show significant activity. Therefore, we consider that a double bond in fatty acid is crucial for the activity; however increasing of the double bond number will decrease the inhibitory activity on α-glucosidase. The investigation of α-glucosidase activity of stearic acid compared to unsaturated fatty acid has not been reported.

This result accorded with previous studies that the presence ofthe double bond in the fatty acid affects the inhibitory potency (Nguyen *et al.*, 2011). The activity of unsaturated fatty acid on α-glucosidase was assumed that the binding of fatty acid may affect the secondary and tertiary structure of proteins because of their detergent effects alone, which was suggested for the effect of fatty alcohol sulfate and palmitic acid binding to bovine serum albumin (Hsia *et al.*, 1984). The resultant alteration in the conformation of the protein molecule may effects its biological activity, which may be one of early effects involved in the inhibition of enzymes by fatty acids (Nguyen *et al.*, 2011). Significant effect of the double bond in the unsaturated fatty acid on α-glucosidase inhibitor, supports our previous results that the prenyl group is important for the inhibitory activity of butyrolactone I (3) compared with butyrolactone II (4) (Dewi *et al.*, 2014). Moreover, Takahashi and Miyazawa (2012) also suggested that the olefin in serotonin derivatives is crucial for the inhibition of α-glucosidase.

In particular compound 7 and 8 showed the highest inhibitory activity on α-glucosidase. The IC_{50} of compound 7 and 8 against *S. cereviceae* α-glucosidase value were 2.7 and 0.79 µg/ml, respectively, which were lower than quercetin as reference compound (IC_{50} 3.3 µg/ml) and butyrolactone I (3) with an IC_{50} 20 µg/ml. Up to our knowledge, this is the first report of inhibitory activity on α-glucosidase of aspulvinone derivatives. Since compound 7 and 8 have identical structure, the activities both of them slightly different, we assume the differences due to stereochemistry an affect to inhibition activity or recognition of active site in α-glucosidase. However this assumes need further study such as, molecular docking approach.

Determination of the Inhibition Pattern on α-glucosidase

The inhibitory mechanisms of all the active compounds were analyzed using Lineweaver-Burk plots. α-Glucosidase solution (0.065 U/mL) was incubated with increasingly higher concentrations of substrate (pNPG) with and without an inhibitor. The results showed various mechanisms of action (Table 2.2).

Tabel 2.2: Inhibition Constant (*Ki* value) and Mode of Active Compounds from *A. terreus* against *S. cereviseae* α-glucosidase

Compound	Inhibition Mode	Ki (µM)
3	Mixed inhibition	70.51
3a	Uncompetitive	235.80
3c	Uncompetitive	127.88
4	Noncompetitive	152.64
5	Mixed inhibition	1.99
7	Noncompetitive	37.16
8	Noncompetitive	10.11
Quercetin	Mixed inhibition	27.13

Antioxidantt Activity of Isolated Compounds

Antioxidants are substance that can prevent, stop or reduce oxidative damage; therefore they able to protect the human body from several diseases, such diabetes and related complication of this disease (Sancheti *et al.*, 2011). An ideal anti-diabetes compound should showed activities of α-glucosidase inhibitors and antioxidants properties (Shibano *et al.*, 2008). Therefore, in the present study the antioxidant activities of isolated compounds from *A. terreus* were studied by DPPH free radical scavenger effect (*See* Table 2.1).

Terreic acid (1) is a quinone epoxide whereas terremutin (2) is a dihydroquinone epoxide. Both are produced by several members of the genus *A. terreus* and have been reported as antibiotic compounds (Cazar *et al.*, 2005). However, the antioxidative properties of these compounds are reported here for the first time. The compounds were able to scavenge DPPH free radicals in a concentration-dependent manner, with an IC_{50} of 0.115±4.02 and 0.114±2.19 mM, respectively. However, the activity of quercetin with IC_{50} of 0.033±1.02 mM was stronger than these compounds.

In this study, terreic acid (1) and terremutin (2) showed potential scavenging activity toward free radicals. The antioxidative activity of these compounds is due to a quinine group which serves as an effective single electron acceptor, acts as a free radical chain-breaker by reacting with the alkyl radical (Liebler and Burr, 2000).This mechanism is similar to that of α-tocopherol hydroquinone (Flora, 2009).

The scavenging activity of butyrolactone I (3) was similar to that of quercetin: the IC_{50} was 21.68 and 11.97 µg/mL respectively. The scavenging mechanism of butyrolactone I was assumed to be the donation of two hydrogen radicals to two molecules of DPPH which produce two molecules of hydrazine DPPH, followed by phenolic conversion into quinine methide, as occurred in Aspernolid A (Sugiyama *et al.*, 2010). Compounds 3, 4, and 3a showed potential scavenging activity while compounds 3b and 3c did not (Table 2.1). The antioxidant activity may originate from the phenolic groups. Compound 4, which lack a prenyl side chain, was the most powerful antioxidant. This result was consistent with the previous reports (Boiko *et al.*, 2006; Osorio *et al.*, 2012) that a prenylated phenolic group in compound 3 decreased antioxidant activity. Cyclization of the prenyl group in compound 3a also decreased the antioxidant activity. In the case of compounds 3b and 3c, replacement of an OH group with an acetyl group dramatically decreased the antioxidant activity (Boiko *et al.*, 2006; Osorio *et al.*, 2012). Hence, it was concluded that the absence of a prenyl side chain increased the activity and acetylation appeared to be detrimental to the antioxidant activity.

Isoaspulvinone E (7) and asvulpinone E(8) have showed significant activity on DPPH free radical, resulted from phenolic group on those compounds. The scavenging DPPH free radicals activity of isoaspulvinone E (7) and aspulvinone (8) have not been reported.

4. CONCLUSIONS

Based on in-vitro study, we recommended that secondary metabolite from *A. terreus* may have beneficial effect in managing the hyperglycemia effects and antioxidant. However further preclinical and clinical studies should be persuade before its pharmaceutical applications. This investigation may also provide as additional information on anti hyperglycemia properties of the secondary metabolites of *A. terreus* which are not reported earlier.

5. ACKNOWLEDGEMENTS

The author are deeply thankful to Prof. Dr. Sanro Tachibana of Faculty of Agriculture, Ehime University, Prof. Dr. LBS Kardono and Dr. Nina Artanti of Research Center for Chemistry –LIPI for support and valuable discussion.

6. REFERENCES

1. Baron AD (1998). Postprandial hyperglycemia and α-glucosidase inhibitors. *Diabetes Res Clin Pract* 40: S51-S55.

2. Boiko MA, Terakh EI, Prosenco AE (2006). Relationship between the electrochemical and antioxidant activities of alkyl-subtituted phenols. *Kinetics and Catalysis* 47: 677-681.

3. Cazar ME, Hirschman SG, and Astudillo L (2005). Antimicrobial butyrolactone I derivates from the Ecuadorian soil fungus *Aspergillus terreus* Thorn. var. *terreus*. *J Micrb and Biotech*, 21: 1067-1075.

4. Cheng AYY and Josse RG (2004). Intestinal absorption inhibitors for type 2 diabetes mellitus: prevention and treatment. *Drug Discovery Today* 192: 201-206.

5. Dewi RT, Iskandar Y, Hanafi M, Kardono, LBS, Angelina M, Dewijanti DI, Banjarnahor, SDS (2007). Inhibitory effect of koji *Aspergillus terreus* on α-glucosidase activity and postprandial hyperglycemia. *Pak J Biol Sci* 10(18): 3131-3135.

6. Dewi RT, Tachibana S, Kazutaka I, Ilyas M (2012). Isolation of antioxidant compounds from *Aspergillus terreus* LS01. *J Microbial Biochem Technol* 4: 010-014.

7. Dewi RT, Tachibana S, Darmawan A (2014a). Effect on α-glucosidase inhibition and antioxidant activities of butyrolactone derivatives from *Aspergillus terreus* MC751. *Med Chem Res* 23: 454-460.

8. Dewi RT, Tachibana S, Kardono LBS, Puspa DNL, and Ilyas M (2014b). Isolation of active compounds from *Aspergillus terreus* LS07. *Indo J Chem* 14: 304-310.

9. Dewi RT, Tachibana S, Fajriah S, and Hanafi M (2015). α-Glucosidase inhibitor compounds from *Aspergillus terreus* RCC1 and their antioxidant activity. *Med Chem Res* 24 (2): 737-743.

10. Du ZY, Liu RR, Shao WY, Mao XP, Ma L, Gu LQ, Huang ZS, and Chan ASC (2006). α-Glucosidase Inhibitions of Natural Curcuminoid and Curcumin Analogs. *Eur J Med Chem* 41(2): 213-218.

11. Evans JL, Goldfine ID, Maddux BA, and Grodsky GM (2002). Oxidative stress and stress-activated signaling pathways: a unifying hypothesis of type 2 diabetes. *Endoctrine Review* 23(5): 599-622.

12. Flora, SJS (2009). Structural, chemical and biological aspects of antioxidant for strategies against metal and metalloid exposure (Review). *Oxidative Medicine and Cellular Longevity* 2 (4): 191-206.

13. Gao H, Guo W, Wang Q, Zhang L, Zhu M, Zhu T, Gu Q, Wang W, and Li D (2013). Aspulvinone from a mangrove rhizophere soil-derived fungus

Aspergillus terreus Gwq-48 with anti-influenza A viral (H1N1). activity. *Bioorg Med Chem Lett* 23: 1776-1778.

14. Gao H, Nishioka T, Kawabata J, and Kasai T (2004). Structure-activity relationships for α-glucosidase inhibition of baicalein, 5,6,7-tryhdroxyflavone: the effect of A-ring substitution. *Biosci Biotechnol Biochem* 68: 369-357.

15. Hsia JC, Wong LT, Tan CT, Er SS, Kharouba S, Balaskas E, Tingker DO, and Feldhoff RC (1984). Bovine serum albumin: characterization of fatty acid binding site on the-N-terminal peptic fragment using a new spin-label. *Biochemistry* 23: 5930-5932.

16. Ingavat N, Dobereiner J, Wiyakrutta S, Mahidol C, Ruchirawat S, and Kittakoop P (2009). Aspergillusol A, an α-glucosidase inhibitor from the marine-derived fungus *Aspergillus aculeatus*. *J Nat Prod* 72: 2409-2052.

17. Kim YM, Jeong YK, Wang MH, Lee WY, and Rhee HI (2005). Inhibitory effect of pine extract on alpha-glucosidase activity and postprandial hyperglycemia. *Nutrion* 21: 756-761.

18. Kim KY, Nam H, and Kim S.M (2008). Potent α-glucosidase inhibitors purified from the red alga *Grateleupia elliptica*. *Phytochemistry* 69: 2820-2825.

19. Krijgsheld P, Bleichrodt R, Veluw GJ, Wang F, *et al.* (2013). Development in Aspergillus. *Mycology* 74: 1-29.

20. Liebler DC, and Burr JA (2000). Antioxidant reactions of α-tocopheroll-hydroquinone. *Lipids* 35 (9): 1045-1047.

21. Meng P and Zhou X (2012). α-Glucosidase inhibitory effect of a bioactivity guided fraction GIB-638 from *Streptomyces fradiae* PWH638 *Med Chem Res* 21: 4422-4429.

22. Nguyen TH, Um BH, and Kim SM (2011). Two unsaturated fatty acid with potent α-glucosidase inhibitory activity purified from the body wall of sea cucumber (*Stichopus japonicas*). *J Food Science* 76 (Nr 9): H208-H214.

23. Parvatkar RR, D'Souza C, Tripathi A, and Naik CG (2009). Aspernolides A and B, butenolides from a marine-derived fungus *Aspergillus terreus*. *Phytochemistry* 70: 128-132.

24. Osorio M, Aravena J, Vergara A, Tabirga L, Baeza E, Catalán K, Gonzáles C, Carvajal M, Carrasco H, and Espinoza L (2012). Synthesis and DPPH radical scavenging activity of prenylated phenol derivatives. *Molecules* 17: 556-570.

25. Ramachandran A, Snehalatha C, Ma RC (2013). Diabetes in South East Asia: an update for the IDF Diabetes atlas. *Diabetes Res Clin Prat*. DOI: 10.1016/j.diabres.2013.11.011.

26. Sancheti S, Sancheti S, Bafna M, and Seo SY (2011). 2,4,6-Trihydroxybenzaldehyde as a potent anti-diabetic agent alleviates postprandial hyperglycemia in normal and diabetic rats. *Med Chem Res* 20: 1181-1187.

27. Shibano M, Kakutani K, Taniguchi M, Yasuda M, and Baba K (2008). Antioxidant constituents in the dayflower (*Commelina communis* L.). and their α-glucosidase inhibitory activity. *J Nat Med* 62: 349-353.

28. Shihabudeen MS, Priscilla H, Thrumurugan K (2011). Cinnamon extract inhibits α-glucosidase activity and dampens postprandial glucose excursian in diabetic rats. *Nutr Metab* 8(46): p.1-p.11.

29. Sugiyama Y, Yoshida K, and Hirota A (2010). Soybean lipoxigenase inhibitory and DPPH radical-scavenging activities of aspernolide A and Butyrolactone I and II. Biosci *Biotechnol Biochem* 74 (4): 881-883.

30. Takahashi T and Miyazawa M (2012a). Synthesis and structure-activity relationships of serotonin derivatives effect on α-glucosidase inhibition. *Med Chem Res* 21: 1762-1770.

31. Tadera K, Minami Y, Takamatsu K, and Matsuoka T (2006). Inhibitor of α-glucosidase and α-amylase by flavonoids. *J Nutr Sci Vitaminol* 52, 149-153.

32. Yen GC, Chang YC, and Su SW (2003). Antioxidant activity and active compounds of rice koji fermented with *Aspergillus candidus*. *Food Chem* 83: 49-54.

33. Yen GC, and Chen HY (1995). Antioxidant activity of various tea extracts in relation to their antimutagenicity. *J Agric Food Chem* 43: 27-32.

34. Zhu PY, Yin LJ, Cheng YQ, Yamaki K, Mori Y, Su YC, and Li LT (2008). Effect of sources of carbon and nitrogen on production of α-glucosidase inhibitor by a newly isolated strain of *Bacillus subtilis* B2. *Food Chemistry* 109: 737-742.

Chapter 3

The Biochemical Study on the Effects of Therapeutic Combination of Herbal Extracts in Diabetic Atherosclerosis

Eman Hussain Abbas[1], Mohamm Khali Mohamme[2],
Ammar Mula Hmood[3] and Saja Ayad Najim[4]

Ministry of Science and Technology,
Director of Division of Sensors for Early Cancer Disease
Baghdad, Iraq
E-mail: [1]emanalzaidi@yahoo.com, [2]mohammed ali2009@yahoo.com
[3]ammur mula@yahoo.com, [4]emanalzaidi@yahoo.com

ABSTRACT

Diabetes mellitus is a group of metabolic disorders with one common manifestation; hyperglycemia. Chronic hyperglycemia causes damage to the eyes, kidneys. Nerves, heart and blood vessels. The etiology and path physiology leading to the hyperglycemia, however, are markedly different among patients with diabetes mellitus, dictating different prevention strategies, diagnostic screening methods and treatments. The adverse impact of hyperglycemia and the rationale for aggressive treatment have recently been reviewed.

This study was under taken to investigate the biochemical changes in 30 diabetic atherosclerosis patient was orally administered daily the therapeutic combination of herbal extract capsules forming of standardized of powder extracts of whit flash pomegranate *Punica granatum*, extract, green tea *Camellia sinensis* extract, olive leaf *Olea europaea* extract and *Trigonella foenum-graecum* extract. The formula was prepared from number of standardized extracts which were studied and analyzed in order to give the optimums of active ingredient. These components was accounted after measuring the toxicity of extracts to be within safe limits. The biochemical changes results of 30 of women with diabetic and atherosclerosis had taken the capsules twice daily for (30) days were compared with 30 volunteers of healthy women.

The results showed a significant ($P0<.05$) decreasing in total glucose close to the normal value, ($p < 0.05$) Glutathion-S-Transferase (G.S.T)enzyme activity, $P<0.05$ total cholesterol, $P<0.05$ malondialdehyde (MDA) level and a significant $P<0.05$ increased in high density lipoprotein (HDL) level.

Keywords: Green tea, Olive leaf, Ellagic acid, Cholesterol, High density-lipoprotein, MDA, Diabetic.

1. INTRODUCTION

Diabetes mellitus is a chronic metabolic disorder characterized by loss of glucose homeostasis caused by defects in insulin secretion and action, resulting in impaired glucose metabolism (Altan, V. 2003). Diabetes is a leading cause of deaths in the world. According to the World Health Organization, at least (171) million people worldwide suffer from diabetes, or (2.8 per cent) of the population. Its incidence is increasing rapidly, and it is estimated that by the year 2030, this number will almost double (Wild, S.G. *et al.*, 2004).

Secondary metabolite (natural product) are substances of low molecular weight, which were not products of the primary metabolic pathway of the producing organism and it is very important to the plant nowadays it is believed that they have vital function.They may act as messenger molecules under specific circumstances or natural pressures in order to protect the producer organism, they also give plants their pigment and odors (Lahlou *et al.*, 2013). Herbal drugs play a role in the diseases most of them speed up the natural healing process. Numerous medicinal plants and their formulations are used for various disorders in ethno medical practices as well as traditional systems of medicines (Banik, G. 2009).

The search for anti oxidants from natural products is on the increase as against antioxidants of synthetic origin. It has been reported that reactive oxygen species (ROS)readily attack and induce oxidatives damage to various biomolecules including proteins, lipids, lipoproteins and DNA (Heikal, *et al.*, 2013).

Diabetes is a disease that causes excessive increase in the production of free radicals which has a key role in the occurrence of complications of diabetes mellitus (Roglic G., *et al.*, 2008, Pritesh Patel *et al.*, 2012). It was observed that the use of herbal plants that contain multiple types of polyphenolic compounds such as ellagic acid in a pomegranate and Chatchens in green tea important in reducing free radicals that cause a number of diseases, including atherosclerosis resulting from the oxidation of light fat and also the ability to effectively re-insulin (Shaik Sameena, *et al.*, 2010, Polina Smirin, *et al.*, 2010, Pitchai Daisy, *et al.*, 2009).

It is found that the Wight flash pomegranate extract that contains the ellagic acid with high efficiency in the inhibition of oxidative stress metadata and reduce risk of atherosclerosis because it has a salicylates groups similar to the installation of penicillin (Karou,D and Dicko, M.H 2005).

The green tea extract contains a number of important compounds catechin, catechin gallate, Epigall- ocatechin, Epigallocatechin-3-gallate which reducing oxidation of light fat that cause of atheroscl- erosis disease, and also re effectiveness of insulin (Xianghong Chen, etal. 2009). The bioactivity of olive tree by product extracts appears to be attributable to antioxidant and phenolic components such as oleuropein, hydroxytyrosol, oleuropein aglycone, and tyrosol(Goulas V etal.2009). Studies indicate that biologically active compounds in olive leaf products are effective in treating disorders (Jemai H, *et al.,* 2008, (Goulas V, *et al.,* 2009) Several studies have shown that oleuropein (up to 6 per cent–9 per cent of dry matter in the leaves), for example, possesses a wide range of pharmacologic and health-promoting properties. (Briante R, *et al.,* 2002) Specifically, oleuropein has been associated with improved glucose metabolism an antihyperglycemic effect in diabetic rats (Gonzalez M. *et al.,* 1992, Jemai H, *et al.,* 2009) The hypoglycemic and antioxidant effects of oleuropein have been reported in alloxan-diabetic rabbits. (Al-Azzawie HF *et al.,* 2006) In streptozotocin (STZ)-induced diabetic rats, olive leaf extract decreases serum concentrations of glucose, lipids, uric acid, creatinine, and liver enzymes, (Eidi A, *et al.,* 2009) implying that olive leaf extract is more effective than glibenclamide and may be of use as an anti-diabetic agent. Rats fed a high-carbohydrate, high-fat diet with olive leaf extract for 16 weeks expressed improved or normalized cardiovascular, hepatic, and metabolic signs than rats fed an identical diet without olive leaf extract. A treatment benefit on blood pressure was not observed (Poudyal H *et al.,* 2010).

Oral administration of 2 and 8 g/kg of plant extract produced dose dependent decrease in the blood glucose levels in both normal as well as diabetic rats. (Chandraprakash Dwivedi, 2013, Xia E. Q. 2010). Fenugreek seeds increased glucose stimulated insulin release by isolated is let cells in both rats and humans. Administration of fenugreek seeds also improved glucose metabolism decreased serum cholesterol level in diabetic rats. and normalized creatinine kinase activity in heart, skeletal muscle and liver of diabetic rats. It also reduced hepatic and renal glucose-6-phosphatase and fructose –1, 6-biphosphatase activity (Manisha Modak, 2007). In recent years, the present study aimed to examine the influence of oral administration of combination therapy of a number of herbal extracts on the levels of some biochemical parameters that are changed by the high level of glucose in serum blood in 30 volunteers of diabetic atherosclerosis patient.

2. MATERIALS AND METHODS

Preparation of Dry Extracts

All herbal plant, leaves, seeds and Fruits were purchased from the local market. All the material that was used in analytical grade.

The extracts of green tea,Wight flash pomegranate and olive leaf dried powder were treated by heating plant with distilled water-ethanol 50 per cent (5 volume to1 weight) of each plant powder in a vacuum soxhlet device connected to a bottle of nitrogen for a period of 6 hours, the fenugreek seeds powder was treated with diethyl ether (1weight to-5 volume) twice and dried in oven at 50C° then treated with distilled water by the same way.

All the extracts was lyophilized to get dried powder.

Preparation of Pharmaceutical Composition

The formula was prepared depending on the amounts of active substances in extracts to be the content per capsule as follows:

Wight flash pomegranates extract *Punica granatum* 0.1 mg

Green tea extract *Camellia sinensis* 0.4 mg

Olive leaf extract *Olea europaea* 0.25 mg

Fenugreek extract *foenum-graecum* 0.25 mg

Experimental Design

1. control group included 30 healthy women aged between 45-55 years old
2. 30 Diabetic patients women with atherosclerosis, aged between 45-55 years old

   A After treatment

   B Before treatment

Treatments

The patients were given capsules of therapeutic combination of herbal Extracts after their consent, according to the provisions convention of Ministry of Health, under the supervision of the attending physician for the purpose of the follow-up of patients and screening level glucose in the blood daily before breakfast and two hours after dinner to see the minimum and maximum level of glucose in the blood and other changes that appear on some biochemical variables

Biochemical Assays

Blood samples were taken before and after oral administration of capsules of the herbal formula, then 5 ml of blood samples were put in test tubes containing anti-clotting EDTA.

0.5 ml of the blood was taken and singled in a test tube for the purpose of measuring the of Glutahtione –S-Trunsferase by colorimetric method using chlorodinitro benzene at 240nm by (Lee, H. *et al.*, 1981).

Plasma samples were obtained by centrifugation of the remaining blood samples at 860rpm for 20 min and stored at –20 °C till measurements.

Stored plasma samples were analyzed for total cholesterol using enzymatic colorimetric method by Richmond (Richmond, W., 1971).

High-density lipoproteins the were determined on a mixture of phosphotengestic acid in calcium ion by colorimetric method (Burstein M. *et al.*, 1970).

Plasma malondialdehyde activity was measured using thiobarbituric acid (TBA) by the colorimetric method (Habing W.H. *et al.*, 1974).

Statistical Analysis

All the data are expressed as Mean±SD and analyzed statistically using one way ANOVA followed Dunnett's test.

3. RESULTS AND DISCUSSION

The selection ratios in combination was done depending on previous research. The active ingredients were measured by high Performance liquid chromatography HPLC technique. These components account after measuring the toxicity of extracts to be within safe limits were given oral twice daily to a group of 30 patients with diabetes and atherosclerosis and 30 of health volunteers, a blood samples Were taken to measure the number of a variable biochemical which described in Table 3.2.

Table 3.1: The Contents of Extracts and the Percentage Yield of Active Ingredients in One Capsule of the Herbal Formula Determined by HPLC Technique

No.	Percentage in Extract per cent	Active Ingredients	Contains in one Capsule	The Plant Extract
1	0.7	ellagic acid	0.1 mg	Wight flash pomegranates
2	23	Catechin Catechin gallate Epigallocatechin	0.4 mg	Green tea
3	30	Oleuropein, Hydroxytyrosol	0.3 mg	Olive leaf
4	13	Trigonelline	0.2 mg	Fenugreek

Table 3.2: The Effect of Herbal Extracts Therapeutic Combination on the Biochemical Parameters Total Glucose, Glutathion-S-Transferase (G.S.T), Cholesterol Level, High-Density Lipoproteins (HDL) and Malondialdehyde Level (MDA) in (30) Patient Women of Diabetic and Atherosclerosis was Compared with (30) Healthy Women

		1	2	3	4	5
Supject	No	Total Glucose mg/dl	G.S.T unit/gHb	Cholesterol Level mg/dl	HDL mg/dl	MDA µmol/gHb
		Mean±SD	Mean±SD	Mean±SD	Mean±SD	Mean±SD
Healthy women Ages 40-45 years	30	90.2±12.1	0.96±0.04	186±9.4	45.3±1.3	0.44±0.17
Patient women Ages 40-45 years	B30A	177.0±3.51	1.45±0.15	271±5.6	37.6±1.9	0.77±0.19
		190±1.22	1.05±0.03	190±1.22	42.4±3.31	0.56±0.07
P value	B	$P<0.001$	$P<0.001$	$P<0.001$	$P<0.001$	$P<0.001$
	A	$P<0.05$	$P<0.05$	$P<0.05$	$P<0.05$	$P<0.05$

B: The patients before treatment; A: The patients after treatment.

Data show in Table 3.2 the biochemical changes of atherosclerosis patient women who suffer from diabetes before and after taking pharmaceutical composition of herbal extract capsules. Each column shows the impact of pharmaceutical composition on one type of biochemical variables on the patients and the healthy women.

Column 1 shows a significant (P<0.001) increase of rate of glucose in the blood plasma of patients women to (177) mg/dl Compared with healthy controls (90.2) mg/dl this rate is relatively high due to insufficient or non-validity of insulin (Edwin,et al.2008) As a significant (P<0.05) decrease in the percentage of glucose in the blood plasma of diabetes and atherosclerosis patients after taken the therapeutic composition capsules to (98.4) mg/dl close to the normal limit. This is due to the presence of polyphenolic compounds such as ellagic acid, cachene, catechin gallate, Epigallocatechin, Epigallocatechin-3-gallate, that help insulin production (Xianghong Chen, etal. 2009,Mackenzie T 2010).

Column 2 showed that the level of effectiveness (G.S.T) enzyme was increased significantly (p <0.001) in blood plasma in patient women to (1.45) unit/gHb as compared to control values (0.96) unit/gHb (Haidara, M, 2006) with the increasing of levels of free radicals and after treatment with pharmaceutical composition capsules the level of effectiveness (G.S.T) enzyme was significantly (p < 0.05) decreased in plasma with mean value of diabetic group (1.05) unit/gHb by the decreasing of free radicals. by the presence of active ingredients inhibit the formation of free radicals. (Mackenzi T 2010, Chandraprakash Dwivedi *et al.*, 2013).

The therapeutic combination of herbal Extracts capsules showed antioxidant activity which is in accordance with the results of previous study significantly (p < 0.05) decreasing the high cholesterol level in diabetes patents women from (271) Mean mg/dl to (190) Mean mg\dl in comparison with normal control (186) Mean mg/dl which showed that the model was successful in reducing cholesterol level because of the effect of anti oxidant combination on the inhibition of some fat decomposition enzymes (Manisha Modak, *et al.*, 2007, Rader, D.J., 2009 Xianghong Chen, *et al.*, 2009). Column 3.

HDL level of diabetic groups was significantly (p < 0.001) decreased by (37.6) Mean mg/dl.

While, after treatment with, pharmaceutical composition capsules the levels was significantly (p < 0.05) increased to (42.4) mg/dl as compared with the mean values of diabetic group (Gennest J. 2011 Rajesham V.2012).column 4.

Diabetic patient women were showed significant (p<0.001) elevation in Plasma malondialdehyde level compared to normal control. The results are shown in the column 5 the treatment with therapeutic composition capsules showed gradual reduction in the Plasma malondialdehyde level in diabetic patients from (0.77) nmol/gHb to (0.56) nmol/gHb on 30 day respectively (Chang C. 2013, Xia E.2010).

4. DISCUSSION

Phyto compound based strategies play a pivotal role in the prevention and treatment of diabetes. Polyphenolic compounds such as flavonoids contribute to increased plasma antioxidant capacity, decreased oxidative stress markers. Growing evidence indicates that various dietary polyphenols may influence carbohydrate metabolism at many levels.

They are bioactive compounds (secondary metabolites) found in plants that works with nutrients and dietary fibers to protect against diseases. The presence of the biologically active ingredients in the fruit, leaves seeds extracts may account for the observed pharmacological actions.

The currently available drugs for management of diabetes mellitus have certain drawbacks and therefore, there is a need to find safer and more effective ant diabetic drugs. Diabetes mellitus of long duration is associated with several complications such as atherosclerosis, myocardial in fraction, nephropathy etc. These complications have long been assumed to be related to chronically elevated glucose.

The present investigation of herbal extracts formula demonstrated the significant anti diabetic activity. The results from the study also indicate that the herbal extracts formula can reduce the levels of serum cholesterol, Glutathion-S-Transferase (G.S.T)enzyme activity, total cholesterol, malondialdehyde (MDA) level and increased the serum high density lipoprotein (HDL).

5. CONCLUSIONS

The present study demonstrated that treatment of diabetic with herbal extracts therapeutic combination could reduce the levels of serum cholesterol, Glutathion-S-Transferase (G.S.T)enzyme activity, total cholesterol, malondialdehyde (MDA) level and increased the serum high density lipoprotein (HDL) be due to its contents generally and in specific to its content of anti oxidants confirms the possibility that the major function of the extracts are the production of vital tissues including pancreas, thereby reducing the causation of diabetes.

6. ACKNOWLEDGEMENTS

The authors gratefully acknowledge the Head of the department of material research and his, assistant, thank the physician Dr. Intesar M.Ali and the biochemist Abeer A.A.Najem for there help in complete this work.

7. RERERENCES

1. Altan, V.M., 2003. The pharmacology of diabetic complications. (10) *Current Medicinal Chemistry*, pp. 1317–132.

2. Al-Azzawie H.F., Alhamdani, MS., 2006. Hypoglycemic and antioxidant effect of oleuropein in alloxan-diabetic rabbits. (78) *Life Sci.* 78, pp. 1371–1377. \

3. Banik, G. 2009. Anti diabetes plants of Southern Assam with special reference to biological screening. Ph.D thesis of Department of Life Science and Bioinformatics, Assam University Silchar. India.

4. Briante, R., Patumi, M., Terenziani, S., Bismuto, E., Febbraio, F. and Nucci, R. 2002. Olea europaea L. leaf extract and derivatives: antioxidant properties. (17) *J Agric Food Chem*, pp.4934– 4940.

5. Burstein, M., Scholinink, H. R.and Morfin, R. J.1970 High density lipoproteins. (583) *J. Lipid Res.* pp. 19.

6.   Chandraprakash Dwivedi., Swarnali Daspaul. 2013. Anti-diabetic Herbal Drugs and Polyherbal Formulation Used For Diabetes. A Review (2) *The Journal of Phytopharmacology*. 3; pp 44-51

7.   Chang CL, Chen, YC., Chen, HM., Yang, NS, and Yang, WC. 2013 Natural cures for type 1 diabetes: a review of phytochemicals, biological actions, and clinical potential. (20) *Curr Med Chem.*7: 899-907.

8.   Eidi, A., Eidi, M. and Darzi, R. 2009 Anti-diabetic effect of *Olea europaea* L. In normal and diabetic rats. (23) *Phytother. Res.* pp. 347–350.

9.   Gennest, J. and Libby, P. 2011. Lipoprotein disorders and cardiovascular disease. In: Heart Disease of Cardiovascular Medicine. 9th ed. Insulin Philadelphia, Pa: Saunders Elsevier 23chap 47.

10. Gonzalez, M., Zarzuelo, A., Ganez, M., Utrilla, MP., Jimenez, J. and Osuna, I.1992 Hypoglycemic activity of olive leaf. (58) *Planta Med.* 513–515.

11. Goulas, V., Exarchou, V., Troganis, AN., Psomiadou, E., Fotsis, T., Briasoulis, E.and Gerothanassis, IP. 2009 Phytochemicals in olive-leaf extracts and their antiproliferative activity against cancer and endothelial cells. (53) *Mol Nutr. Food Res.* Pp. 600–608.

12. Habing, W.H., Pabst, M.H. and Jakob, Y., 1974 MAD and lipid peroxidation. (249) *J Bio Chem.*,pp. 7130.

13. Haidara, M., Yassin, HZ., Rateb, M., Ammar, H. and Zorkana, MA.,2006 Role of oxidative stress in development of cardiovascular complications in diabetes mellitus. (4) *Curr Vasc Pharmacol.*, pp.215-227.

14. Heikal,T.M., Mossa,A,T,H., Marei, G,I,K. and Rasoul, M,A. 2012 Cyromazine and chlorpyrifos induced renal toxicity in rats. The Amelioration effects of green tea extract. (2) *J. Environ.Anal Toxicol.* 146

15. Jemai, H., El Feki, A. and Sayadi, S. 2009 Anti-diabetic and antioxidant effects of hydroxytyrosol and oleuropein from olive leaves in alloxan-diabetic rats. (57) *J Agric Food Chem*, pp.8798– 8804.

16. Jemai, H., Bouaziz, M., Fki, I., El Feki, A. and Sayadi, S. 2008 Hypolipidimic and antioxidant activities of oleuropein and its hydrolysis derivative-rich extracts from Chemlali olive leaves. (176) *Chem Biol Interact* pp. 88–98.

17. Karou, D., Dicko, M.H., Simpore, J.and Traore, A.S. 2005 Antioxidant and anti bacteria activities of polyphenols from ethno medicinal plants of Burkina Faso. African(4) *Journal of Biotechnology*, pp. 823–828.

18. Lahlou, M. 2013 The success of natural products in drug discovery. (4) *Pharmacology and Pharmacy*. pp. 17-31.

19. Lee, H.H., Layman, D.K. and Bell, R.R. 1981. Determination of Glutahtione peroxides. (111) *J Nutr.* pp. 194.

20. Mackenzie, T., Leary, L. and Brooks, WB. 2010. The effect of an extract of green and black tea on glucose control with type 2 diabetes mellitus. (5) *Metabolism.* pp.1340– 1344.

21. Manisha Modak, Priyanjali Dixit, Jayant Londhe, Saroj Ghaskadbi, and Thomas Paul A. 2007. Indian Herbs and Herbal Drugs Used for the Treatment of Diabetes. (40) *J. Clin. Biochem. Nutr.* pp 163–173.

22. Poudyal, H., Campbell, F. and Brown, L. 2010 Olive leaf extract attenuates cardiac, hepatic, and metabolic changes in high carbohydrate-fed, high fat-fed rats. (140) *J Nutr.* pp. 946–953.

23. Pritesh Patel, Pinal Harde, Jagath Pillai, Nilesh Darji, Bhagirath Patel,and Sat Kaival. 2012. College of Pharmacy Pharmacophore (An International Research Journal Anti-diabetic Herbal Drugs A Review. (3) *Pharmacophore.* pp. 18-29.

24. Pitchai, Daisy., James, and Eliza., Khanzan Abdul Majeed Mohamed.,2009. A novel di hydroxyl gymic triacetate isolated from *Gymnemasylvestre* glycemic,hypo lipidemic activity on induced diabetic rats. (126) *Journal of Ethnopharmacology* pp.339–344.

25. Polina, Smirin., Dvir, Taler., Guila, Abitbol., Tamar, Brutman- Barazani., Zohar, Kerem.,Sanford, R. and Sampson, Tovit., 2010. Rosenzweig. *Sarcopoterium spinosum* extract as an anti diabetic agent: *In vitro* and *in vivo* study. (129) *Journal Ethnopharmacology.* pp. 10–17

26. Rader, D.J., 2009. Cholesterol actyltransferase and atherosclerosis. (120) American Heart Assiociation. pp.552-599.

27. Rajesham. V. V., Ravindernath. A, D. and Bikshapathi V.R.N. 2012. A review on medicinal plant and herbal drug formulation used diabetes mellitus, Indo American. (2) *Journal of Pharmaceutical Research.* pp. 10.

28. Richmond, W.,1971 total cholesterol. (19) *Clim. Chem.* pp.1350.

29. Shaik Sameena Fatima., Maddirala Dilip Rajasekhar., Kondeti Vinay Kumar., Mekala Thur Sampa Kumar., Kasetti Rame Babu.and Chippa Appa Rao. 2010. Ant diabetic and anti hyperlipidemic activity of *Vernoni anthelmintica* seeds in Streptozotocin induced diabetic rats. (48) *Food and Chemical Toxicology.* pp. 495-501.

30. Wild SG. Roglic, A., Green, R.and King, H., 2004. Global prevalence of diabetes. Estimated for the year 2000 and projection for 2030. (27) *Diabetes Care.* pp. 1047–105.

31. Xia E. Q., Deng, G. F. and Guo, Y. 2010. Review of biological activities of polyphenols from seeds (11), *Int J MOD Sci.* pp. 2; 622-662.

32. Xianghon Chen, Xue Bai, Yihui Liu, Luanyuan Tian, Jianqiu Zhou, Qun Zhou. Jinbo Fang. and Jiachun Chen. 2009 Anti-diabetic effects of water extract and crude green tea In mice. (122) *Journal of Ethnopharmacology.* pp205-209

33. Chandraprakash Dwivedi, and Swarnali Daspaul, 2013. Anti-diabetic Herbal Drugs and Polyherbal Formulation Used For Diabetes. A Review(2) *The Journal of Phytopharmacology.* 3; pp. 44-51.

34. Chang, CL., Chen, YC., Chen, HM., Yang, NS., andYang, WC., 2013 Natural cures for type 1 diabetes: a review of phytochemicals, biological actions, and clinical potential. (20) *Curr Med Chem.* 7; pp. 899-907.

Appendix 1:
This tables is a combined parameters for each individual patient before and after treatment with pharmaceutical composition capsules

Patient	Age	MDA nmol/gHb Control 0.44±0.17		HDL mg/dl Control 45.3±1.3		Cholesterol mg/dl Control 186±9.4		GST unit/gHb Control 0.96±0.04		Total Glucose mg/dl Control 9 0.2±12.1	
		A	B	A	B	A	B	A	B	A	B
1	40	0.56	0.77	42.4	37.6	191.0	271.0	1.04	1.4	98.4	177.0
2	41	0.52	0.78	43.6	36.5	191.1	265.4	1.03	1.3	97.5	175.1
3	40	0.53	0.79	39.1	38.7	188.8	271.1	1.05	1.45	98.4	173.2
4	40	0.56	0.74	42.7	37.6	189.0	271.0	1.08	1.51	99.7	178.1
5	40	0.58	0.75	42.8	37.6	190.0	276.6	1.07	1.46	99.1	177.1
6	41	0.63	0.96	42.7	38.9	188.8	265.4	1.06	1.48	97.1	177.1
7	40	0.54	0.78	43.8	37.2	191.1	269.0	1.05	1.51	98.4	180.2
8	42	0.56	0.72	43.2	36.7	191.1	265.4	1.03	1.36	97.5	179.1
9	42	0.56	0.73	42.1	37.6	190.0	268.2	1.05	1.43	98.41	178.2
10	43	0.56	0.75	41.8	37.6	199.0	269.0	1.04	1.6	99.5	175.0
11	43	0.58	0.96	42.4	37.6	191.1	271.0	1.02	1.45	98.8	174.1
12	45	0.63	0.78	42.7	35.8	188.8	270.0	1.03	1.6	99.2	180.5
13	41	0.59	0.77	43.2	37.5	190.2	277.6	1.04	1.48	98.5	177.5
14	42	0.52	0.76	43.3	39.5	191.1	268.1	1.08	1.4	99.5	177.1
15	43	0.56	0.73	42.4	37.7	190.0	261.3	1.04	1.4	98.1	177.1
16	44	0.51	0.73	42.5	37.3	189.0	276.6	1.08	1.4	97.1	175.6
17	45	0.55	0.74	39.8	37.7	190.1	271.0	1.04	1.47	99.1	177.0
18	41	0.56	0.73	42.4	38.4	189.8	272.1	1.06	1.44	97.8	176.0
19	42	0.56	0,77	43.1	36.4	190.0	271.1	1.04	1.45	97.1	177.0
20	44	0.54	0.71	42.2	36.1	190.5	272.2	1.06	1.46	98.8	177.6
21	40	0.56	0.77	43.4	35.6	191.0	274.1	1.06	1.39	98.4	177.3
22	40	0.59	0.78	43.7	36.6	188.0	270.2	1.08	1.41	99.7	178.1
23	41	0.56	0.79	42.7	37.7	189.0	272.0	1.04	1.5	97.1	174.9
24	44	0.58	0.76	42.6	38.6	188.0	270.3	1.06	1.44	98.2	178.6
25	42	0.54	0.74	42.7	37.8	191.1	271.4	1.05	1.45	98.5	177,0
26	43	0.49	0.76	45.2	39.5	190.0	277.6	1.06	1.5	97.4	174.0
27	45	0.56	0.58	39.1	36.8	191.0	268.0	1.03	1.46	98.5	178.1
28	40	0.57	0.71	43.1	38.6	191.0	265.4	1.04	1.45	98.9	178.5
29	42	0.58	0.77	39.1	37.4	188.8	270.1	1.05	1.44	98.41	177.0
30	41	0.57	0.96	42.4	37.6	190.0	272.3	1.06	1.46	98.9	177.2
Mean	B	0.56	0.77	42.4	37.6	190.0	271.0	1.05	1.45	98.4	177.0

Chapter 4

Prophylactic Potential and Medicinal Properties of Protein Sweeteners from Underutilised Plant Species

Gabrial Gbolagade Awolehin[1] and H.D. Ibrahim[2]

[1]Deputy Director and Head,
[2]Director General and Chief Executive Officer
Policy Analysis and Development,
Raw Materials Research and Development Council [RMRDC], Nigeria
E-mail: [1]awogab2010@yahoo.com

ABSTRACT

Diabetes is a prevalent disease all over the world and people that have the disease at global level is estimated to have risen by about 133 per cent in the recent years to approximately 350 million. Diabetes and elevated glucose have been reported to cause about 3 million deaths, worldwide annually, and it is on the increase.

Prevention, as it is often said, is better than cure. The costs of diabetes could be enormous, often beyond the ability of moderate family. Therefore, effective prevention of diabetes leads to cost effective health care. Prevention of the onset of diabetes which is called primary prevention or the prevention of its immediate and long term consequences which is called secondary prevention can partly be effected significantly by reduction in sugar intake. Continued intake of sugar via confectionery, food and beverages will aggravate diabetic condition in patient, as high sugar consumption is linked to type 2 diabetes.

Some plant species have substances that have intense sweetness, some up to 5,000 times sweeter than sugar but the sweet substances in them are proteins which range from moderately soluble to highly soluble proteins that have high potentials of replacing sugar in the food, beverages and confectionery industries. Some of such plants namely *T. danielli,*

D. cumminsi, P. brazzeana, S. dulcificum, C. masaikai are presented, giving characteristics of their protein sweeteners, which can be used to prevent, ameliorate or serves as palliative for diabetic patient. The protein sweeteners have versatile applications, in food, beverages, pharmaceuticals, and in alternative medicine. The protein sweetener will help to ameliorate or lesson incidence of type 2 diabetes in patients that administer them rather than sugar. Other uses of the plants in alternative medicine are also discussed.

Keywords: Diabetes, Protein sweeteners, Proteins-thaumatin, Monellin.

1. INTRODUCTION

There are two approaches to finding solutions to diseases. They can either be prevented from establishment or cured, after establishment. To prevent diseases is however a better alternative, as cost of curative approaches [if curable at all] could be scaring and unaffordable in some cases. One of such cases is diabetes. Diabetes is a prevalent disease all over the world. The International diabetes Federation estimates that 285 million people around the world have diabetes. This total is expected to rise to 438 million within 20 years, as 7 million people are said to develop diabetes each year and people that have the disease at global level is estimated to have risen by about 133 per cent in the recent years. Diabetes and elevated glucose have been reported to cause about 3 million deaths, worldwide annually, and it is on the increase.

Prevention, as it is often said, is better than cure. The costs of diabetes could be enormous, often beyond the ability of moderate family. Therefore, effective prevention of diabetes leads to cost effective health care. Prevention of the onset of diabetes which is called primary prevention or the prevention of its immediate and long term consequences which is called secondary prevention can partly be effected significantly by reduction in sugar intake. Continued intake of sugar via confectionery, food and beverages will aggravate diabetic condition in patient, as high sugar consumption is linked to type 2 diabetes.

Type 1 diabetes, according to the International Diabetes Federation, called juvenile-onset diabetes. It is usually caused by an auto-immune reaction where the body's defence system attacks the cells that produce insulin. The reason this occurs is not fully understood. People with type 1 diabetes produce very little or no insulin. The disease may affect people of any age, but usually develops in children or young adults. People with this form of diabetes need injections of insulin every day in order to control the levels of glucose in their blood. If people with type 1 diabetes do not have access to insulin, they will die.

Type 2 diabetes used to be called non-insulin dependent diabetes or adult-onset diabetes, and accounts for at least 90 per cent of all cases of diabetes. It is characterised by insulin resistance and relative insulin deficiency, either or both of which may be present at the time diabetes is diagnosed. The diagnosis of type 2 diabetes can occur at any age. Type 2 diabetes may remain undetected for many years and the diagnosis is often made when a complication appears or a routine blood or urine glucose test is done. It is often, but not always, associated with overweight or obesity, which itself can cause insulin resistance and lead to high blood glucose levels. People with type 2 diabetes can often initially manage their condition through exercise and diet. However, over time most people will require oral drugs and or insulin.

Diabetic condition is often aggravated by sugar intake. But some plant species have substances that have intense sweetness, some up to 5,000 times sweeter than sugar but the sweet substances in them are proteins which range from moderately soluble to highly soluble proteins that have high potentials of replacing sugar in the food, beverages and confectionery industries. Some of such plants are: *Thaumatococcus danielli* (Ewe eran), *Dioscoreophyllum cumminsii* (serenpidity berry), *Pentadiplandra brazzeana* Baillon, *Synsepalum dulcificum* (Agbayun), *Curculigo latifolia*, and *Capparis masaikai*.

Therefore, exploration of possible [proteinous]alternatives to sugar is very strategic in that it performs dual role of addressing industrial as well as medical problems in alternative medicine.

Background Information on Herbal Medicine in Nigeria

The contribution of natural value added products to orthodox medicine cannot be overlooked. This is in view of the fact that African Traditional Medicine has over time been the mainstay of primary health care for majority of Nigerian living in rural areas.

In some African countries such as Nigeria, mechanism has been put in place for registration regulatory frame work and institutional instruments for developing African Traditional Medicine and locally produced commercial quantities of standardized Traditional medicines.

Okujagu (2010) declared that Research works on Traditional Medicine have been documented in line with the procedures of the World Health Organisation [WHO] as well as with an African Union [AU] directing that Research and Development on African Traditional Medicine should be made a priority in recognition of its vast potentials and immense contribution to health and poverty alleviation.

Herbal medicine, otherwise called traditional medicine has become entrenched in the medical practice in Nigeria and there are calls that Government should really integrate herbal medicine into Nigerian health care delivery system.

Institutional Framework and Synergy on Herbal Medicine in Nigeria

The federal Ministry of health has a desk for traditional medicine. Similarly, the relevant Research Institutes such as Nigeria Institute of Pharmaceutical Research and Development [NIPRD] has section dealing with herbal practice and packaging.

The body overseeing Food and drug administration and control called National Agency for Food and Drug Administration and control (NAFDAC) now recognises the herbal medicine practice.

An agency called Nigerian Natural Medicine Development Agency (NNMDA) has been established and recognized by the government to among other roles, ensure development of natural medicine. The NNMDA was established in 1997 by Nigerian Federal Ministry of Science and Technology order in accordance with the National Science and Technology Act of 1980. The NNMDA's mission is to promote traditional medicine through research. To accomplish its mission, the NNMDA is establishing and maintaining a virtual/digital library on traditional Nigerian medicine and indigenous health healing systems; researching, collating, documenting, and disseminating all published and unpublished research works on all aspects of traditional medicine; and establishing a college and zonal training centers for the study of natural medicine. Some institutions of higher learning having pharmacy course have also pharmacognosy. Pharmaconosy is the vital link between traditional and orthodox system of medicine. There is Nigerian Medicinal plants Development Company (NMPDC) a well known Green Pharmacy, having done extensive works on several medicinal plants.

A very cordial relationship by way of collaborative efforts exist among universities, Research institutes such as Nigeria Institute of Pharmaceutical Research and Development (NIPRD), Bioresearches Development and conservation programme, Raw Materials Research and Development Council, National stored Products Research Institute, institute of public policy analysis of Nigeria etc.

Harnessing Nigerian Medicinal Plants

To fully harness medicinal plants, the following recommendations are made:

1. Need for increased sensitisation of Nigerians to appreciate the value and enormous potentials that abound with their use.
2. The Federal Government on its part must initiate and facilitate policies through the Federal Ministry of health by passing the traditional medicine policy, to regulate the practice.
3. There is also the need to facilitate the integration of herbal medicine into the health care delivery system. This will stimulate research and development as well as processing and packaging of herbal products.
4. Need to discourage importation of such products that can be produced locally.
5. There is need for reduced interest in the quest for chemically synthesised drugs and desire to promote Green pharmacy.

2. IDENTIFICATION OF PROTEINOUS ALTERNATIVES TO SUGAR

Some plants have been identified that produce fruits containing proteinous sweeteners of various degree of sweetness and which are sweeter than sugar by 200-5000 times. These plants grow well in our local environment.

Six of the plants on which some serious works have been done are:-

1. *Thaumatococcus danielli* (Ewe eran) - which produces protein —**thaumatin**
2. *Dioscoreophyllum cumminsii* (serenpidity berry) – which produces protein **monelin**

3. *Pentadiplandra brazzeana* Baillon – which produces **brazzein**
4. *Synsepalum dulcificum* (Agbayun) – which produces protein **miraculin**
5. *Curculigo latifolia* – which produces protein **curculin**
6. *Capparis masaikai* – which produces **mabinlin**

Thaumatin

Thaumatin is a low-calorie (virtually calorie-free) protein sweetener and flavor modifier. The substance is also used for its flavor modifying properties and not exclusively as a sweetener. The thaumatins were first found as a mixture of proteins isolated from the *T. danielli* fruit (*Thaumatococcus daniellii* Bennett) of West Africa.

Sweetness Properties

T. danielli is a tropical rain forest shrub in the family of Marantaceae that produces reddish trigonal fruit with 2-3 seeds that contain the sweet fluid in the mucilage around the seed. The fluid is essentially proteinous and it is 5000 times sweeter than sugar. Although very sweet, thaumatin's taste is markedly different from sugar's. The sweetness of thaumatin builds up very slowly. Perception lasts a long time leaving a liquorice-like aftertaste at high usage levels. Thaumatin is highly water-soluble, and stable to heat and stable under acidic conditions.

As a sweetener

Within west Africa, the T. danielli fruit has been locally cultivated and used to flavor foods and beverages for some time. The fruit's seeds are encased in a membranous sac, or aril, that is the source of thaumatin. In the 1970s, Tate and Lyle began extracting thaumatin from the fruit. In 1990, researchers at Unilever reported the isolation and sequencing of the two principal proteins found in thaumatin, which they dubbed thaumatin 1 and thaumatin II.

Thaumatin has been approved as a sweetener in **the European Union** (E957), **Israel**, and **Japan**. In the **United States**, it is a generally recognized as safe flavouring agent (FEMA GRAS 3732).

Uses of the Fruits

The most popular use of *T. daniellii* is as a sweetener. The aril contains a non-toxic, intensely sweet protein named thaumatin, which is at least 3000 times as sweet as sucrose. In West Africa, the aril is traditionally used for sweetening bread, over-fermented palm-wine and sour food. When the seeds are chewed, for up to an hour afterwards they cause sour materials eaten or drunk to taste very sweet. Since the mid-1990s, thaumatin is used as sweetener and flavour enhancer by the food and confectionary industry. Substituting synthetic sweeteners, it is used as a non-caloric natural sweetener. The seeds of *T. daniellii* also produce a jelly that swells to 10 times its own weight and hence provides a substitute for agar

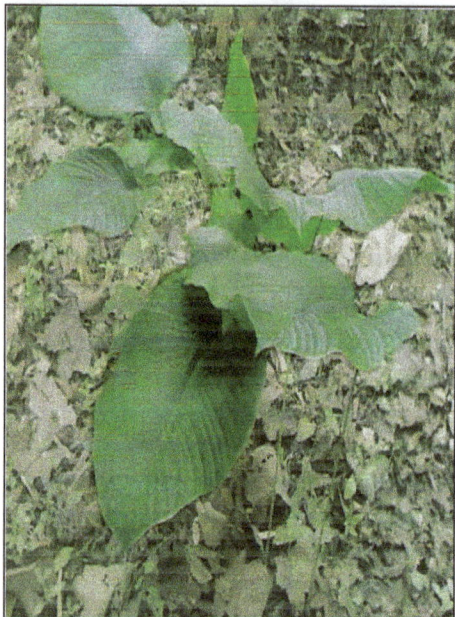

Figure 4.1: A Growing *T. danielli* Plant.

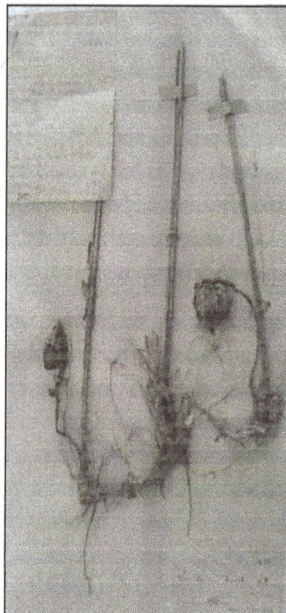

Figure 4.2: A *T. danielli* Plant Showing its Fruiting Point.

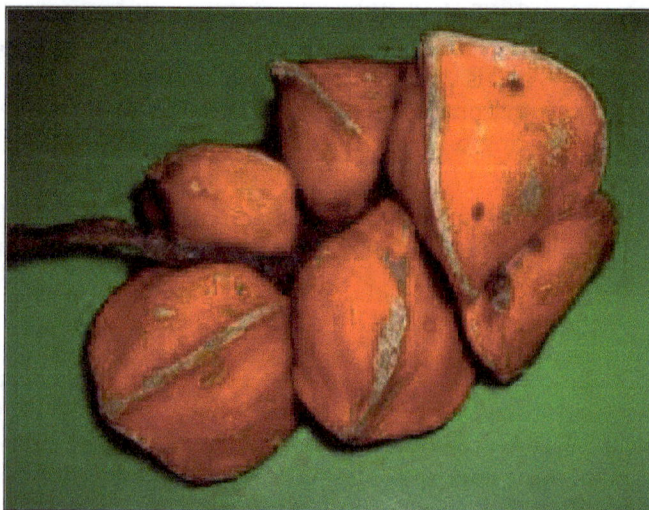

Figure 4.3: Matured Trigonal Fruits of *T. danielli*.

Uses of Leaves

In West Africa, *T. danielli* is mostly cultivated for the leaves. The lamina of the leaves is used for wrapping foods. The petiole is used to weave mats and as tools and building materials. The entire leaf is also used for roofing.

Traditional Medicinal Use

Thaumatin is not a carbohydrate, thus it is an ideal sweetener for diabetics.

T. daniellii is also used in traditional medicinal uses in the Ivory Coast and Congo. The fruit is used as a laxative and the seed for pulmonary problems. In traditional medicinal use, the leaf sap is used as antidote against venoms, stings and bites. The Leaf and root sap are used as sedative and for treating insanity.

Propagation

T. daniellii is propagated by rhizome fragments bearing one or two stools each. The seed has low percentage of germination and the slow growth of the seedling. Therefore, for commercial propagation, the plant should be established from rhizomes and not from seed

Dioscoreophyllum cumminsii (Serendipity Berry)

D. cumminsii is a tropical rain forest vine belonging to the family Menispermacea. Its berries have a water soluble proteinous substance of intense sweetness called **monellin**, that is **3000 times** sweeter that sucrose.

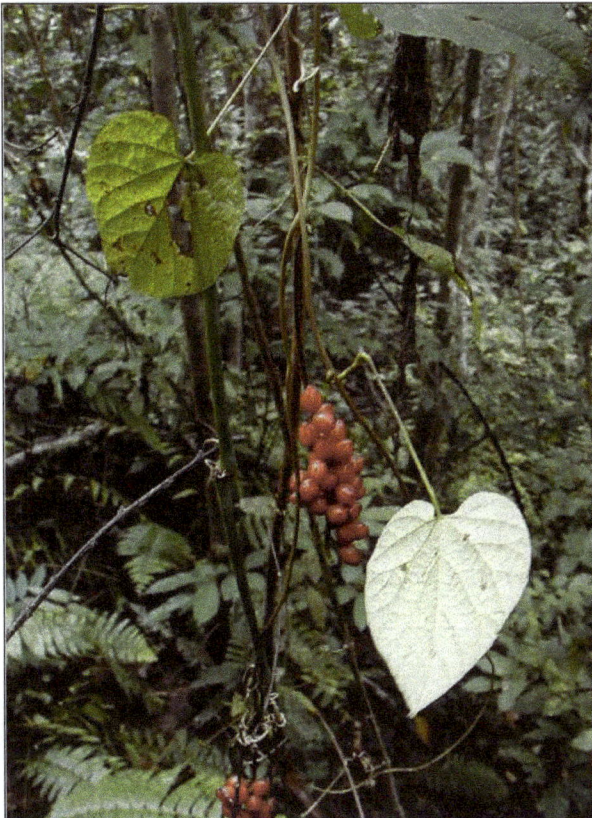

Figure 4.4: Fresh Plant of *D. cumminsii* Showing the Fruit Cluster.

Monellin

Monellin is a sweet protein which was discovered in 1969 in the fruit of the West African shrub known as serendipity berry (*Dioscoreophyllum cumminsii*).

The plant is a tropical rain forest vine belonging to the family Menispermacea. Its berries have a water soluble proteinous substance of intense sweetness called monellin, that is 3000 times sweeter that sucrose.

The protein was named in 1972 after the Monell Chemical Senses Center in Philadelphia, USA., where it was isolated and characterized.

Sweetness Properties

The relative sweetness of monellin is about 3000 times sweeter than sucrose, Monellin has a slow onset of sweetness and lingering aftertaste. Monellin's sweetness is pH dependent; the protein is tasteless below pH 2 and above pH 9. Blending the sweet protein with bulk and or intense sweeteners reduces the persistent sweetness and shows a synergictic sweet effect. Heat over 50°C at low pH denatures monellin protein causing a loss of the sweetness.

As a sweetener

Monellin can be useful for sweetening some foods and drinks as it is a protein readily soluble in water due to its hydrophilic properties. However it may have limited application because it denatures under high temperature conditions, which makes it unsuitable for processed food. It may be more relevant as noncarbohydrate tabletop sweetener, especially for individuals such as diabetics who must control their sugar intake.

Pentadiplandra brazzeana Baillon

It is an African plant in the family of Pentadiplandraceae. Brazzein was originally isolated from the fruit of the plant.

Brazzein

Brazzein is the smallest, heat-stable and pH-stable member of the set of proteins known to have intrinsic sweetness.

Sweetness Properties

The protein, consisting of 54 amino acid residues, is reported to be about 2000 times sweeter than sucrose and represents an excellent alternative to available low calorie sweeteners.

As a Sweetener

The red pulp of the fruits is eaten as a snack, or sometimes used to sweeten maize porridge. The protein brazzein, originally extracted from the fruit pulp, is being developed into a low-calorie sweetener for the food Industry.

Figure 4.5: A Fruiting Branch of *P. brazzeana* Baillon.

Medicinal Uses of the Plant

The roots are used throughout Central Africa against several problems related to giving birth. The roots of P. brazzeana also have laxative purgative and carhartic properties, and are applied to the abdomen to treat oedema.

In Cameroon, the root bark is one of more than 20 constituents of a sauce that is given to mothers who have just given birth, to stimulate milk production. Among the Mezime people of Cameroon a root decoction is given orally or applied as an enema to facilitate the expulsion of the placenta; it also helps in reducing pain caused by hernia. In Cameroon and DR Congo the macerated roots, alone or mixed with other ingredients, are taken orally or applied as an enema against malaria. The Monzombo people in Cameroon drink a decoction of the tuber, as well as the juice from macerated roots, mixed with pounded leaves as an anthelmintic. A root decoction is taken to treat pneumonia and serious bronchitis, but administration to pregnant women is avoided owing to risk of miscarriage. It is even misused to induce abortion in some cases. Together with leaves of *Kalanchoe crenata* (Andrews) haw, a root preparation is used as nose drops to stop epileptic crises. The roots smell of aspirin and are hung over the doorway or are placed inside the roof to keep away snakes.

In the Central African Republic a tuber decoction of the plant is said to prevent haemorrhages after parturition. The fresh root is pulped, or the dry root pounded and mixed with palm oil, to make an ointment for topical application to prevent infections of the navel in newborn babies. Because the plant is vesicant, the duration of treatment must be limited to avoid blistering.

The crushed root or root bark is applied as an infusion drunk to soothe chest pain, toothache, lumbago, rheumatism and haemorrhoids. Powder of dried root bark is applied to scarifications to treat intercostal and abdominal pains. In the Central African Republic *Capsicum* pepper is added to macerated roots to prepare a drink that soothes cough

In Nigeria the crushed root is used to treat several skin infections and in south-western Cameroon a leaf decoction is used to wash the skin against scabies.

In **Congo** pulped roots are applied externally against itch and as an antiseptic, and to treat wounds, sores and ulcers. A decoction of the bark mixed with bark and roots of other plants is taken against stiffness or weakness of the limbs and back. Roots and tubers are also commonly used in the treatment of intestinal problems such as dysentery, colic, urethritis, gonorrhoea and other urino-genital infections.

Synsepalum dulcificum

Synsepalum dulcificum, also known as the miracle fruit, is a plant of the family Sapotaceae, which originated in Ghana (Tropical West Africa) with a berry that when eaten, causes sour foods subsequently consumed to taste sweet. This effect is due to miraculin, which is used commercially as a sugar substitute. [Wikipedia]

Figure 4.6: A *S. dulcificum* Plant Showing its Fruiting Points.

Miraculin

Miraculin is a glycoprotein extracted from the fruit of the miracle fruit plant (*Synsepalum dulcificum* or *Richadella dulcifica*). It helps to improve insulin resistance and thus useful in managing Type 2 diabetes.

Miraculin itself is not sweet. However, after the taste buds are exposed to miraculin, ordinarily sour foods, such as citrus, are perceived as sweet thus acting as taste modifier. This effect lasts up to an hour.

The reactive substance, isolated by Prof. Kenzo Kurihara (Kurihara Kenzo) a Japanese scientist, was named miraculin after the miracle fruit when kurihara published his work in Science in 1968.

As a Sweetener

As miraculin is a readily soluble protein and relatively heat stable, it is a potential sweetener in acidic food (*e.g.* soft drinks).

Figure 4.7: A Matured *S. dulcificum* Berry.

Propagation

Propagated by seed following de-pulping and drying.

Figure 4.8: Seeds of *S. dulcificum*.

Medicinal Uses

According to Wikipedia, attempts have been made to create a commercial sweetener from the fruit, with an idea of developing this for patients with diabetes. Fruit cultivators also report a small demand from cancer patients, because the fruit allegedly counteracts a metallic taste in the mouth that may be one of the many side effects of chemotherapy. This claim has not been researched scientifically, though in late 2008, an oncologist at Mount Sinai Medical Center in Miami, Florida, began a study, and by March 2009, had filed an investigational new drug application with the U.S. Food and Drug Administration.

In Japan, miracle fruit is popular among patients with diabetes and dieters. It is however unfortunate that high heat destroys the active principle, so that canning, jams, preserves, baking, drying, etc. are impossible. The fruits can be held for a short period of time by refrigeration or freezing.

Curculigo latifolia

C. latifolia is a plant from Malaysia in the family Hypoxidaceae, from which Curculin, a sweet protein was discovered and isolated in 1990 from its fruit.

Curculin

Like miraculin, curculin exhibits taste- modifying activity; however, unlike miraculin, it also exhibits a sweet taste by itself. After consumption of curculin, water and sour solutions taste sweet.

Sweetness Properties

Curculin is considered to be a high-intensity sweetener, with a reported relative sweetness of 430-2070 times sweeter than sucrose on a weight basis.

As a Sweetener

Like most protein, curculin is susceptible to heat. At a temperature of 50°C the protein starts to degrade and lose its "sweet-tasting" and "taste-modifying" properties, so it is not a good candidate for use in hot or processed foods. However, below this temperature both properties of curculin are unaffected in basic and acidic solutions, so it has potential for use in fresh foods and as a table-top sweetener.

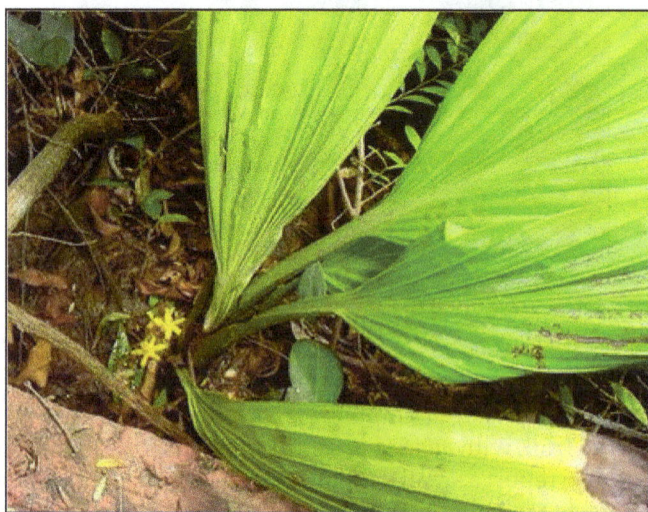

Figure 4.9: A *Curculigo latifolia* Plant Showing Flowering Part.

Figure 4.10: A *Curculigo latifolia* Plant Showing its Fruits.

Propagation

Popularly by rhizomes and corms, other propagules are in vitro plantlets regeneration from shoot tips and seed but seed has low germination percentage.

Traditional Medical Uses

Rhizome is used for menorrhagia and Ophthalnia when concocted with other plants.

Roots are used as internal medicine for fever.

Flowers and roots are used against stomach ache and diuretic in genito-urinary disorders.

Other Uses

The tough high-weight leaf fibres are used for production of finishing nets, mopes, twines, rice bags, garments and fabrics leaves are used for wrapping fruits, vegetables and food in Indonesia while they are used in magical healing ceremonies in Borneo.

Curculigo latifolia var latifolia was found to contain arbutin (1.10mg/g) used as ingredient in skin whitening.

Capparis masaikai

Capparis masaikai grows in the subtropical region and bear fruits of tennis-ball size. The mature seeds are used in traditional Chinese medicine.

C. masaikai is found in China. The protein mabinlin is isolated from it.

Figure 4.11: A *Capparis masaikai* Plant Showing Fruiting Points.

Mabinlin

Mabinlin is a sweet protein with the highest known thermostability. It is derived from Capparis masaikai which is and its sweetness was estimated to be around **400** times that of sucrose on weight basis. The sweetness of Mabinlin are unchanged after 48hr incubation at boiling point and after 1hr at 80°C.

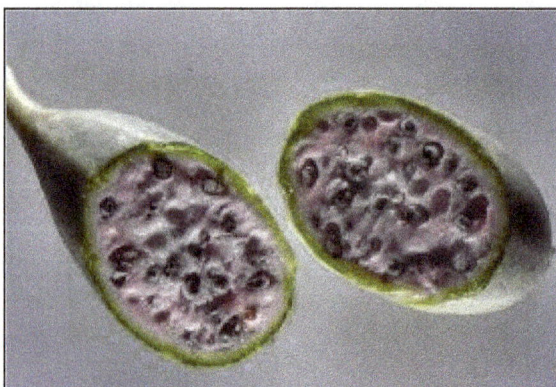

Figure 4.12: A *Capparis masaikai* Fruit Showing the Seeds.

Propagation

Propagation is by seed, tissue culture and rapid multiplication

Health and Other Challenges of Sugar Production in Nigeria

The importance of proteinous alternatives to sugar will better be perceived and appreciated when we know the challenges facing sugar production in the Country.

Sugar production in Nigeria is faced with myriads of challenges. Some of the challenges are enumerated below:

1. The shortfall in local production: Due to the gross shortfall in sugar production in the Country, Government imports 90 per cent of the Country's sugar requirements. The National Sugar Development Council showed that above 500,000 hectares of low land suitable for sugar cane cultivation existed nationwide which remained uncultivated in about 40 locations.

   Importation of a large chunk of sugar needs in any Country has both direct and indirect implications on the economy, namely:

   a. Importation depletes foreign reserves

   b. The Country is at the mercy of the exporting Countries.

   c. Massive importation shrinks and eventually paralyses the home made efforts to produce locally as the local efforts cannot compete with Countries who over the years have gained cheaper/economic scale of production.

   d. The exporting Countries have the prerogatives of price determination, and so we have no choice.

2. Problems of diversity in use of sugar cane: sugar cane which is the primary raw materials for the production of sugar has several applications for which it can be put into. Recently, use of sugar cane for production of Bio-ethanol had gained prominence, thus implying competition for the already scarce raw materials base.

3. Problems of bulkiness of sugar: The bulkiness of sugar implies that huge amount is spent for international shipment and local haulage to destination of use.

4. Problems of diabetics aggravation: Where there are tendencies or traits of diabetes, uncontrolled intake of sugar will aggravate the diabetic condition either for Type 1 or Type 2 diabetes.

   a. Type 1 diabetes – called insulin dependent diabetes, occur when body fails to produce insulin, the hormone required for controlling blood sugar level.

   b. Type 2 diabetes – Develops slowly, much more common than Type 1 accounting for 75 per cent of cases. It is strongly related to being overweight.

5. Problems of obesity: since Type 2 diabetes is strongly related to being overweight, intake of sugar needs to be monitored and moderated and if there are alternatives that are not sugar based, it is safer.

Prospects of Proteinous Alternatives to Sugar

If the potentials of these plants could be developed and harnessed, an inestimable array of prospects and benefits stand to be accrued.

1. The protein sweeteners have versatile applications, in food, beverages, pharmaceuticals, and in alternative medicine

2. The shortfall in production of sugar will be made up with. with the use of proteinous sweetener.

3. Importation of sugar could completely cease as the protein sweetener takes the place of the imported sugar, thus conserving a huge amount of foreign reserves.

4. These sweeteners have potentials of becoming an export commodity to Countries all over the world thus becoming a major foreign exchange earner.

5. Home grown industries for processing of the sweetener will spring up leading to generation of employments, increase in Gross domestic products, wealth creation and enhanced economy.

6. The protein sweeteners are not bulky since its sweetness is intense, a pinch of final product will be sufficient for a large chunk of food products and thus cost of transportation to destination of use is infinitesimal. A few kilogram of the processed sweetener will do the work of thousands of tonnes of sugar.

7. Similarly, the protein sweetener will help to ameliorate or lesson incidence of type 2 diabetes in patients that administer them rather than sugar.

8. The plants grow well in the rainforest belt of the country and so there is no problem of adaptation.

3. SUGGESTIONS/RECOMMENDATIONS

The following suggestions/recommendations on this subject matter for stakeholders in the economic development of the Country, are made:

1. Exploration and development of the full potentials of these proteinous plants should be embarked upon.

2. There is urgent need to domesticate these plants, in view of deforestation less they are extinct.

3. An establishment of a pilot plant in the first instance is important to perfect the developed technologies and then eventual commercialization

4. Public Private Partnership in the development of this material: The private component will fast track the development while the public component will provide conducive environment, favourable policy formulation and infrastructural frame work.

5. All the relevant Research Agencies should also start to work on other possible sources of proteinous alternatives to sugar.

6. National Sugar Development Council should have a unit for exploring the proteinous alternatives to sugar.

7. Since the protein sweeteners grow in the wild, there is need for a multidisciplinary approach involving Botanists, Agronomists, Farmers, Relevant aspect of Pharmacy, Industrial Chemists *etc.* to see to the propagation, popularization, cultivation, processing and extraction of the

sweetening substances in the plant for commercialization. Massive seed multiplication is the starting point.

8. Most of the protein sweeteners have seed dormancy which can be broken by use of some physical or chemical methods. But vegetative propagation or tissue culture technique is recommended.

4. CONCLUSIONS

If the potentials of these plants are exploited and harnessed, the hitherto problem of aggravation of diabetic condition of diabetic patients resulting from sugar consumption will be alleviated, shortage of sugar for food, beverage and pharmaceutical industries would be a thing of the past, Sugar importation will be reduced, foreign reserves enhanced and employment will be generated.

Therefore, all efforts must be made to fully tap the great treasure inherent in these natural resource potentials we are endowed with.

For stakeholders – Government, Farmers, Researchers, Industrialists, all hands must be on deck to give the needed attention and urgency to the development of these proteinous alternatives to sugar. The advantages are numerous and timeless.

5. REFERENCES

1. Awolehin, G. G. (2009) Strategic Exploration of Proteinous alternatives to sugar for a double remedial applications: In Industry and in Medicine – A paper presented at 1st RMRDC International Seminar on Natural Resources Development and Utilization, Abuja, Feb. 2009.

2. Capparis masaikai : From Wikipedia, the free encyclopedia.

3. HU Qi-min; HUANG Yun-feng; LAI Mao-xiang [2013] Tissue culture and rapid propagation technique of Capparis masaikai. Journal of Southern Agriculture Year, Issue 9, p.1431-1434.

4. Ravi Kant (2005) Sweet Proteins– Potential replacement for artificial low calorie sweeteners. *Nutritional Journal* 4:5.

5. Shariff, Z [2014]. Herbal medicine should be integrated into Nigerian health care. Herbal Drug Development interviews of Pharm [Hajia] Zainab Shariff, CEO, Nigerian Medicinal Plants Development Company[NMPDC] by Temitope Obayendo. Fri, April 25.

6. Thaumatococcus danielli (Bennett) Collections of University of Connecticut Accession Data 2002.

7. Thaumatococcus daniellii : From Wikipedia, the free encyclopedia.

8. Obioh (nee G.U. Ahuama), G.I.B., Isichei, A.O [2000] A population viability analysis of serendipity berry Dioscoreophyllum cumminsii in a semi-deciduous forest in Nigeria. *Ecological Modelling* 201(3-4): 558-562.

9. Ogunsola Kayode (2008) Miracle Berry: A Potential Industrial Sweetner. *In Quarantine News, A quarterly news bulletin of Nigeria Agricultural Quarantine Service*, Vol. 4 No. 2.

10. Okujagu, T [2010]. Evaluating traditional medicine research in Nigeria. *Business Day Newspaper*. August 20.

11. Pentadiplandra brazzeana Baill. Prota 11(1): Medicinal plants/Plantes médicinales 1Prota 11(1): *Medicinal Protologue Bull. Mens. Soc. Linn.* Paris 1: 611 (1886).[htttp://www.prota.org]

12. *Synsepalum dulcificum*: Everyday Miracle - Grow the Dream!. www.TopTropicals. com

Chapter 5

Effect of Aqueous Extract of Black Seed (*Nigella sativa*) on Glucose, Insulin, Lipid Profile and Various Related Parameters in Type 2 Diabetic Individuals

Ahmad Bilal[1], Tariq Masud[2], Arshad Mahmood Uppal[3], Abdul Khaliq Naveed[4] and Muhammad Siddique Afridi[5]

[1]*Pakistan Council of Scientific and Industrial Research, Pakistan*
E-mail: dr.ahmadbilal@yahoo.com
[2]*Department of Food Technology,*
PMAS Arid Agriculture University Rawalpindi, Pakistan
E-mail: drmasud_tariq@hotmail.com
[3]*District Headquarter Teaching Hospital, Rawalpindi, Pakistan*
E-mail: amuppal@live.com
[4]*Department of Biochemistry and Molecular Biology,*
Rifah Medical College, Rawalpindi, Pakistan
E-mail: khaliqnaveed2001@yahoo.com
[5]*Medicinal Botanic Center,*
PCSIR Laboratories Complex, Peshawar, Pakistan
E-mail: afridi_mdk@hotmail.com

ABSTRACT

Type 2 Diabetes mellitus (T2DM) is rapidly spreading disease because of changes in our lifestyle and dietary habits. There is a true concern about its treatment in wake of increased side effects of pharmaceutical preparations coupled their price hike. This study included

44 T2DM individuals, fulfilling especially laid down criteria. All the patients consumed Aqua's Extract of Black Seed (*Nigella sativa*) (NsE) for 40 days followed by a placebo for another 40 days. Fasting blood sample of each subject was collected on 0, 40th and 80th days of the study. Glucose, Insulin, Lipid profile, total leukocyte count, platelet count, ALT, AST and Urea were analyzed. SPSS version 22 was used for statistical analysis. A decrease in glucose and increase in insulin level was observed after treatment that reversed in placebo phase. A significant fall in total cholesterol level was seen after treatment (p=0.019), which reversed in placebo phase. There was no effect on LDL and HDL cholesterol. TG levels fell significantly (p=0.007) and their reversal was seen after placebo (p=0.020). There was no significant change in cellular blood components like leukocytes and platelets. The levels of ALT and AST reduced significantly after the treatment, which was carried to the placebo phase. No significant change was observed in blood urea levels in any phase. Effect of NsE on subjective feelings of the patients suggest that it did not harm any of these, but improved aches and pains, burning feet, tingling sensations and blurred vision to some extent, though, these changes were statistically not significant. The study suggests that NsE reasonably lower the glucose, improves the lipid profile without affecting the leukocytes, platelets, liver enzymes and renal profile, thus highlighting its safe use.

Keywords: Nigella sativa, Aqueous extract, Diabetes mellitus, Insulin, Glucose, Lipids.

1. INTRODUCTION

Since man was crowned for the kingdom of earth, he had been suffering from multiple diseases. As he lived by water courses and so his life was dependent on the vegetation's grown nearby. These were used for food, wearing, safety and treatment of his ailments. As he progressed he separated active ingredients useful for his diseases, which were subsequently processed artificially and there was birth of allopathic medicine. The synthetic compounds are prone to lead to toxicity. Since their side/toxic effects were highlighted he was much concerned about these and has to be returned to the original source. Since the advent of this need, much has been done in herbal medicine. In fact this resurge is reconnaissance of herbal medicine. These have proved to be cheaper as compared to the allopathic medicines and less toxic. Especially this is true for the diseases that result from unhealthy dietary habits and lifestyle.

Type 2 diabetes mellitus (T2DM) is one of these disorders and has been documented since the time of Hippocrates. This is also mentioned in the Egyptian papyruses. Increasing prevalence of T2DM and its costly treatment is a matter of concern on the globe (Kanter, M., 2003), which has aroused the need for cost effective and safer alternatives. Herbs are once again ready to serve the purpose. Moreover it is very satisfying for the patients to get their ailments treated with traditional recipes *i.e.* herbs. These are thought to be safe and easy to administer provided their pharmacological properties are established. Out of these *Nigella sativa* (Ns) is famous for the treatment of various ailments for the last 2000 years (Malcolm,1987; John,1984; Felix,1968). Avicenna referred black seeds in his book "The Canon of Medicine", as it stimulates the body's energy and helps recovery from fatigue and dispiritedness. Black seeds and their oil have a long history of folklore usage in Indian and Arabian civilization as food and medicine (Warrier, 2004 and Yarnell, 2011).

Nigella sativa commonly known as black seed is herbaceous annual plant, from family Ranunculaceae (Rchid, 2004) **and is** half a meter tall, with blue flowers and pungent triangular black seeds (Malcolm, 1987). These contain considerable amount of fixed and volatile oils (Nickavar, B., 2003), proteins, alkaloids and saponins (Ali,2003; Al-Ghamdi,2003; Turkdogan,2003; Kalus,2003; Michelitsch,2003). It's high LD50 (28.8 mL/kg orally) (Zaoui,2002), signifies its lower toxicity. Its ability to stabilize hepatic enzymes, organ integrity, and antioxidant activity adds on its safety, but the fall in leukocyte and platelet count must be taken into consideration (Zaoui, 2002). Ns has been tried as anti diabetic in animal type 2 like models with considerable efficacy, (Fararh, 2004; Kanter,2004) but very little research has been reported in humans especially diabetic patients. However, Ns in combination with various other herbs and Ns oil have been studied on human individuals for the treatment of diabetes (Al-Rowais, 2002; Bilal,2009). So this study on human subjects was designed, with the objectives of investigating the effects of Aqua's Extract of Black Seed (*Nigella sativa*) on fasting blood glucose, insulin level and lipids in T2DM individuals; in consideration with its possible side effects *i.e.* the fall in leukocyte and platelet count, in addition to ALT, AST and blood urea.

2. MATERIAL AND METHODS

Study Design

This was a placebo control changeover clinical trial, where 75 T2DM patients were initially registered according to specified criteria. The study comprised of two phases *i.e.* the treatment phase (40 days), where patients were given NsE and placebo phase (40 days), where they were given a placebo. Default was 31, where in 15 didn't report back after taking first treatment, 11 couldn't take treatment regularly, 5 did not come fasting on the day of blood sample. Forty-four (20 male and 24 female) completed the study successfully. All subjects were treated with Aqueous Extract of Ns (NsE) 2ml/day for 40 days and then with placebo for another 40 days. Fasting blood sample of each individual was collected at base line (0 day), after NsE treatment (40th day) and after placebo administration (80th day) of the study. The baseline readings served as concurrent control. Total leukocyte count and platelet count were analyzed on the day of collection of blood samples, while serum and plasma were stored at a temperature of -20°C till completion of study, when glucose, insulin level, total cholesterol, HDL cholesterol, LDL cholesterol, triglycerides, ALT, AST and blood urea were determined.

Selection Criteria

The criteria used for the selection of T2DM individuals, were patients with; a) known T2DM history and of either sex; b) elevated fasting blood glucose levels and taking usual diabetic medicine; c) age 30-60 years; d) no insulin therapy; e) no other chronic disease; f) no medication, other than that for diabetes; g) no herbal treatment for any disease, where the recipe contains black seed in any form and h) no pregnancy.

Registration, Briefing and Consent

All the selected cases from twin cities of Rawalpindi/Islamabad, Pakistan were registered and briefed about the study. They were told that this study will not only guide us for a better alternate treatment for their sufferings, but they will also get an opportunity to have all their tests done free of cost. They were explained that they will receive two treatments phases each lasting for 40 days. Their fasting blood samples will be collected on 0, 40th and 80th day of the study. They will keep taking their usual diabetic medicine throughout the study period and intimate any alteration in the treatment. They will also keep taking their usual diet and continue with their usual activities. Their written consent was obtained.

Nigella sativa Seeds

Nigella sativa seeds were purchased from a herborist from the local market (Rawalpindi, Pakistan) after careful selection. The seeds were identified at "Herbarium Medicinal Botanic Centre; Pakistan Council of Scientific and Industrial Research (P.C.S.I.R.) Laboratories Complex, Peshawar, Pakistan". A voucher specimen No. (PES): 9747 was deposited there for future reference.

Preparation of *Nigella sativa* Aqueous Extract

Nigella sativa seeds were manually cleaned, washed with distilled water and dried in an oven at 40°C. Dried seeds were then grinded into powder using electric grinder in small amounts avoiding heating. Aqueous Extract was prepared using soxlet apparatus. One kg grinded seeds were placed in the thimble and 2.5 liters H_2O was added in the round bottom flask as solvent; at 100°C for 24 hrs. The extract obtained was further concentrated up to 1 lit. Aqueous Extract was first collected in 1000 ml sterilize bottle and dispensed in 30ml brown sterile glass bottles having a dropper and cap at their mouth. Each dropper was checked for the uniform size and quantity of droplet and was rejected otherwise.

Preparation of Placebo

Placebo capsules were filled with ready to use wheat bran which was also made available from the local market.

Collection of Blood Samples

Seven mL of fasting blood samples was collected from each patient under sterile conditions in Medical Center of PMAS Arid Agriculture University, Rawalpindi. Each blood sample was collected in three different tubes *i.e.*, i) Plain tube (4 mL), ii) Glucose tube (2 mL), iii) CP tube (1 mL). Plane tubes are without any additive, glucose tubes contain Potassium Fluoride and CP tubes contained potassium editate. All the tubes were properly labeled to afford an identity of the patient.

Preparation and Storage of Serum and Plasma

Serum and plasma were separated in plain tube through centrifugation at 85 G for nine minutes. These were stored in labeled sterilized plain tubes at -20°C in the Department of Food Technology for assay at a proper time.

Tests for Total Leukocyte Count (TLC) and Platelet Count

Blood samples in CP tubes were processed on the day of collection for TLC and platelet count on auto analyzer (Sysmex KX-21, Made in Japan) in the Department of Pathology.

Analysis of Plasma for Glucose and Serum for Insulin and Lipid Profile

Analysis of plasma for glucose and serum for total cholesterol, HDL cholesterol, LDL cholesterol and triglycerides were performed on auto analyzer (VITALAB Selectra E, Made in Netherlands). Kit used for the purpose was (Linear chemicals, Spain). Serum insulin level was determined on (IMMULITE–1000 Hormone Analyzer, Made in USA). Kit used for this purpose was (Diagnostic Products Corporation, PILINC-23, USA). Analyses were executed in the Department of Pathology.

Statistical Analysis

Data collected from all the patients was analyzed statistically using SPSS (Statistical Package for Social Sciences) Version 22. The results are expressed as Mean±SEM. For each parameter, mean values were compared by paired sample test.

Type 2 Diabetes Mellitus Patients

Forty four T2DM patients successfully completed the study. Their mean age was 47.814±1.053 years falling in forties, 45.45 per cent were males (20) and 54.54 per cent females (24).

Parameters

The main parameters studied were glucose, insulin and lipid profile. However total leukocyte, platelet count, ALT, AST and blood urea levels were taken into consideration as safety parameters where as subjective feelings of the patients like defective vision, tingling sensations, burning feet, aches and pains, diabetic foot, visual acuity, diabetic retinopathy, polyuria and polydipsia were also taken under observation.

Some factor on which diabetes depends directly or indirectly, such as activity, stress and body mass index (BMI) were also taken into account. Similarly dietary intake (protein, carbohydrates, fats and calories) also play an important role in diabetes (Samaha, 2003; Yancy, 2004). A high-starch, high-carbohydrate diet excessively stimulates appetite and may disturb energy level in diabetic patients. A reduction of carbohydrates normalizes the balance, reduces insulin concentrations and favours utilization of stored fat as fuel as well as significantly reducing insulin resistance (Boden, 2005). Weight loss in overweight persons is improved by a higher proportion of protein, presumably due to protein's effect on satiety and/or metabolic efficiency (Krieger, 2006; Skov, 1999; Stubbs, 1996). A reduction in carbohydrates for patients with T2DM effectively reduces both fasting and postprandial glucose. These effects can be independent of weight loss (Boden, 2005; Gannon and Nuttall 2004). Similarly activity level and stress status also plays important role in diabetes mellitus. So keeping in view the above mentioned facts, these factors were also studied at base line, after forty days of treatment with NsE and then after another forty days of placebo.

3. RESULTS

Glucose Profile

The mean fasting glucose level±SEM at base line (concurrent control) was 190.272±7.653 mg/dL, which fell to 181.022±6.126 mg/dL after 40 days treatment with NsE. That again rose to 189.022±7.655 mg/dL after treatment with placebo for a similar period (Figure 5.1). When pretreatment and after treatment results were subjected to paired t test, the fall in glucose level after treatment was non significant (p=0.051), which increased after the placebo non significantly (p=0.054) (Appendix 01). Fall in fasting blood glucose level was seen in 65.90 per cent cases (n=44) after NsE treatment and a rise was observed in 93.18 per cent of the cases after the placebo.

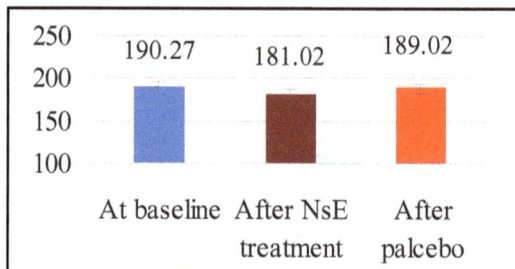

Figure 5.1: Mean±SEM of Fasting Glucose at Various Phases of Treatment.

Insulin Level

The mean insulin level±SEM at base line was 7.453±0.780 ulU/mL, which increased to 8.113±0.728 ulU/mL after treatment with NsE and again decreased to 7.866±0.728 ulU/mL after placebo for a period similar to the treatment period (Figure 5.2). The rise in insulin level after treatment was non significant (p=0.099), which non significantly fell after the placebo (p=0.584) (Appendix 02). Rise in fasting blood insulin level was seen in 58.139 per cent cases (n=44) after NsE treatment and a fall was observed 52.27 per cent of the cases after the placebo.

Activity Level

The value labels allotted for calculation of activity/exercise level were (0) - does not exercise, (1) - exercises for 1 hour weekly, (2) - exercises for 2 hours weekly, (3) - exercises for 3 hours weekly, (4) - exercises for 4 hours weekly, (5) - exercises for ≥5 hours weekly. The mean activity level±SEM was 2.568±0.190 hrs/week at base line, 2.636±0.175 hrs/week after treatment with NsE and 2.613±0.175 hrs/week after the placebo. However these changes were almost negligible and statistically non significant *i.e.* change after treatment with NsE (p=0.498) and after the placebo (p=0.838) (Appendix 03). Activity levels of the patients at base Line, after NsE treatment and after placebo are expressed in Table 5.1:

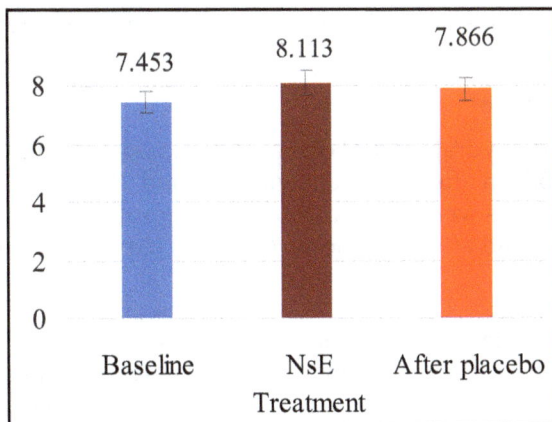

Figure 5.2: Mean±SEM of Fasting Insuline at Various Phases of Treatment.

Table 5.1: Activity Level of Patients

	No Activity	I hr/Week	2 hrs/Week	3 hrs/Week	4 hrs/Week	5 hrs/Week
Patients at Base Line	3	5	13	14	7	2
Patients after NsE	2	5	12	14	10	1
Patients after Placebo	1	8	9	17	7	2

Stress Level

The value labels allotted for calculation of stress level were (0) for No Stress, (1) for Mild Stress, (2) for Moderate Stress and (3) for Severe Stress. The mean stress level±SEM of the patients was 1.931±0.075 at base line, 1.818±0.074 after NsE treatment and 1.886±0.066 after the placebo. These changes were negligible and statistically non significant *i.e.* change after treatment with NsE (p=0.133) and after the placebo (p=0.183) (Appendix 04). Stress status of the patients at base Line, after NsE treatment and after placebo are expressed in Table 5.2:

Table 5.2: Stress Level of Patients

	No Stress	Mild Stress	Moderate Stress	Severe Stress
Patients at Base Line	Nil	7	33	4
Patients after NsE	Nil	10	32	2
Patients after Placebo	Nil	7	35	2

Dietary Intake

Protein Intake

The mean Protein intake±SEM of the patients was 87.767±3.947 gm at base line 86.720±3.985 gm after treatment with NsE and 84.023±4.831 gm after the placebo.

These changes were statistically non significant *i.e.* change after treatment with NsE (p=0.117) and after the placebo (p=0.224) (Appendix 05).

Carbohydrates Intake

The mean Carbohydrates±SEM of the patients was 321.976±14.988 gm at base line, 319.767±15.058 gm after treatment with NsE and 313.837±15.876 gm after the placebo. These changes were statistically not significant *i.e.* change after treatment with NsE (p=0.058) and after the placebo (p=0.183) (Appendix 06).

Fats Intake

The mean Fats intake±SEM of the patients was 48.372±2.607 gm at base line, 48.790±2.561 gm after treatment with NsE and 49.604±2.419 gm after the placebo. These changes were statistically non significant *i.e.* change after treatment with NsE (p=0.299) and after the placebo (p=0.236) (Appendix 07).

Calories Consumption

The mean Calories±SEM of the patients was 2070.558±87.766 at base line, 2063.116±88.892 after treatment with NsE and 2039.302±91.644 after the placebo. Statistically these changes were non significant *i.e.* change after treatment with NsE (p=0.175) and after the placebo (p=0.277) (Appendix 08).

Body Mass Index (BMI)

The mean level of BMI±SEM of the patients was 28.681±0.697 at base line, 28.582±0.689 after treatment with NsE and 28.549±0.690 after the placebo. Statistically the changes were not significant *i.e.* change after treatment with NsE (p=0.061) and after the placebo (p=0.518) (Appendix 09).

Lipids Profile

Total Cholesterol (TC)

The mean total cholesterol±SEM at base line was 200.409±6.818 mg/dL, which fell to 187.795±5.210 mg/dL after treatment with NsE and again rose to 194.477±6.376 mg/dL after the placebo (Figure 5.3). The fall in total cholesterol level after treatment was significant (p=0.019), which rose after the placebo non significantly (p=0.168) (Appendix 10). Fall in TC level was recorded in 52.27 per cent of the patients, 40.90 per cent experienced a rise while 45.45 per cent experienced no change after NsE treatment. The TC level increased again in 54.44 per cent of the cases after the placebo.

HDL Cholesterol

The mean HDL cholesterol level±SEM at base line was 40.386±0.986mg/dL, it was 40.022±0.879mg/dL after treatment with NsE and 40.704±1.274 mg/dL at the end of placebo (Figure 5.3). No significant change was noticed in HDL cholesterol level after NsE treatment (p=0.713), or after the placebo (p=0.549) (Appendix 11).

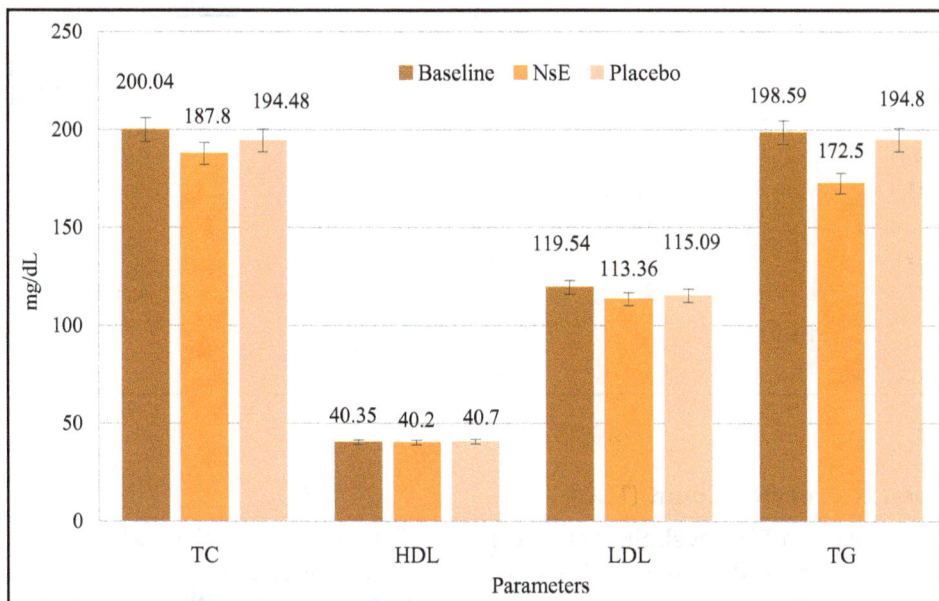

Figure 5.3: Comparison of Total Cholesterol (TC), HDL Cholesterol (HDL), LDL Cholestrol (LDL) and Trigycerides (TG).

LDL Cholesterol

The mean LDL cholesterol level±SEM at base line was 120.409±4.618 mg/dL, which fell to 113.363±3.998 mg/dL after treatment with NsE and rose to 115.090±3.851 mg/dL after the placebo (Figure 5.3). The fall in LDL cholesterol level after treatment was not significant (p=0.119) and the same trend was noted after the placebo (p=0.609) (Appendix 12).

Triglycerides (TG)

The mean TG level±SEM at base line was 200.227±13.667 mg/dL, which fell to 172.522±12.715 mg/dL after treatment with NsE and again rose to 194.818±14.989 mg/dL after placebo (Figure 5.3). Statistically the fall in TG level after treatment was significant (p=0.007), and same was the rise after the placebo (p=0.020) (Appendix 13). Fall in TG level was recorded in 56.81 per cent of the patients, 36.36 per cent experienced a rise while 6.8 per cent experienced no change after NsE treatment. After the placebo a rise was observed in 65.9 per cent of the cases.

Safety Parameters

Platelet Count

The mean platelet count±SEM was 23.459±1.176 ×10^{10}/l, at base line, which changed to 22.284±1.025 ×10^{10}/l after treatment with NsE and changed to 22.588±0.943 ×10^{10}/l after the placebo (Figure 5.4). The change after treatment with NsE was non significant (p=0.237) and change after the placebo was also non significant (p=0.524) (Appendix 14).

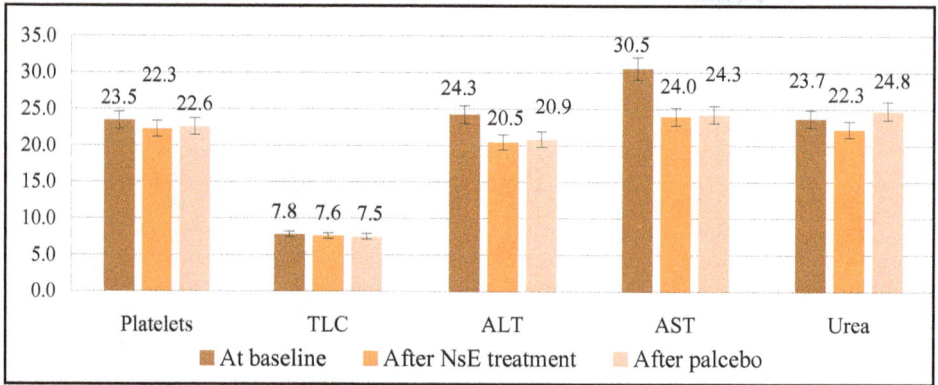

Figure 5.4: Safety Parameters.

Total Leucocytes Count (TLC)

The mean TLC level±SEM at base line was 7.829±0.302 ×10^9/l, which changed to 7.620±0.256 ×10^9/l after treatment with NsE and changed to 7.522±0.252×10^9/l after the placebo (Figure 5.4). The change after treatment with NsE was non significant (p=0.206) and change after the placebo was also non significant (p=0.214) (Appendix 15).

ALT Level

The mean ALT level±SEM at base line was 24.250±1.504 which reduced to 20.454±1.196 after treatment with NsE and changed to 20.863±1.440 after the placebo (Figure 5.4). The change after treatment with NsE was significant (p=0.002) however the change after placebo was non significant (p=0.722) (Appendix 16).

AST Level

The mean AST level±SEM at base line was 30.545±1.620 which reduced to 23.977±1.471 after treatment with NsE and increased to 24.272±1.660 after the placebo (Figure 5.4). The change after treatment with NsE was significant (p=0.000) but the change after placebo was non significant (p=0.812) (Appendix 17).

Blood Urea Level

The mean Blood Urea Level±SEM at base line was 23.704±0.935 which changed to 22.250±0.827 after treatment with NsE and further changed to 24.772±1.046 after the placebo (Figure 5.4). However the change after treatment with NsE was non significant (p=0.088) but the change after placebo was found significant (p=0.001) (Appendix 18).

Subjective Feelings

Subjective feelings of the DM2 patients like Defective Vision, Tingling Sensations, Burning Feet, Aches and Pains, Diabetic Foot, Visual Acuity, Diabetic Retinopathy, Polyurea and Polydypsia were taken into consideration to see the effect of treatment with *Nigella sativa* Extract (NsE). All the patients were clinically examined and interviewed at base line (concurrent control), after NsE treatment of forty days and after the placebo for another forty days. All the parameters except Visual Acuity and Diabetic Retinopathy were recorded as Yes or No. To analyze the collected data the value labels allotted to (Yes) was (01) and (No) was (02). However for Visual Acuity (00) for (No), (01) for (Yes Right Eye), (02) for (Yes left Eye) and (03) for (Yes Right and Left Eye) while for Diabetic Retinopathy (00) for (No Signs of Diabetic Retinopathy), (01) for (Background Diabetic Retinopathy), (02) for (Pre-proliferative Diabetic Retinopathy) and (03) for (Poliferative Diabetic Retinopathy). Results are as under:

Defective Vision

The mean of Defective Vision level±SEM at the base line in the patients was 1.431±0.075 which changed to 1.409±0.074 after treatment with Ns Extract and changed to 1.454±0.075 after the placebo. The change after treatment with NsE was significant (p=0.000) and the change after placebo was also significant (p=0.000) (Appendix 19). At the base line, 40.90 per cent patients (n=44) were not suffering with defective vision, this percentage changed to 38.63 per cent after NsE treatment and changed to 43.18 per cent after the placebo. The decreased value after NsE treatment reveals the less number of patients replying (No) when examined for the defective vision as compared to the base line.

Tingling Sensations

The mean of Tingling Sensations level±SEM at the base line in the patients was 1.545±0.075 which changed to 1.568±0.075 after treatment with Ns Extract and remain unchanged to 1.568±0.075 after the placebo. The change after treatment with NsE was non significant (p=0.323) and no change occurred after the placebo hence the t could not be computed because the standard error of the difference was 0 (Appendix 20). At the base line, 54.54 per cent patients (n=44) replied No, when investigated for Tingling Sensations, this percentage increase to 56.81 per cent after NsE treatment while it remained the same after the placebo.

Burning Feet

The mean of Burning Feet level±SEM at the base line in the patients was 1.545±0.075 which changed to 1.590±0.074 after treatment with Ns Extract and changed to 1.477±0.076 after the placebo. The change after treatment with NsE was non significant (p=0.323) and the change after placebo was also non significant (p=0.058) Appendix 21). At the base line, 52.27 per cent patients (n=44) replied No, when investigated for Burning Feet, this percentage increase to 58.13 per cent after NsE treatment and reduced to 45.45 per cent after the placebo. The increased value after NsE treatment indicates that more number of patients replied (No) when inquired about the Burning Feet.

Aches and Pains

The mean level for Aches and Pains±SEM at the base line in DM2 patients was 1.250±0.066 which changed to 1.318±0.071 after treatment with Ns Extract and changed to 1.204±0.061 after the placebo. The change after treatment with NsE was non significant (p=0.083) and the change after placebo was also non significant (p=0.058) (Appendix 22). At the base line, 25 per cent patients (n=44) replied No, when investigated for Aches and Pains, this percentage increase to 31.81 per cent after NsE treatment and reduced to 20.45 per cent after the placebo. The increased value after NsE treatment indicates that more number of patients replied (No) when asked about the Aches and Pains.

Diabetic Foot

The mean Level of Diabetic Foot±SEM at base line was 1.909±0.043 which remain unchanged to 1.909±0.043 after treatment with NsE and also remained unchanged to 1.909±0.043 after the placebo. The correlation and t could not be computed because the standard error of the difference was 0 (Appendix 23). At the base line, 9.09 per cent patients (n=44) were suffering from Diabetic Foot, this percentage remained the same after NsE treatment and after the placebo.

Visual Acuity

The mean Level of Visual Acuity±SEM at base line was 1.159±0.217 which remained 1.159±0.217 after treatment with NsE and remained unchanged to 1.159±0.217 after the placebo. The correlation and t could not be computed because the standard error of the difference was found 0 (Appendix 24). At the base line, 38.63 per cent patients (n=44) were found in the distress of Visual Acuity. Out of these 2.27 per cent with Right, 2.27 per cent with Left and 34.09 per cent with both Right and Left eyes. This percentage however remained the same after NsE treatment and after the placebo.

Diabetic Retinopathy

The mean Level of Diabetic Retinopathy±SEM at base line was 0.113±0.074 which remained 0.113±0.074 after treatment with NsE and remained unchanged to 0.113±0.074 after the placebo. The correlation and t could not be computed because the standard error of the difference was found 0 (Appendix 25). At the base line, 6.81 per cent patients (n=44) were found in the misery of Diabetic Retinopathy. Out of these 4.54 per cent had background diabetic retinopathy while 2.27 per cent had proliferative diabetic retinopathy. This percentage remained unmoved after NsE treatment and after the placebo.

Poly-Urea

The mean level of Poly-Urea±SEM at the base line in DM2 patients was 1.522±0.076 which changed to 1.590±0.074 after treatment with NsE and changed to 1477±0.076 after the placebo. The change after treatment with NsE was non significant (p=0.323) and the change after placebo was also non significant (p=0.096) (Appendix 26). At the base line, 52.27 per cent patients (n=44) replied No, when investigated about Poly-urea, this percentage increase to 56.81 per cent after NsE treatment and dropped to 45.45 per cent after the placebo. The increased value after NsE treatment indicates that more number of patients replied (No) when asked about Poly-urea.

Poly-Dypsia

The mean level of Poly-dypsia±SEM at the base line in DM2 patients was 1.431±0.075 which changed to 1.500±0.076 after treatment with Ns Extract and changed to 1.386±0.074 after the placebo. The change after treatment with NsE was not significant (p=0.262) and the change after placebo was also non significant (p=0.058) (Appendix 27). At the base line, 40.90 per cent patients (n=44) replied No, when investigated about Poly-dypsia, this percentage increase to 47.72 per cent after NsE treatment and dropped to 36.36 per cent after the placebo. The increased value after NsE treatment indicates that more number of patients replied (No) when inquired about Poly-dypsia.

4. DISCUSSION

Glucose and Insulin Levels with Reference to Activity Level, Stress Level, BMI and Dietary Intake

A noticeable decrease in glucose and increase in insulin levels was observed after the treatment with NsE, which reversed after the placebo, but these changes were not significant. The factors that possibly can affect the insulin or glucose level such as activity, stress or the dietary profile (protein, carbohydrates, fats and calories intake), remained unchanged and only very minor and statistically non-significant changes were noticed. So it can reasonably be assumed that whatever the improvement was observed it definitely was due to NsE. This fall can be attributed to rise in insulin due to this treatment (Halima, 2004). Similar results have been obtained in streptozotocin-induced diabetic rats, with plant and seed extracts (Labhal, 1997; Al Awadi, 1991) and alloxan induced diabetic rats with seed extract (Meral, 2001). However they concluded that this effect is due to increased insulin release from the pancreas. It has been proved that Ns in rats decreases the oxidative stress and preserves pancreatic β-cell integrity and subsequently leads to better insulin production (Benhaddou-Andaloussi, 2011; Abdelmeguid, 2010) in diabetic Meriones shawi. The researcher also assessed lipids, insulin, leptin, and adiponectin levels. ACC phosphorylation and Glut4 protein content were determined in liver and skeletal muscle.

Result of this study regarding glucose and insulin levels seem to be almost in line with the previous studies. A possible reason for non significant improvement in insulin and glucose levels might be the low dose of NsE 2ml/day which we used as compared to the doses used by weight of the animal models. Probably the increased dose would have yielded better results but safety was the concern. Another reason which cannot be neglected is thymoqunine an active component of Ns present in volatile and fixed oils was missing here, as here aqueous extract was used. Hence added advantage of thymoqunine was missing. Our previous study on effect of Ns oil on type 2 diabetic patients (Bilal, *et al.*, 2009) yielded much better insulin increase and glucose fall. The volatile oil, a pale yellow liquid with a characteristic aromatic odor and taste is readily soluble in organic solvents such as ether, chloroform and ethanol but only sparingly soluble in water (E1-Alfy, 1975). Further investigations are therefore recommended in human subjects.

Lipid Profile

A significant fall was observed in total cholesterol (TC) and triglycerides (TG) levels after treatment phase. In placebo phase this fall reversed but didn't revert to the baseline (see Figure 5.3). These changes can reasonably be attributed to NsE as their activity, stress level and dietary habits remained almost constant throughout the study. The individual TC pattern of the most patients after the treatment revealed that majority had TC < 190 mg/dL. An important thing to notice is that TG reduction was in all patients except one, who had baseline level of above 300 mg/dL. Though statistically not significant a noticeable decrease was also seen in LDL cholesterol after treatment, which can be attributed to fall in TC. Their individual LDL pattern shows that NsE treatment provided relief to the patients specially with a level of 175 mg/dL or above. HDL cholesterol remained almost unchanged showing no effect of NsE on it. Our findings are almost in line with previous studies where a significant decrease in lipid levels was observed in diabetes induced sand rats and rabbits treated with 0.5 g/kg of NsE (Faruk, 2014; Labhal, 1997). The carry over effect *i.e.* fall in levels (as the placebo level was less than the base line level) both in TC and TG is favorable for T2DM patients. Previously no studies have been reported on the effect of NsE on lipid profile in human subjects. However, in our previous study using Ns oil in T2DM patients showed an undesirable significant increase in TG (Bilal, 2009). Studies with different dosage schedules in T2DM patients are recommended, which may bring forth more positive outcomes.

Safety Profile

Though Ns has been used by the humans since ages and thought beneficial to the health. It's LD50 in animals is 0.50 ml/kg of body weight. Even then this is a valid concern and to look into this aspect hepatic, renal and hematological parameters were included in the study. The total leukocyte and platelet counts of these patients dropped slightly after the treatment with NsE as compared to the base line levels, which were not statistically significant, declaring it safe as regards hematological parameters are concerned. Similar observations were made by (Zaoui, 2002), who stated that Ns is safe and beneficial but total leukocyte and platelet count must be taken into consideration. This small decrease in these cellular counts is feared to exaggerate in case larger doses of NsE are used; larger studies are suggested to alley this anxiety. However its platelet lowering effect may be used to protect against vascular events among patients with stable angina, intermittent claudication, and atrial fibrillation (Trialists' Collaboration, A., 2002).

Adverse drug reactions affecting the liver represent an important challenge for safety in drug development. Drugs can reproduce practically the whole spectrum of liver diseases, but acute hepatitis is the most common syndrome. Most hepatic drug reactions occur in only a small proportion of individuals, making them difficult to detect at the time of drug development (Labhal, 1997). The markers of hepatic safety ALT and AST were studied in detail, which remained in normal ranges in all phases of study. The drop in both ALT and AST was significant (p<0.050) after treatment and placebo phase. This announces the hepatic safety of the NsE. No significant change was observed after the placebo period, showing a beneficial effect of NsE and no effect of the placebo. Here it is difficult to say that NsE can be used to lower the elevated ALT and AST levels; for which large scale studies are a requirement.

As for renal safety profile of NsE is concerned, it proved considerably safe. The minor and not statistically significant changes were seen in the blood urea level of the patients after treatment and placebo phases of the study. If we look at the data a statistically non significant change from 23.7045 with SD 6.204 to 22.25 SD (p=0.088), but after the treatment was replaced with inert place, it crossed the baseline value *i.e.* 24.7727±6.944. This shows that initially there was a change in urea level, which may be highlighted, if larger dose schedule is applied.

These findings are suggestive of beneficial effect of NsE at a dose of 2ml/day in T2DM patients.

Subjective Feelings

As for subjective feelings are concerned, no harmful effect of NsE was seen on any of these, rather it improved aches and pains, burning feet, tingling sensations and blurred vision to some extent though these improvements or changes were statistically non-significant. Improvement in some of these parameters may be attributed to the reduction in glucose level and lipid profile of the patients. No previous studies have been reported to the best of our knowledge on the effect of NsE on subjective feelings of T2DM patients. As diabetes mellitus is a chronic metabolic syndrome and these subjective feelings develop in the patients slowly and gradually so a period of forty days treatment may not be sufficient to ameliorate these sufferings. Moreover sometimes it is not possible for the depressed and hopeless chronic sufferers of this disease to perceive the minor changes. These patients have pathetic attitude towards their ailment and may ignore the minor improvements.

5. CONCLUSIONS

There are sufficient evidences to believe or conclude that NsE reasonably lowers the fasting blood glucose, improves the lipid profile to some extent without affecting the leukocytes, platelets, liver enzymes and renal profile, thus highlighting its safe use as anti-diabetic. On the other hand in this study we also assume that a bit higher dose of *Nigella sativa* aqueous extract (>2ml/day) for a longer period (more than 40 days) may result in more encouraging and prominent findings regarding T2DM patients. Safety margin is such that it does not seem harmful in human beings. For this, more human studies are recommended.

6. RECOMMENDATIONS

On the basis of findings of the study it is strongly recommended that:-

1. NsE should be advocated as useful solo or in combination with other diabetic herbal treatments.
2. Large scale human studies on *Nigella sativa* should be planned to establish;
    a. Its efficacy
    b. Its safety
    c. The best dosage schedule.

7. ACKNOWLEDGMENTS

We highly acknowledge the financial support provided by Higher Education Commission (HEC) of Pakistan for this study in the form of HEC Indigenous PhD Scholarship. We are thankful to Mr. Chaudhary Abdur Shakoor, Head, Statistics Department, PMAS Arid Agriculture University Rawalpindi, for providing us guidance in statistical analysis of data.

8. REFERENCES

1. Abdelmeguid, N.E., Fakhoury R., Kamal S.M., Wafai R.J. 2010. Effects of *Nigella sativa* and thymoquinone on biochemical and subcellular changes in pancreatic β-cells of streptozotocin-induced diabetic rats. *J Diabetes*; 2 (4): 256-266.

2. Al-Awadi, F., H. Fatania, U. Shamte. 1991. The effect of a plants mixture extract on liver gluconeogenesis in streptozotocin induced diabetic rats. *Diabetes Res.*, 18 (4): 163-168.

3. Al-Ghamdi, M. S. 2003. Protective effect of *Nigella sativa* seeds against carbon tetrachloride- induced liver damage. *Am. J. Chin. Med.*, 31 (5): 721-728.

4. Ali, B. H., and G. Blunden. 2003. Pharmacological and toxicological properties of *Nigella sativa*. *Phytother Res.*, 17 (4): 299-305.

5. Al-Rowais, N. A. 2002. Herbal medicine in the treatment of diabetes mellitus. *Saudi Med. J.*, 23 (11): 1327-1331.

6. Benhaddou-Andaloussi A, Martineau L, Vuong T, Meddah B, Madiraju P, Settaf A. 2011. The *in vivo* anti-diabetic activity of *Nigella sativa* is mediated through activation of the AMPK pathway and increased muscle glut4 content. Evid Based Complement Alternat Med. Published online 2011 Apr 14. doi: 10.1155/2011:538671.

7. Bilal, A., Masud T., Uppal A.M., Naveed A.K. 2009. Effect of *Nigella sativa* oil on some blood parameters in type 2 diabetes mellitus patients. *Asian Journal of Chemistry* 21 (7): 5373-5381.

8. Boden G., 2005. Effects of a low-carbohydrate diet on appetite, blood glucose levels, and insulin resistance in obese patients with type 2 diabetes. *Ann Intern Med.*, 142 (6): 403-411.

9. El-Alfy, T.S., El-Fattatry, H.M. and Toama, M.A. 1975. *Pharmazie* 30 : 109-111.

10. El-Dakhakhny, M., Mady N. I., Lembert N. and Ammon H. P. 2002. The hypoglycemic effect of *Nigella sativa* oil is mediated by extrapancreatic actions. *Planta Med.*, 68 (5): 465-466.

11. El-Missiry, M. A. and A. M. El-Gindy. 2000. Amelioration of alloxan induced diabetes mellitus and stress in rats by oil of *Eruca sativa* seeds. *Ann. Nutr. Metab.*, 44 (2): 97-100.

12. Fararh, K. M., Atoji Y., Shimizu Y. and Takewaki T. 2002. Isulinotropic properties of *Nigella sativa* oil in Streptozotocin plus Nicotinamide diabetic hamster. *Res. Vet. Sci.*, 73 (3): 279-82.

13. Fararh, K. M., Atoji Y., Shimizu Y., Shiina T., Nikami H. and Takewaki T. 2004. Mechanisms of the hypoglycaemic and immunopotentiating effects of *Nigella sativa* L. oil in streptozotocin-induced diabetic hamsters. *Res. Vet. Sci.*, 77 (2): 123-129.

14. Faruk H., Al-jawad, Hashim M. Hashim, Batool A. Al-khafaji, 2014. Effect of Aqueous Extract of some medicinal plants on plasma lipid profile in diabetes induced rabbits. *Medical Journal of Babylon* 2:4 doi:1812-156X-2-4.

15. Felix, G., *New Larousse Encyclopedia of Mythology*. 1968. The Hamlyn Publishing Group Ltd, Middlesex, England.

16. Gannon, M. C. and Nuttall F. Q., 2004. Effect of a high-protein, low-carbohydrate diet on blood glucose control in people with type 2 diabetes. *Diabetes*, 53: 2375-2382.

17. Godin, D. V., Saleh A. W., Maureen E. G. and Goumeniouk A. D. 1998. Anti-oxidant enzyme alterations in experimental and clinical diabetes. *Mol. Cell. Biochem.*, 84: 223-231.

18. Halima, R., 2004. *Nigella sativa* seed extracts enhance glucose-induced insulin release from rat-isolated Langerhans islets. *Fundamental and Clinical Pharmacology* 18 (5): p 525.

19. Hawsawi, Z. A., Ali B. A., Bamosa A. O. 2001. Effect of *Nigella sativa* (black seed) and thymoquinone on blood glucose in albino rats. *Annals of Saudi Medicine*, 21 (5): 3-4.

20. John Lust, 1984. *The Herb Book*. Bantam Books, New York, USA.

21. Kaleem, M., Kirmani D., Asif M., Ahmed B. 2006. Biochemical effects of *Nigella sativa* L seeds in diabetic rats. *Indian J. Exp. Biol.*, 44 (9): 745-748.

22. Kalus, U., Pruss A., and Bystron J. 2003. Effect of *Nigella sativa* (black seed) on subjective feeling in patients with allergic diseases. *Phytother Res.*, 17 (10): 1209-1214.

23. Kamimura, S., 1992. Increased 4-hydroxynonenal levels in experimental alcoholic liver disease; association of lipid peroxidation with liver fibrogenesis. *Hepatology*, 16: 93-98.

24. Kanter, M., Coskun O., Korkmaz A. and Oter S. 2004. Effects of *Nigella sativa* on oxidative stress and beta-cell damage in streptozotocin-induced diabetic rats. *Anat. Rec. A. Discov. Mol. Cell. Evol. Biol.*, 279 (1): 685-691.

25. Kanter, M., Meral I., Yener Z., Ozbek H. and Demir H. 2003. Partial regeneration/proliferation of the beta-cells in the islets of Langerhans by *Nigella sativa* L. in streptozotocin-induced diabetic rats. *Tohoku J. Exp. Med.*, 201: 213-219.

26. Krieger, J. W., H. S. Sitren, M. J. Daniels, B. Langkamp-Henken. 2006. Effects of variation in protein and carbohydrate intake on body mass and composition during energy restriction. *Am. J. Clin. Nutr.*, 83 (2): 260-274.

27. Labhal, A. 1997. Action anti-obesity. Hypocholestrolemiante et hypotrigly-ceridemiante de *Nigella sativa* chez le *Psammomys obesus*. *Caducee*, 27: 26-28.

28. Larrey, D. (2001, December). Epidemiology and individual susceptibility to adverse drug reactions affecting the liver. In *Seminars in liver disease* (Vol. 22, No. 2, pp. 145-155).

29. Malcolm, S., 1987. *The Encyclopedia of Herbs and Herbalism*. Macdonald and Co, Turin, Italy.

30. Meral, I., Yener Z., Kahraman T. and Mert N. 2001. Effect of *Nigella sativa* on glucose concentration, lipid peroxidation, anti-oxidant defence system and liver damage in experimentally-induced diabetic rabbits. *J. Vet. Med. A. Physiol. Pathol. Clin. Med.*, 48 (10): 593-599.

31. Michelitsch, A., Rittmannsberger A., 2003. A simple differential pulse polarographic method for the determination of thymoquinone in black seed oil. *Phytochem. Anal.*, 14 (4): 224-227.

32. Nickavar, B., Mojab F., Javidnia K. 2003. Chemical composition of the fixed and volatile oils of *Nigella sativa* L. from Iran. *Z Naturforsch, [C]*. 58 (9): 629-31.

33. Rchid H., Chevassus H., Nmila R., Guiral C., Petit P., Chokairi M. and Sauvaire Y. 2004. *Nigella sativa* seed extracts enhance glucose-induced insulin release from rat-isolated Langerhans islets. *Fundamental and Chinical Pharmacology*, 18 (5); 525-529.

34. Samaha, F.F., Iqbal N., Seshadri P., 2003. A low-carbohydrate as compared with a low-fat diet in severe obesity. *New Eng. J. Med.*, 348: 2074-2081.

35. Skov, A.R., Toubro S., Ronn B., Holm L. Astrup A., 1999. Randomized trial on protein vs carbohydrates in ad libitum fat reduced diet for the treatment of obesity. *Int J Obes Relat Metab Disord.*, 23 (5): 528-536.

36. Stubbs, R.J., M.C. vanwyk, A.M. Johnstone, C. G. Harbron. 1996. Breakfast high protein, fat or carbohydrates: effect on within-day appetite and energy balance. *Eur. J. Clin. Nutr.*, 50 (7): 409-417.

37. Trialists'Collaboration, A. (2002). Collaborative meta-analysis of randomised trials of antiplatelet therapy for prevention of death, myocardial infarction, and stroke in high risk patients. *Bmj*, 324(7329), 71-86.

38. Turkdogan, M. K., H., Ozbek. 2003. The role of Urtica dioica and *Nigella sativa* in the prevention of carbon tetrachloride-induced hepatotoxicity in rats. *Phytother Res.*, 17 (8): 942-6.

39. Warrier PK, Nambiar VPK, Ramankutty. Chennai: Orient Longman Pvt Ltd; 2004. Indian medicinal plants-a compendium of 500 species; pp. 139-142.

40. Yancy W. S., M. K. Olsen, J. R. Guyton, R. P. Bakst, E. C. Westman. 2004. A low-carbohydrate ketogenic diet versus a low-fat diet to treat obesity and hyperlipidemia. *Ann. Int. Med.*, 140: 769-777.

41. Yarnell E, Abascal K. 2011. *Nigella sativa*: holy herb of the middle East. *Altern Compl Therap.*, 17 (2): 99-105.

42. Zaoui, A., Cherrah Y., and Mahassini N. 2002. Acute and chronic toxicity of *Nigella sativa* fixed oil. *Phytomedicine*, 9 (1): 69-74.

Appendix 1: Paired Samples Statistics (Glucose Level)

		Mean	N	SD	SEM	t	df	Sig. (2-tailed)
Pair 1	At Base Line	190.2727	44	50.7691	7.65374	2.006	43	0.051
	After NS Extract	181.0227	44	40.6387	6.12653			
Pair 2	After NS Extract	181.0227	44	40.6387	6.12653	−1.977	43	0.054
	After placebo	189.0227	44	50.7792	7.65526			
Pair 3	At Base Line	190.2727	44	50.7691	7.65374	0.241	43	0.810
	After placebo	189.0227	44	50.7792	7.65526			

Appendix 2: Paired Samples Statistics (Insulin Level)

		Mean	N	SD	SEM	t	df	Sig. (2-tailed)
Pair 1	At Base Line	7.4530	44	5.17740	0.78052	−1.686	43	0.099
	After NS Extract	8.1130	44	4.83448	0.72883			
Pair 2	After NS Extract	8.1130	44	4.83448	0.72883	0.551	43	0.584
	After placebo	7.8664	44	4.82993	0.72814			
Pair 3	At Base Line	7.4530	44	5.17740	0.78052	−1.013	43	0.317
	After placebo	7.8664	44	4.82993	0.72814			

Appendix 3: Paired Samples Statistics (Activity Level)

		Mean	N	SD	SEM	t	df	Sig. (2-tailed)
Pair 1	At Base Line	2.5682	44	1.26487	0.19069	−.684	43	0.498
	After NS Extract	2.6364	44	1.16321	0.17536			
Pair 2	After NS Extract	2.6364	44	1.16321	0.17536	0.206	43	0.838
	After placebo	2.6136	44	1.16571	0.17574			
Pair 3	At Base Line	2.5682	44	1.26487	0.19069	−.298	43	0.767
	After placebo	2.6136	44	1.16571	0.17574			

Appendix 4: Paired Samples Statistics (Stress Level)

		Mean	N	SD	SEM	t	df	Sig. (2-tailed)
Pair 1	At Base Line	1.9318	44	0.50106	0.07554	1.530	43	0.133
	After NS Extract	1.8182	44	0.49522	0.07466			
Pair 2	After NS Extract	1.8182	44	0.49522	0.07466	−1.354	43	0.183
	After placebo	1.8864	44	0.44282	0.06676			
Pair 3	At Base Line	1.9318	44	0.50106	0.07554	0.628	43	0.533
	After placebo	1.8864	44	0.44282	0.06676			

Appendix 5: Paired Samples Statistics (Protein intake)

		Mean	N	SD	SEM	t	df	Sig. (2-tailed)
Pair 1	At Base Line	87.7674	44	25.8878	3.94787	1.601	43	0.117
	After NS Extract	86.7209	44	26.1373	3.98590			
Pair 2	After NS Extract	86.7209	44	26.1373	3.98590	1.233	43	0.224
	After placebo	84.0233	44	31.6848	4.83189			
Pair 3	At Base Line	87.7674	44	25.8878	3.94787	1.506	43	0.140
	After placebo	84.0233	44	31.6848	4.83189			

Appendix 6: Paired Samples Statistics (Carbohydrates intake)

		Mean	N	SD	SEM	t	df	Sig. (2-tailed)
Pair 1	At Base Line	321.976	44	98.2883	14.9888	1.949	43	0.058
	After NS Extract	319.767	44	98.7480	15.0589			
Pair 2	After NS Extract	319.767	44	98.7480	15.0589	1.354	43	0.183
	After placebo	313.837	44	104.108	15.8763			
Pair 3	At Base Line	321.976	44	98.2883	14.9888	1.725	43	0.092
	After placebo	313.837	44	104.108	15.8763			

Appendix 7: Paired Samples Statistics (Consumption of Fats)

		Mean	N	SD	SEM	t	df	Sig. (2-tailed)
Pair 1	At Base Line	48.3721	44	17.0964	2.6071	−1.051	43	0.299
	After NS Extract	48.7907	44	16.7993	2.5618			
Pair 2	After NS Extract	48.7907	44	16.7993	2.5618	−1.202	43	0.236
	After placebo	49.6047	44	15.8657	2.4195			
Pair 3	At Base Line	48.3721	44	17.0964	2.6071	−1.667	43	0.103
	After placebo	49.6047	44	15.8657	2.4195			

Appendix 8: Paired Samples Statistics (Calories intake)

		Mean	N	SD	SEM	t	df	Sig. (2-tailed)
Pair 1	At Base Line	2071.5581	44	575.521	87.7661	1.381	43	0.175
	After NS Extract	2063.1163	44	582.906	88.8923			
Pair 2	After NS Extract	2063.1163	44	582.906	88.8923	1.102	43	0.277
	After placebo	2039.3023	44	600.952	91.6444			
Pair 3	At Base Line	2071.5581	44	575.521	87.7661	1.378	43	0.175
	After placebo	2039.3023	44	600.952	91.6444			

Appendix 9: Paired Samples Statistics (BMI)

		Mean	N	SD	SEM	t	df	Sig. (2-tailed)
Pair 1	At Base Line	28.6816	44	4.62851	0.69777	1.923	43	0.061
	After NS Extract	28.5825	44	4.57089	0.68909			
Pair 2	After NS Extract	28.5825	44	4.57089	0.68909	0.651	43	0.518
	After placebo	28.5490	44	4.57828	0.69020			
Pair 3	At Base Line	28.6816	44	4.62851	0.69777	1.834	43	0.074
	After placebo	28.5490	44	4.57828	0.69020			

Appendix 10: Paired Samples Statistics (Total cholesterol)

		Mean	N	SD	SEM	t	df	Sig. (2-tailed)
Pair 1	At Base Line	200.4091	44	45.2293	6.81857	2.445	43	0.019
	After NS Extract	187.7955	44	34.5627	5.21053			
Pair 2	After NS Extract	187.7955	44	34.5627	5.21053	−1.402	43	0.168
	After placebo	194.4773	44	42.2943	6.37612			
Pair 3	At Base Line	200.4091	44	45.2293	6.81857	1.661	43	0.104
	After placebo	194.4773	44	42.2943	6.37612			

Appendix 11: Paired Samples Statistics (HDL Cholesterol)

		Mean	N	SD	SEM	t	df	Sig. (2-tailed)
Pair 1	At Base Line	40.3864	44	6.54578	0.98681	0.370	43	0.713
	After NS Extract	40.0227	44	5.83290	0.87934			
Pair 2	After NS Extract	40.0227	44	5.83290	0.87934	−.604	43	0.549
	After placebo	40.7045	44	8.45117	1.27406			
Pair 3	At Base Line	40.3864	44	6.54578	0.98681	−.349	43	0.729
	After placebo	40.7045	44	8.45117	1.27406			

Appendix 12: Paired Samples Statistics (LDL Cholesterol)

		Mean	N	SD	SEM	t	df	Sig. (2-tailed)
Pair 1	At Base Line	120.4091	44	30.6331	4.61811	1.591	43	0.119
	After NS Extract	113.3636	44	26.5199	3.99803			
Pair 2	After NS Extract	113.3636	44	26.5199	3.99803	−.515	43	0.609
	After placebo	115.0909	44	25.5477	3.85147			
Pair 3	At Base Line	120.4091	44	30.6331	4.61811	1.755	43	0.086
	After placebo	115.0909	44	25.5477	3.85147			

Appendix 13: Paired Samples Statistics (Triglyceride)

		Mean	N	SD	SEM	t	df	Sig. (2-tailed)
Pair 1	At Base Line	200.2273	44	90.6582	13.6672	2.842	43	0.007
	After NS Extract	172.5227	44	84.3427	12.7151			
Pair 2	After NS Extract	172.5227	44	84.3427	12.7151	−2.414	43	0.020
	After placebo	194.8182	44	99.4284	14.9894			
Pair 3	At Base Line	200.2273	44	90.6582	13.6672	0.526	43	0.602
	After placebo	194.8182	44	99.4284	14.9894			

Appendix 14: Paired Samples Statistics (Platelet Count)

		Mean	N	SD	SEM	t	df	Sig. (2-tailed)
Pair 1	At Base Line	234.5909	44	78.0492	11.7663	1.199	43	0.237
	After NS Extract	222.8409	44	68.0174	10.2540			
Pair 2	After NS Extract	222.8409	44	68.0174	10.2540	−.642	43	0.524
	After placebo	225.8864	44	62.5977	9.43696			
Pair 3	At Base Line	234.5909	44	78.0492	11.7663	1.141	43	0.260
	After placebo	225.8864	44	62.5977	9.43696			

Appendix 15: Paired Samples Statistics (Total Leucocyte Count)

		Mean	N	SD	SEM	t	df	Sig. (2-tailed)
Pair 1	At Base Line	7.8295	44	2.00518	0.30229	1.283	43	0.206
	After NS Extract	7.6205	44	1.69987	0.25627			
Pair 2	After NS Extract	7.6205	44	1.69987	0.25627	1.262	43	0.214
	After placebo	7.5227	44	1.67566	0.25262			
Pair 3	At Base Line	7.8295	44	2.00518	0.30229	1.994	43	0.053
	After placebo	7.5227	44	1.67566	0.25262			

Appendix 16: Paired Samples Statistics (ALT Level)

		Mean	N	SD	SEM	t	df	Sig. (2-tailed)
Pair 1	At Base Line	24.2500	44	9.97701	1.50409	3.260	43	0.002
	After NS Extract	20.4545	44	7.93712	1.19657			
Pair 2	After NS Extract	20.4545	44	7.93712	1.19657	−.358	43	0.722
	After placebo	20.8636	44	7.98717	1.20411			
Pair 3	At Base Line	24.2500	44	9.97701	1.50409	2.124	43	0.039
	After placebo	20.8636	44	7.98717	1.20411			

Appendix 17: Paired Samples Statistics (AST Level)

		Mean	N	SD	SEM	t	df	Sig. (2-tailed)
Pair 1	At Base Line	30.5455	44	10.7475	1.62025	6.121	43	0.000
	After NS Extract	23.9773	44	9.76345	1.47190			
Pair 2	After NS Extract	23.9773	44	9.76345	1.47190	−.240	43	0.812
	After placebo	24.2727	44	11.0123	1.66018			
Pair 3	At Base Line	30.5455	44	10.7475	1.62025	4.196	43	0.000
	After placebo	24.2727	44	11.0123	1.66018			

Appendix 18; Paired Samples Statistics (Blood Urea Level)

		Mean	N	SD	SEM	t	df	Sig. (2-tailed)
Pair 1	At Base Line	23.7045	44	6.20420	0.93532	1.743	43	0.088
	After NS Extract	22.2500	44	5.49048	0.82772			
Pair 2	After NS Extract	22.2500	44	5.49048	0.82772	−3.511	43	0.001
	After placebo	24.7727	44	6.94451	1.04692			
Pair 3	At Base Line	23.7045	44	6.20420	0.93532	−1.157	43	0.254
	After placebo	24.7727	44	6.94451	1.04692			

Appendix 19: Paired Samples Statistics (Defective Vision)

		Mean	N	SD	SEM	t	df	Sig. (2-tailed)
Pair 1	At Base Line	1.4318	44	0.50106	0.07554	0.573	43	0.570
	After NS Extract	1.4091	44	0.49735	0.07498			
Pair 2	After NS Extract	1.4091	44	0.49735	0.07498	−1.431	43	0.160
	After placebo	1.4545	44	0.50369	0.07593			
Pair 3	At Base Line	1.4318	44	0.50106	0.07554	−.443	43	0.660
	After placebo	1.4545	44	0.50369	0.07593			

Appendix 20: Paired Samples Statistics (Tingling Sensations)

		Mean	N	SD	SEM	t	df	Sig. (2-tailed)
Pair 1	At Base Line	1.5455	44	0.50369	0.07593	−1.000	43	0.323
	After NS Extract	1.5682	44	0.50106	0.07554			
Pair 2	After NS Extract	1.5682(a)	44	0.50106	0.07554	−	−	−
	After placebo	1.5682(a)	44	0.50106	0.07554			
Pair 3	At Base Line	1.5455	44	0.50369	0.07593	−1.000	43	0.323
	After placebo	1.5682	44	0.50106	0.07554			

(a) The t cannot be computed because the standard error of the difference is 0.

Appendix 21: Paired Samples Statistics (Burning Feet)

		Mean	N	SD	SEM	t	df	Sig. (2-tailed)
Pair 1	At Base Line	1.5455	44	0.50369	0.07593	−1.000	43	0.323
	After NS Extract	1.5909	44	0.49735	0.07498			
Pair 2	After NS Extract	1.5909	44	0.49735	0.07498	1.949	43	0.058
	After placebo	1.4773	44	0.50526	0.07617			
Pair 3	At Base Line	1.5455	44	0.50369	0.07593	1.774	43	0.083
	After placebo	1.4773	44	0.50526	0.07617			

Appendix 22: Paired Samples Statistics (Aches and Pains)

		Mean	N	SD	SEM	t	df	Sig. (2-tailed)
Pair 1	At Base Line	1.2500	44	0.43802	0.06603	−1.774	43	0.083
	After NS Extract	1.3182	44	0.47116	0.07103			
Pair 2	After NS Extract	1.3182	44	0.47116	0.07103	1.949	43	0.058
	After placebo	1.2045	44	0.40803	0.06151			
Pair 3	At Base Line	1.2500	44	0.43802	0.06603	1.000	43	0.323
	After placebo	1.2045	44	0.40803	0.06151			

Appendix 23: Paired Samples Statistics (Diabetic Foot)

		Mean	N	SD	SEM
Pair 1	At Base Line	1.9091(a)	44	0.29080	0.04384
	After NS Extract	1.9091(a)	44	0.29080	0.04384
Pair 2	After NS Extract	1.9091(a)	44	0.29080	0.04384
	After placebo	1.9091(a)	44	0.29080	0.04384
Pair 3	At Base Line	1.9091(a)	44	0.29080	0.04384
	After placebo	1.9091(a)	44	0.29080	0.04384

(a) The t cannot be computed because the standard error of the difference is 0.

Appendix 24: Paired Samples Statistics (Visual Acuity)

		Mean	N	SD	SEM
Pair 1	At Base Line	1.1591(a)	44	1.44581	0.21796
	After NS Extract	1.1591(a)	44	1.44581	0.21796
Pair 2	After NS Extract	1.1591(a)	44	1.44581	0.21796
	After placebo	1.1591(a)	44	1.44581	0.21796
Pair 3	At Base Line	1.1591(a)	44	1.44581	0.21796
	After placebo	1.1591(a)	44	1.44581	0.21796

(a) The t cannot be computed because the standard error of the difference is 0.

Appendix 25: Paired Samples Statistics (Diabetic Retinopathy)

		Mean	N	SD	SEM
Pair 1	At Base Line	0.1136(a)	44	0.49254	0.07425
	After NS Extract	0.1136(a)	44	0.49254	0.07425
Pair 2	After NS Extract	0.1136(a)	44	0.49254	0.07425
	After placebo	0.1136(a)	44	0.49254	0.07425
Pair 3	At Base Line	0.1136(a)	44	0.49254	0.07425
	After placebo	0.1136(a)	44	0.49254	0.07425

(a) The t cannot be computed because the standard error of the difference is 0.

Appendix 26: Paired Samples Statistics (Poly-Urea)

		Mean	N	SD	SEM	t	df	Sig. (2-tailed)
Pair 1	At Base Line	1.5227	44	0.50526	0.07617	−1.000	43	0.323
	After NS Extract	1.5909	44	0.49735	0.07498			
Pair 2	After NS Extract	1.5909	44	0.49735	0.07498	1.702	43	0.096
	After placebo	1.4773	44	0.50526	0.07617			
Pair 3	At Base Line	1.5227	44	0.50526	0.07617	1.431	43	0.160
	After placebo	1.4773	44	0.50526	0.07617			

Appendix 27: Paired Samples Statistics (Poly-Dypsia)

		Mean	N	SD	SEM	t	df	Sig. (2-tailed)
Pair 1	At Base Line	1.4318	44	0.50106	0.07554	−1.138	43	0.262
	After NS Extract	1.5000	44	0.50578	0.07625			
Pair 2	After NS Extract	1.5000	44	0.50578	0.07625	1.949	43	0.058
	After placebo	1.3864	44	0.49254	0.07425			
Pair 3	At Base Line	1.4318	44	0.50106	0.07554	1.431	43	0.160
	After placebo	1.3864	44	0.49254	0.07425			

Chapter 6

The Need for an Integrated, Multi-target *In vitro* Anti-diabetic Screening Platform

Maryna van de Venter

Nelson Mandela Metropolitan University,
Port Elizabeth, 6031, South Africa
E-mail: maryna.vandeventer@nmmu.ac.za

ABSTRACT

Medicinal plants have attracted much interest as potentially inexpensive and culturally acceptable therapies for the treatment of diabetes in developing countries; however most have not been scientifically evaluated for safety and efficacy. The high cost of clinical studies unfortunately reside beyond the reach of most academic and research institutions and subsequently do not represent a viable option for exploring the many medicinal plants reported to treat this disease. While animal models afford an accepted alternative, the multifactor nature of diabetes together with the lack of suitable animal models which simultaneously encompasses the entire pathological spectrum of human diabetes, poses a significant challenge as it is not possible to accurately select appropriate animal models without some knowledge of the mechanism through which complex plant extracts elicit their therapeutic effect. To bridge this gap between ethnobotany and ethnopharmacology, we investigated target directed *in vitro* screening as a possible tool to identify precise therapeutic mechanisms and subsequently allow prioritisation of samples toward specific *in vivo* models. Currently *in vitro* assays are regarded by many to have little value as a predictive technique; however evaluation of the published literature suggests that the extensive abuse and over interpretation of *in vitro* data is a primary factor contributing to the eroded confidence of this technique in the field of ethnopharmacology. Where appropriate data exists, it is evident that many of the *in vitro* identified activities have been confirmed in animal models and, albeit somewhat limited, also in preliminary clinical studies. This review proposes an integral role for *in vitro* assays in the search for successful anti-diabetic therapies and attempts to restore some enthusiasm toward the acceptance of *in vitro* assays in the field of ethnopharmacology, but more importantly, to encourage its proper use so as to ultimately accelerate the pace of medicinal plant based research.

Keywords: Anti-diabetic, Screening, In vitro, Multi-target, Diabetes, Drug targets.

1. INTRODUCTION

The incidence of diabetes is rapidly escalating and expected to become a major health problem especially in developing countries. Subsequently there is an urgent need to identify safe and effective treatment solutions. Although medicinal plants are well recognised as a potential therapeutic solution, clinical acceptance understandably demands meaningful scientific validation of documented anecdotal evidence. The main problems associated with clinical evaluation of medicinal plants with reported anti-diabetic properties reside both with the complexity of constituents of the plant material and with the lack of appropriate information as to the precise pharmacological mechanism by which such plant extracts provide therapeutic relief. While animal models afford an accepted strategy to investigate the therapeutic potential of new remedies, the multifactor nature of diabetes poses a significant challenge in identifying the anti-diabetic activity of unknown and often complex test samples as it is not possible to accurately select appropriate animal models and experimental designs without some knowledge of the mechanism through which these elicit a therapeutic effect. To address this problem multi-target *in vitro* anti-diabetic screening can be used to prioritize test samples according to pathogenic target(s) and consequently allow the selection of appropriate animal models.

In this review we aim to emphasise the value of *in vitro* screening as an integral component in the validation of anti-diabetic medicinal plants. A thorough understanding of diabetes and the metabolic perturbations that characterize this complex disease is essential for any researcher working in this field. There are many excellent reviews on this topic and due to limited space this manuscript only covers the most essential background information.

2. DIABETES MELLITUS

Prevalence of Diabetes Mellitus in Developing Countries

Diabetes mellitus is now recognized as a significant global health problem affecting more than 300 million people in the world (Rawal *et al.*, 2011; IDF). In the past diabetes mellitus was perceived as a disease of the affluent community, people of developed countries. However, non-communicable diseases such as diabetes mellitus in developing countries are evolving rapidly. In 2014, the International Diabetes Federation published data stating that 77 per cent of people with diabetes live in low- to middle-income countries (www.idf.org). These countries include China, India, Russia, Brazil, Pakistan, Indonesia and Bangladesh (Rawal *et al.*, 2012).

Globally, the prevalence of diabetes is 8.3 per cent with the highest percentage being in North America and Caribbean, with a value of 11.4 per cent. The Middle

East and North Africa have the second highest percentage of 9.7 per cent. The global region with the lowest prevalence percentage is Africa, with an estimated value of 5.1 per cent. The top ten countries globally are shown in Figure 6.1.

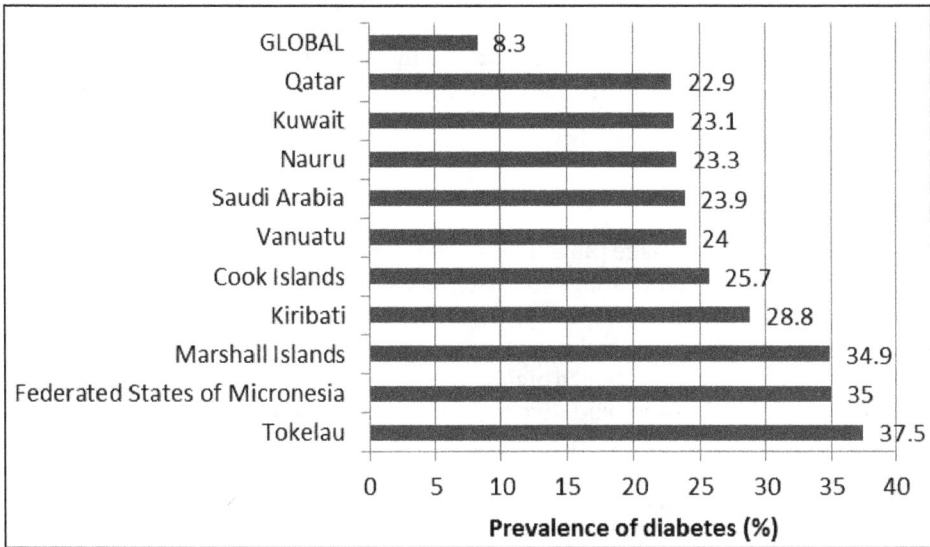

Figure 6.1: The Top Ten Countries with High Prevalence of Diabetes (Adapted from www.idf.org).

The geographic distribution of the prevalence of diabetes is diverse. However, 7 of the top 10 countries are located in the Western Pacific Region, where it is estimated that 138 million individuals live with diabetes and has a prevalence of 8.5 per cent. The country with the lowest prevalence of diabetes is Mali with a value of 1.6 per cent, a country of Africa, where 22 million people have been estimated to be living with diabetes (www.idf.org; Martinez, 2013).

Many studies have projected that there will be a significant increase in the prevalence of diabetes worldwide, but particularly in developing countries, due to urbanization and lifestyle changes (WHO, 2005).

Types of Diabetes

Diabetes mellitus is a chronic, metabolic disorder characterised by hyperglycaemia (*i.e.* high blood glucose levels) due to defects in insulin secretion, insulin action or both. Diabetes is mainly classified into type 1, type 2 and gestational diabetes. Type 1 diabetes (syn. insulin-dependent/juvenile-onset) is associated with cellular-mediated autoimmune destruction of the pancreas' β-cells, causing a deficiency in insulin secretion. About 5-10 per cent of diabetes is affected by type 1 diabetes. Type 2 diabetes (syn. non-insulin-dependent/adult-onset) is associated with a combination of resistance to insulin action and inadequate compensatory insulin secretory response. Overweight and obesity can lead to insulin resistance and type 2 diabetes but the two do not always occur together. About 90-95 per cent of diabetes is affected by type 2 diabetes (American Diabetes Association,

2014). Gestational diabetes mellitus, defined as diabetes with initial onset during pregnancy, may have implications for both the mother and child. Women with gestational diabetes have an increased risk of developing type 2 diabetes later in life. Pregnancy influences maternal metabolism with a reduction in insulin sensitivity as the pregnancy reaches the final stages (Lindsay, 2009). Less commonly found types of diabetes include neonatal diabetes mellitus, maturity-onset diabetes of the young, and diabetes insipidus (Siddiqui *et al.*, 2015, Bichet, 2012). Diabetes insipidus is characterised by the excretion of abnormally large volumes of diluted urine (Bichet, 2012).

Causes of Diabetes

Hyperglycemic damage plays an important role in the initial onset of diabetes. Hyperglycemic-induced damage of tissue is associated with the over activation of four main biochemical pathways, namely the hexosamine-, polyol-, protein kinase C- and advanced glycation end products' formation pathways. Superoxide overproduction is the link between high glucose and the activation of these pathways. Cellular redox potential and oxidative stress play an important role in hyperglycemic damage (Teodoro *et al.*, 2013, Rolo and Palmeira, 2006). Mitochondrial functioning impairment – mitochondrial DNA mutations and decrease in mitochondrial DNA copy numbers – is related to type 2 diabetes (Rolo and Palmeira, 2006). More than 80 per cent of diabetes arises spontaneously with no family history (Greene, 2002).

Causes of diabetes can be divided into genetic-, environmental- and lifestyle factors.

Genetic Factors

Genetic defects of β-cells and insulin action, and genetic syndromes are associated with diabetes (American Diabetes Association, 2014). Mutations in the insulin gene, *INS*, lead to folding impairments of proinsulin in the pancreatic β-cells' endoplasmic reticulum (Weiss, 2013). Cysteine to tyrosine substitution at position 96 of the *INS* gene prevents the formation of one of the three disulfide bonds between chains A and B of mature insulin (Fonseca *et al.*, 2011). Correct folding contributes towards a protein's stability and biological activity. Impaired insulin secretion by the β-cells is linked to ER stress, organelle architecture distortion and cell death (Weiss, 2013). Apart from genetic mutations, other causes of ER stress in β-cells include high glucose, long-chain free fatty acids, inflammatory cytokines, and hypoxia and misfolded islet amyloid polypeptide (hIAPP) (Fonseca *et al.*, 2011). Most *INS* gene mutations are associated with proinsulin misfolding and insulin biosynthesis, whereas other mutations engage different molecular and cellular mechanisms resulting in β-cell failure and diabetes.

Environmental Factors

Virus infections – Coxsackie B, enterovirus, rubella virus, cytomegalovirus, retrovirus and others – might induce diabetes by a direct cytolytic effect or triggering an autoimmune response leading to β-cell destruction (Åkerblom *et al.*, 1997, American Diabetes Association, 2014). Environmental toxins, organic pollutants (*i.e.* pesticides, plastics and industrial petrochemicals) and heavy metals (*i.e.* arsenic, lead and mercury) play an important role in obesity and diabetes development. Environmental toxins interfere with glucose and cholesterol metabolism and induce insulin resistance. The mechanisms of action include oxidative stress, inflammation and mitochondrial injury to name a few (Hyman, 2010). High dietary intake of N-nitroso compounds, nitrates and nitrites is a risk factor for diabetes mellitus (Åkerblom *et al.*, 1997, Longnecker and Daniels, 2001).

Lifestyle Factors

An inactive and high stress lifestyle, in combination with an unhealthy diet leading to obesity, contributes towards a higher risk for diabetes. Unhealthy diets consist of high amounts of sugar, trans- and saturated fatty acids, and low amounts of fiber and phytonutrients (Hyman, 2010). Obesity lowers insulin sensitivity in peripheral tissues, whereby the β-cells compensate for insulin sensitivity by upregulating insulin secretion. This again is influenced by an individual's genetic constitution. Lipids affect β-cell function, while free fatty acids suppress glucose-induced insulin secretion in the long run (Ashcroft and Rorsman, 2012).

Affected Organs

Ketoacidosis and coma, due to hyperglycemia and hypoglycemia, respectively, are acute metabolic complications associated with mortality in diabetes. The chronic elevation of blood glucose levels lead to angiopathy – damage to blood vessels. Diseases associated with angiopathy are divided into microvascular and macrovascular diseases. Microvascular diseases – damage to small blood vessels – include retinopathy, nephropathy and neuropathy. Macrovascular diseases – damage to arteries – include cardiovascular (myocardial infarction) and cerebrovascular (strokes) diseases (Forbes and Cooper, 2013). Microvascular complications are attributed to the biochemical and structural changes in basement membrane proteins of different organ systems (Forbes and Cooper, 2013).

Nephropathy

Nephropathy is the major cause of renal failure in diabetes (American Diabetes Association, 2014). It is characterised by the development of proteinuria and a decline in the glomerular filtration rate over a long period of time (Forbes and Cooper, 2013, Sen *et al.*, 2012). Nephropathy is associated with hypertension, resulting in heart attacks and strokes (Forbes and Cooper, 2013).

Retinopathy

Retinopathy is the major cause of blindness in diabetes (American Diabetes Association, 2014, Ford *et al.*, 2012). It is characterised by lesions of the retina due to the degeneration or occlusion of retinal capillaries, resulting in vision impairment (Forbes and Cooper, 2013, Pescosolido *et al.*, 2014). Thirty percent of diabetes is affected by retinopathy (Pescosolido *et al.*, 2014).

Neuropathy

Neuropathy is characterised by a loss of neurons and/or damage of nerves in the lower extremities or organs (Ahamed and Banji, 2012). It affects more than half of diabetes and is the major cause of lower extremity amputations (Forbes and Cooper, 2013). It affects autonomic, motor and sensory neurons of the peripheral nervous system (Hamada and Debrky, 2014). Neuropathy is the major cause of erectile dysfunction, wound healing impairment, gastrointestinal-, genitourinary- and cardiovascular dysfunction (American Diabetes Association, 2014, Søfteland *et al.*, 2014, Forbes and Cooper, 2013)

Cardiovascular Complications

Cardiovascular complications in diabetes are characterised by alterations in cardiac contractile functions (Sen *et al.*, 2012). This includes atherosclerosis, which lead to strokes, myocardial infarction and impaired myocardial function disorders. Cardiovascular complications account for half of the mortalities seen in the diabetic population, with a threefold risk in myocardial infarctions compared to the general population (Forbes and Cooper, 2013).

Diabetes mellitus sufferers are more prone to infections, including urinary tract infections, skin and soft tissue infections, osteomyelitis, lower respiratory tract infection, and others (Cooke, 2014).

Current Therapies and their Molecular Targets

The main aim of any diabetes therapy is to reduce the blood glucose level, while minimising side effects like hypoglycaemia (American Diabetes Association, 2015). A complex practical management programme needs to be carefully explained; hence education plays an important role in diabetes treatment. Blood glucose needs to be monitored and maintained at normal levels (Greene, 2002).

Type 1 Diabetes

Most people suffering from Type 1 diabetes are treated with multiple-dose insulin injections, or continuous subcutaneous insulin infusion (insulin pump therapy) to address the main cause of the disease, namely insufficient insulin production. The biggest risk in insulin treatment is hypoglycaemia, which can be reduced by using insulin analogs. The use of metformin, incretin-based therapies and sodium-glucose co-transporter 2 inhibitors for type 1 diabetes are being investigated (American Diabetes Association, 2015).

Type 2 Diabetes

In type 2 diabetes, the first step in glycaemic control is lifestyle changes which can improve the main metabolic cause namely insulin resistance. Lifestyle changes include healthy eating, weight control, increased physical activity and diabetes education. It improves glycaemic control, but is rarely sufficient to achieve target glycaemic control; hence it is combined with an oral anti-diabetic drug (Wallia and Molitch, 2014, American Diabetes Association, 2014, Lebovitz, 2011). Type 2 diabetes is controlled by mono-, dual- or triple therapy of orally administered anti-diabetic drugs. The biguanide metformin is mainly administered as soon as type-2 diabetes is diagnosed, unless there are contraindications or intolerance towards metformin. Combination therapy involves the addition of one or more noninsulin agent/s to metformin, if glycaemic control is not achieved within a certain timeframe. These include sulfonylurea, thiazolidinediones, dipeptidyl peptidase 4 inhibitors, sodium-glucose co-transporter inhibitors and glucagon-like peptide 1 receptor agonists (American Diabetes Association, 2015).

Available oral anti-diabetic drugs exert their effects on one or more of the following molecular targets: they (1) enhance insulin secretion by β-cells (*i.e.* sulfonylureas, non-sulfonylureas, incretins and glucagon-like peptide analogs and agonists), (2) reduce glucose absorption in the gastrointestinal tract (*i.e.* α-glucosidase inhibitors, sodium-glucose co-transporter inhibitors), (3) decrease glucose release from the liver (*i.e.* biguanides), (4) improve the disposal of peripheral glucose (*i.e.* biguanides and thiazolidinediones) and (5) increase sensitivity of target tissues to insulin (*i.e.* biguanides and thiazolidinediones) (Prabhakar *et al.*, 2014, Campbell-Tofte *et al.*, 2012, Kavishankar *et al.*, 2011).

Oral hypoglycaemic agents are effective in reducing insulin resistance or facilitating insulin secretion in the early stages of type 2 diabetes. Prolonged type 2 diabetes is associated with a decline in β-cell function. Insulin therapy is prescribed to type 2 diabetes patients if glycaemic control was not achieved by two or more oral hypoglycaemic agents, and when severe hyperglycaemia is indicated by fasting plasma glucose (Wallia and Molitch, 2014, Lebovitz, 2011). Morbidly obese patients with type 2 diabetes, not responding to medical therapy, may choose to undergo interventional gastric surgery known as bariatric surgical procedures. This improves hyperglycaemia and decreases the risk for obesity-related cardiovascular diseases (Lebovitz, 2011).

Future Treatment

Future diabetes treatment, especially for type 1 diabetes mellitus, might include immunotherapy (*i.e.* tolerance to β-cell autoantigens, anti-T-cell monoclonal antibodies, cytokine-based immunotherapy and non-antigen-specific immunostimulation) (Bach, 2001), artificial pancreas (Breton *et al.*, 2012) and stem cells (Soria *et al.*, 2001).

3. MEDICINAL PLANTS AS ANTI-DIABETIC THERAPY

In developing countries and especially in Africa and Asia, it has been estimated that about 80 per cent of the population depends on traditional medicine for the treatment of diseases. Proximity of medical facilities, availability and cost of drugs play an important role in disease treatment (Campbell-Tofte *et al.*, 2012). Commercially available anti-diabetic drugs have various side effects. Natural products, taken with synthetic drugs, may reduce side effects and may have additive and/or synergistic properties (Prabhakar *et al.*, 2014). According to world ethnobotanical information reports and ethnopharmacological surveys, 800-1200 medicinal plants are used in traditional medicine for their anti-diabetic activity (Prabhakar *et al.*, 2014, Rupeshkumar *et al.*, 2014). Experimentally, about 450 plants have shown hypoglycaemic activity, but the complete mechanisms of action of only 109 of these plants are known. Mechanisms of action include improving glucose uptake and utilization, controlling insulin release, manipulating carbohydrate metabolism, preventing and restoring β-cell integrity and function, and antioxidant properties (Prabhakar *et al.*, 2014).

Plants contain high concentrations of detoxifying, antioxidant and anti-inflammatory secondary metabolites. These protect plants against pests, infections and harsh climate conditions (Hyman, 2010, Rupeshkumar *et al.*, 2014). Secondary metabolites with anti-diabetic activities include flavonoids, alkaloids, terpenoids, phenolics and others (Bahmani *et al.*, 2014). The complexity of plant material components and incomplete understanding of diabetes disease etiology contribute towards problems with the clinical evaluation of medicinal plants (Campbell-Tofte *et al.*, 2012). Some herbs may provide symptomatic relief of diabetes and assist in the prevention of secondary complications of the disease (Rupeshkumar *et al.*, 2014).

It is important to note that the most commonly used oral hypoglycaemic agent for type 2 diabetes, metformin, has its origins from *Galega officinalis* (syn. French lilac). The syntheses of biguanides are based on the guanidine derivative, galegine, which is an active component (Campbell-Tofte *et al.*, 2012). Metformin is administered for the treatment of overweight and obese type 2 diabetes with normal kidney function, where its effectiveness depends on the presence of insulin. It does not directly stimulate insulin secretion, but increases insulin's action (Kavishankar *et al.*, 2011).

The Sequence from Ethnobotanical Claims to Clinical Approval

In a recent review by Newman and Cragg (2012), it is indicated that more than half the approved clinical drugs in the U.S.A are of natural sources and thus, the importance of ethnomedicine is absolutely clear. Unlike conventional chemically-defined drugs, herbal or traditional medicines have a substantial history of human use. This is a major advantage. However, when considering a natural product as a candidate for clinical trials, a few important factors need to be taken into account that is not necessary for synthetic drugs. When a traditional formulation is used, it is important to achieve a chemical fingerprint of the total ingredients as well as adequate quantification of the constituents. This is because variation of content from batch to batch may be an issue. It is also important that levels of herbicides, pesticides and other toxic contaminants are addressed. Information on plants and extracts, such as description of the plant, plant processing, analytical procedures and storage must be well documented. Traditional formulations that are intended for human administration must follow GMP (WHO, 2005).

The sequence from ethnobotanical claims to clinical approval involves basic research where knowledge of the disease and potential targets for treatment is gained, drug discovery and pre-clinical testing, and thereafter the candidate enters clinical trials (Figure 6.2).

The first important pre-clinical step in the sequence from traditional medicine to clinical approval is evaluation of safety of use. Thus, the first series of tests to be conducted is toxicity evaluation. This is done in a number of ways. Cytotoxicity is evaluated using an *in vitro* system, making use of primary cells or permanent cell lines (Li, 2005; WHO, 2005). *In vitro* studies using target cells are then conducted to initially characterise the extracts mechanism of action. The next step is to conduct *in vivo* studies using an animal model to assess single-dose acute, sub-chronic and

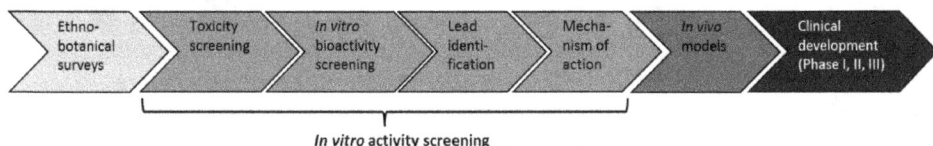

In vitro activity screening

Figure 6.2: The Sequence from Ethnobotanical Claims to Clinical Approval. Many potential anti-diabetic drug leads are never identified because of the complexity of the disease and the limited approach to *in vitro* screening.

chronic toxicity and maximum tolerated dose. Repeat dose toxicological effects in a rodent and non-rodent model should be investigated. Reproductive toxicity is also recommended using a rodent or rabbit model. Carcinogenicity assessment also needs to be performed *in vivo* (WHO, 2005).

The next step is to determine the efficacy of traditional products. This needs to be demonstrated using *in vitro* and *in vivo* methods. The last pre-clinical assessment for a drug candidate is pharmacokinetics, *i.e.* absorption, distribution of the active ingredient around the body and the rate of metabolism and the nature of the resultant metabolites. This is evaluated using animal models (WHO, 2005).

Crude plant extracts are often subjected to bioassay-guided fractionation in order to isolate and characterise the active compound responsible for bioactivity of the plant extract. This is done to identify a new active compound which can lead to a new drug candidate for a disease or to identify a new lead compound that can be altered, synthetically. Bioassay-guided fractionation can either be performed before crude plant extract toxicity evaluation or after. The active fraction and compound needs to undergo the same evaluation process as mentioned above.

Once *in vitro* and animal testing has been performed, the drug candidate or extract can enter clinical trials, which is divided into three phases. Basically, the WHO describes Phase I studies determine the safety associated with increasing the dose of the product. Phase II determines the efficacy of a range of doses in individuals with the disease and Phase III expands trials of safety and efficacy of a potential therapy as well as evaluate the benefit-risk ratio. Statistical comparison of a new therapy to used standards is also done in Phase III (For detailed descriptions of Clinical Trials, visit www.who.int).

The major disadvantage of plants as a source of new drugs is the time it takes from plant collection to the drugs' clinical approval. It has been estimated that this process takes 10 to 20 years (Fabricant and Farnsworth, 2001). Thus, development and improvement of methods currently employed is ongoing to allow for quick identification of lead compounds from natural sources and plants.

4. TESTING FOR ANTI-DIABETIC ACTIVITY

In vivo Models of Diabetes Mellitus

Experimental animal models of diabetes mellitus have been used in pre-clinical studies to study the pathophysiology of diabetes and to investigate *in vivo* efficacy and mode of action of potential anti-diabetic candidates. These models provide valuable information about molecular pathways that contribute to the induction of diabetes in humans (Chatzigeorgiou *et al.*, 2009). Diabetes is a complex disease with abnormalities that cannot be investigated in a single model, therefore, a variety of animal models in different species are employed in order to study different aspects of the disease. Diabetic experiments are carried out in rodents, dogs and non-human primates induced by several methods that include chemical, surgical and genetic (immunological) manipulations (Anunciado, 2000; Kumar *et al.*, 2012). Rodents are the most commonly used animal models to mimic human diabetes. They are less expensive to maintain than larger animals and their diabetic condition manifests rapidly (Rees and Alcolado, 2005). Examples of experimental animal diabetic models are:

Experimentally Induced Models

Surgical, chemical or diet induced methods are used to study the consequences of hyperglycaemia.

☆ Surgical models: Pancreatectomy is conducted mainly in dogs, pigs, rabbits and rats, which results in hyperglycaemia (Kumar *et al.*, 2012).

☆ Chemically induced diabetes: Alloxan and streptozotocin (STZ) are toxic glucose analogues that accumulate in pancreatic β-cells and are widely used to induce experimental diabetes in animals. They enter the β-cell via a glucose transporter, GLUT2, and causes DNA damage (Lenzen, 2008, Skudelski, 2001). This treatment leads to insulin deficient hyperglycemic models without insulin resistance (Eddouks *et al.*, 2012).

☆ Diet induced obesity and insulin resistance or diabetes: Rats or mice are fed a high fat diet (HFD) containing between 32 and 60 per cent of calories from fat, usually with a high content of saturated fats. These rodents tend to become obese and insulin resistant but not hyperglycemic (Eddouks *et al.*, 2012) while chronic inflammation may also be absent.

Genetically Induced or Inbred Diabetic Models

Genetically modified rodents are deficient in factors involved in pancreas development or in β-cell growth. Other genetically diabetic animal models, such as *obese rodent models*, exhibit mutations in the leptin gene (ob/ob mouse) or receptors (db/db mouse), resulting in leptin deficiency (Chatzigeorgiou *et al.*, 2009; Kumar *et al.*, 2012). These animals are expensive and require sophisticated maintenance (Eddouks *et al.*, 2012).

Hormone Induced Diabetic Models

Administration of growth hormone in dogs and cats can induce intensive diabetes. Cortisone is also used to induce hyperglycaemia and glycosuria in treated rats (Kumar *et al.*, 2012).

Surgical or chemical pancreatectomy models give a fair representation of type 1 diabetes. However, not one of the above models truly represents the type 2 diabetic state in humans. To address this, combinations of treatments are sometimes employed to more accurately reflect type 2 diabetes as it occurs in human subjects.

One such example is the combination of HFD and STZ and although the dual model is an improvement, it still has limitations (Skovsø, 2014). The fact is that there is no perfect animal model for type 2 diabetes. Further complicating factors when working with animals include ethical considerations, the need for specialized animal housing facilities, the requirement of large amounts of test samples and high cost. Genetically modified strains are often very vulnerable and special care has to be taken when working with them.

Existing *In vitro* Anti-diabetic Assays

As discussed previously, diabetes is a complex disease affecting various organs in different ways. Although this could be seen as an advantage because of the potential for multiple drug targets, it also introduces complications in the search for drug candidates. Medium to high throughput screening cannot be achieved with *in vivo* models, while each *in vitro* anti-diabetic screening model only includes a single target or cell type.

Existing *in vitro* anti-diabetic assays can be grouped according to their therapeutic targets (Table 6.1). Restoring glucose homeostasis is the primary therapeutic target while some anti-diabetic drugs address adipose dysfunction or secondary diabetic complications. A number of molecular or metabolic targets exist through which each of the therapeutic effects can be achieved. The list in Table 6.1 is not exhaustive but includes all the most commonly used assays that directly target the metabolic processes involved in diabetes or complications associated with the disease.

5. INTEGRATION OF *IN VITRO* AND *IN VIVO* RESULTS

The Need for an Integrated, Multi-Target *In vitro* Screening Platform

Most publications on *in vitro* anti-diabetic activity of plant extracts only report results of one or two bioassays. In studies involving screening of a list of plants used for the treatment of diabetes, only a small percentage of the extracts are usually active in the selected *in vitro* assay. This is not surprising considering the great number of possible targets that are not considered in such an investigation. If the purpose of screening is identification of extracts with potential anti-diabetic activity, more assays should be included to ensure the highest possible 'hit rate'. This was one of the points highlighted by Houghton *et al.*, in 2007 but since then not much seems to have changed. One plausible explanation for this is the facilities and expertise required to perform a large number of bioassays ranging across the fields of enzymology, cell biology, signal transduction and more. *In vitro* assays are also very often over interpreted or abused by using inappropriate concentrations (Houghton *et al.*, 2007). As a result, many researchers have lost faith in them. Considering the slow progress in anti-diabetic drug discovery and the large number of unexplored plants known for their use in diabetes, the need for an integrated, multi-target *in vitro* screening platform should be obvious. Such a platform will enable researchers to perform more meaningful screening of plant extracts to increase the success rate in identifying those with potential for further studies. Strict guidelines should be applied to prevent a repetition of mistakes made in the past that have led to the

Table 6.1: Existing *In vitro* Anti-diabetic Screening Assays Grouped According to Therapeutic Targets

Molecular/ Metabolic Target	In vitro Model(s)	Principle	Examples of Appropriate In vivo Model(s)
Therapeutic target: Glucose homeostasis			
Intestinal glucose absorption	Inhibition of α-glucosidase and α-amylase	Inhibition of carbohydrate digestion to delay post-prandial glucose absorption	Normal or any diabetic rodents
	Inhibition of glucose transport across Caco2 intestinal epithelial cell monolayer	Inhibition of intestinal sodium dependent glucose transporter to prevent postprandial hyperglycaemia	Normal or any diabetic rodents
Gluconeo-genesis	Inhibition of glucose-6-phosphatase; Inhibition of fructose 1,6 bisphosphatase; Inhibition of hepatocyte glucose production using hepatocyte cell lines or primary hepatocytes	Reduction of glucose released from hepatocytes into the circulation	Zucker diabetic fatty (ZDF) -diabetic rats (hyperinsulin-aemic)
Hepatic glucose utilisation	Removal of extracellular glucose by cultured hepatocytes (cell lines or primary hepatocytes); Increased glycogen storage in cultured hepatocyte cell lines or primary hepatocytes	Increased hepatic glucose utilization and storage leads to reduced blood glucose levels	ZDF-diabetic rat (hyperinsulin-aemic)
Peripheral glucose metabolism	Increased glucose utilization by cultured myocyte or adipocyte cell lines; Increased uptake of [3H]deoxyglucose by cultured myocyte or adipocyte cell lines; Stimulation of GLUT4 translocation and/or activation of insulin signaling intermediates in cultured myocyte or adipocyte cell lines	Increased uptake and utilization of glucose into peripheral tissues (especially skeletal muscle and adipose) leads to reduction of postprandial blood glucose levels; Improvement of insulin sensitivity to stimulate glucose uptake and metabolism in myocytes and adipocytes	High fat diet (HFD) induced insulin resistant (non-obese) rodents
Pancreatic β-cell function	Stimulation of β-cell proliferation (cultured primary β-cells or cell lines); Increased insulin secretion by cultured primary β-cells or cell lines	Improved postprandial glucose clearance in response to increased insulin levels	*STZ or alloxan induced hyperglycemic rodents; Goto-Kakizaki rats
Potentiation of incretin receptor signaling	Inhibition of dipeptidyl peptidase 4 (DPP-IV)	Prolonged incretin half-life, contributing to increased insulin secretion and reduced postprandial blood glucose levels	HFD rodents ob/ob mice db/db mice

Contd...

Table 6.1–Contd...

Molecular/Metabolic Target	In vitro Model(s)	Principle	Examples of Appropriate In vivo Model(s)
Therapeutic target: Adipose tissue function			
Adipo-genesis	Triglyceride accumulation in differentiating cultured primary preadipocytes or cell lines	PPARγ agonists improve insulin sensitivity	Any insulin resistant model
Inflam-mation	Adipokine production in differentiated cultured primary adipocytes or cell lines	Improved insulin sensitivity	Obese rodent model
Therapeutic target: Secondary anti-diabetic activity			
Protein glycation	BSA glycation assay	Prevent or reverse protein glycation as result of hyperglycaemia	HFD + STZ rodents (prolonged hyperglycaemia)
Wound healing	Scratch assay using cultured fibroblasts	Promote wound healing in diabetic conditions	HFD + STZ rodents (prolonged hyperglycaemia)
Chronic inflam-mation	Inhibition of reactive oxygen species (ROS) and pro-inflammatory cytokine production in cultured peripheral blood mononuclear cells or RAW 264.7 cells	Inflammation contributes to pancreatic β-cell destruction, insulin resistance, diabetes and associated complications	HFD induced obese rodents
	Inhibition of cyclo-oxygenase and lipoxygenase	Enzymes involved in production of pro-inflammatory cytokines	HFD induced obese rodents
Anti-oxidant assays	Reducing capacity (DPPH, FRAP) Radical scavenging activity (NO, O⁻, OH⁻, H_2O_2)	Reduce oxidative stress induced by hyperglycaemia	Models with hyperglycaemia and/or inflammation

*Dosage of STZ or alloxan will determine the characteristics of the model: complete or partial destruction of β-cells can be achieved.

bad reputation they currently have.

Careful selection of assays for an integrated, multi-target *in vitro* anti-diabetic drug screening platform holds further advantages. Unlike *in vivo* models, most of the listed *in vitro* assays can be adapted and automated for medium- to high throughput screening. By analyzing the *in vitro* activities across a range of assays, it will be possible to select the most appropriate *in vivo* models to validate the *in vitro* activities. In this way the number of unnecessary animal experiments and the chances of 'missing' an activity because of selection of the wrong animal model will be greatly reduced.

Integration of *In vitro* Screening Data

Screening of plant extracts or compounds should be performed using a carefully selected range of *in vitro* assays that include all the major anti-diabetic targets. Results from all the bioassays can then be considered together to group the samples according to their mode of action and a scoring system can be implemented to prioritise them for further investigation. Grouping according to mode of action will also greatly assist in selection of the most appropriate animal model(s) for confirmation of the observed *in vitro* effects. Table 6.1 includes a column indicating the most appropriate models for confirmation of each *in vitro* activity. This approach should greatly increase the success rate during screening and identify extracts acting on more than one anti-diabetic target.

Does an Adequate *In vitro*/*In vivo* Relationship Exist to Justify Multi-target *In vitro* Screening?

Diabetes is now clearly established as a multi-factor based disease, consequently it follows that the concurrent modulation of several targets can theoretically provide a superior therapeutic effect compared to the action of a single selective agent. Indeed, contemporary therapeutic strategies employed in the treatment of diabetes now often include combinations of various existing anti-diabetic drugs aimed to influence multiple biological pathways simultaneously, and thereby elicit a more effective clinical outcome. Considering that nature based compounds are synthesised in a biological/enzymatic environment, the propensity for such compounds to interact with various molecular targets is expected to be much greater than that of any synthetic drug. In this regard medicinal plants possess a clear advantage in that it may not only contain several active constituents, but that each individual compound has the potential to interact with multiple modes of anti-diabetic action, thus providing multifaceted benefits. The challenge however is to correctly identify the full domain of this therapeutic potential using currently available research tools and to do so in a manner that allows accurate extrapolation to the clinical setting.

Multiple targeting may be considered a double-edged sword in diabetes research. On the one hand it is more probable that such an approach will provide superior therapeutic efficacy, however on the other hand multiple targets make *in vivo* studies infinitely more complicated. Since no single animal model that encompasses the entire spectrum of anti-diabetic targets exists, *in vitro* target directed screening provides the only meaningful approach to pre-identify potential anti-diabetic mechanisms and thereby allow selection of appropriate animal models and experimental designs.

Examples from Literature

Despite much scepticism with regards to the use of *in vitro* assays as a predictive tool for clinical efficacy assessment, there is no shortage of literature to demonstrate how *in vitro* studies have contributed to the advancement of drug development. In order to illustrate the value of comprehensive *in vitro* target directed screening as an integral component in the approach to identify the full therapeutic potential of natural products, we may consider berberine and oleanolic acid as two well-studied examples of natural compounds for which *in vitro* screening has played a fundamental role to establish these molecules as bona fide anti-diabetic agents. Their chemical structures can be seen in Figure 6.3. The documented multi-target nature of these compounds substantiates the remarkable capacity of natural products to interact with a varied array of anti-diabetic mechanisms, many of which would probably not have been identified in animal models unless specifically designed to do so. Where appropriate data exists, it is evident that many of the *in vitro* identified activities have been confirmed in animal models and, albeit somewhat limited, also in preliminary clinical studies.

Berberine Targets Adipose Dysfunction

Berberine, a plant alkaloid well known in Chinese and Ayurvedic medicine, has been demonstrated to have potential as a treatment for diabetes and obesity. Several studies implicate adipose tissue as a primary target in the therapeutic potential of berberine, suggesting that improvement in insulin sensitivity and attenuation of dyslipidaemia are secondary to restoration of adipocyte function. Although the precise molecular mechanism remains to be elucidated, *in vitro* studies have confirmed a strong lipocentric mode of action. Lipid accumulation in both human and 3T3-L1 preadipocytes is inhibited upon berberine treatment which may involve transcription factors such as PPARγ and GATA (Hu and Davies, 2009). In addition, berberine treatment significantly altered adipokine mRNA levels in differentiating human preadipocytes (Hu and Davies, 2009). Extrapolating these findings to an

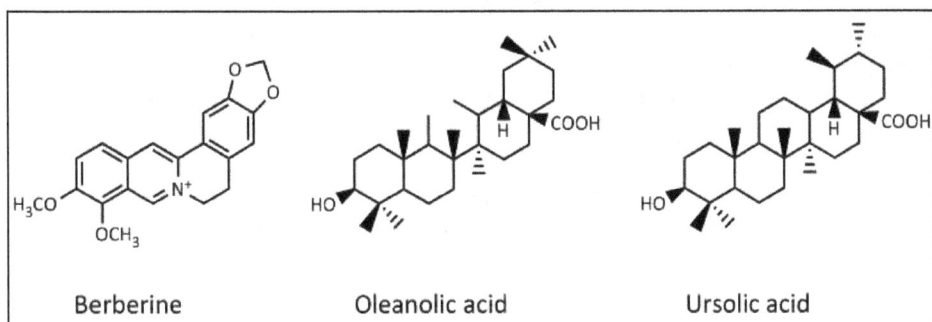

Figure 6.3: Chemical Structures of the Natural Products Berberine, a Plant Alkaloid, and the Two Triterpenoids Oleanolic Acid and Ursolic Acid. Their anti-diabetic potential is discussed in the text.

animal model, Hu and Davies, (2010) demonstrated that in high fat diet fed mice, berberine reduced body weight gain, and serum glucose, triglyceride, and total cholesterol levels. Furthermore these changes were accompanied by a down-regulation in PPARγ expression and an up-regulation of GATA-3 expression in epididymal adipose tissues, however not all studies concur that PPARγ expression is attenuated by berberine. The difference may be related to the concentrations used, since high *in vitro* concentrations of berberine directly inhibit PPARγ activity in adipocytes, whereas it shows an opposite effect at low concentrations as in *in vivo* studies (Yin *et al.*, 2012).

Oleanolic Acid Inhibits Carbohydrate Digestion

Oleanolic acid and the structurally related ursolic acid (Figure 6.3) are ubiquitous triterpenoids often occurring in plants used in traditional medicine for the treatment of diabetes and inflammatory diseases. Both these triterpenoids are reported to be strong *in vitro* inhibitors of enzymes involved in carbohydrate digestion. Oleanolic acid inhibits α-glucosidase and α-amylase with IC_{50} values of 10-15 µM and 100 µM respectively (Castelano *et al.*, 2013). Using non-diabetic male Wistar rats, Tiwari *et al.* (2010) measured a 22 per cent decrease in the blood glucose levels following a starch tolerance test. A similar hypoglycaemic effect is reported in GK/Jcl rats treated with oleanolic acid (Komaki *et al.*, 2003). At a dose of 1 mg/kg, oleanolic acid reduced blood glucose levels after 30 min by 23 per cent, producing a hypoglycaemic effect in prediabetic patients fed cooked rice (Komaki *et al.*, 2003). In contrast, Matsuda *et al.* (1998) reported that pre-treatment of rats with oleanolic acid had no effect on serum glucose level in their glucose tolerance test even when using a dose of 100 mg/kg body weight. This discrepancy is explained by the prerequisite for starch digestion prior to glucose absorption and highlights the importance of using appropriate experimental design when attempting to extrapolate *in vitro* findings to animal models.

Triterpenoids Protect against Inflammation

Inflammation is well recognised as a contributory factor in the development of diabetes. Obesity, which is often associated with type 2 diabetes, is characterised by chronic adipose tissue inflammation resulting in adipocyte dysfunction and insulin resistance. Pancreatic β-cells are extremely sensitive to inflammatory cytokines and oxidative stress, which lead to islet degeneration and a concomitant decline in insulin secretion. Although chronic inflammation is currently not included as a primary therapeutic target in diabetes treatment, it is recognised to have anti-diabetic potential. Many natural occurring triperpenoids, such as ursolic acid have been shown to attenuate nitric oxide production in activated mouse macrophages (RAW 264.7 cells) suggesting possible anti-inflammatory activity (Shu *et al.*, 1998). The anti-inflammatory activity has been studied *in vivo* and demonstrated to involve the suppression of NF-κβ (Checker *et al.*, 2012). In STZ diabetic mice, oleanolic acid prolongs survival of transplanted islets by altering macrophage cytokine production (Nataraju *et al.*, 2009).

Although these above described examples represent only the tip of the iceberg, it is clear that *in vitro* identification of putative molecular targets provide an added dimension when attempting to characterise the therapeutic potential of medicinal plants. There exists adequate scientific evidence to support *in vitro* screening as an integral component in the design of animal and clinical studies. Furthermore, *in vitro* data can provide meaningful information to assist the selection of appropriate animal models and end point analysis to ensure that the full anti-diabetic potential of medicinal plants are exposed. However it is mandatory that *in vitro* data be evaluated within the limitations of the technique and within the appropriate experimental context.

6. CONCLUSIONS

Natural products hold great potential as anti-diabetic drugs but progress has been slow due to the complexity of the disease. A thorough knowledge and understanding of the causes and metabolic perturbations involved in diabetes is essential for meaningful research in this field. Although many *in vivo* models of diabetes are available, there is no perfect model that exhibits all the characteristics of diabetes (especially type 2 diabetes) as it occurs in humans. Ethical, practical and cost implications further limit the potential of *in vivo* studies for screening of large numbers of samples. On the other hand, many *in vitro* anti-diabetic assays have the potential for medium to high throughput screening applications, they are less expensive and can yield results in a much shorter time than most *in vivo* models. The limitations should, however be realised and acknowledged to restore trust in them. Progress in screening for anti-diabetic activity will be much faster and more successful if researchers include a wider range of assays to cover as many of the molecular targets as possible. A further benefit will be the identification of samples with multiple targets, with the potential of improved therapeutic efficacy. This approach will reduce the amount of *in vivo* testing by eliminating inactive samples and providing insight into the mechanism of action to assist in selection of the most appropriate *in vivo* models for further investigations.

7. ACKNOWLEDGEMENTS

The authors would like to acknowledge Anli Hattingh's contribution in drawing the chemical structures for this manuscript.

8. REFERENCES

1. Åkerblom, H.K., Knip, M., Hyöty, H., Reijonen, H., Virtanen, S., Savilahti, E., Ilonen, J. 1997. Interaction of genetic and environmental factors in the pathogenesis of insulin-dependent diabetes mellitus. *Clinica Chimica Acta*, **257**: 143-156.

2. American Diabetes Association. 2014. Diagnosis and classification of diabetes mellitus. *Diabetes Care*, **37** (Suppl. 1): S81-S90.

3. American Diabetes Association. 2015. Approaches to glycaemic treatment. *Diabetes Care*, **38** (Suppl. 1): S41–S48.

4. Anunciado, R. V. P., Imamura, T., Ohno, T., Horio, F., Namikawa, T. 2000. Developing a new model for non-insulin dependent diabetes mellitus (NIDDM)

by using the Philippine wild mouse, Mus musculus castaneus. *Experimental Animals,* **49**(1): 1-8.

5. Ashcroft, F.M., Rorsman, P. 2012. Diabetes mellitus and the β cell: the last ten years. *Cell,* **148**: 1160-1171.

6. Bach, J-F. 2001. Immunotherapy of insulin-dependent diabetes mellitus. *Current Opinion of Immunotherapy,* **13**: 601-605.

7. Bahmani, M., Golshahi, H., Saki, K., Rafieian-Kopaei, M., Delfan, B., Mohammadi, T. 2014. Medicinal plants and secondary metabolites for diabetes mellitus control. *Asian Pacific Journal of Tropical Disease,* **4**: S687-S692.

8. Bichet, D.G. 2012. Genetics and diagnosis of central diabetes insipidus. *Annales d'Endocrinologie,* **72**: 117-127.

9. Breton, M., Farret, A., Bruttomesso, D., Anderson, S., Magni, L., Patek, S., Dall Man, C., Place, J., Demartini, S., Del Favero, S., Toffanin, C., Hughes-Karvetski, C., Dassau, E., Zisser, H., Doyle III, F.J., De Nicolao, G., Avogaro, A., Cobelli, C., Renard, E., Kovatchev, B. 2012. Fully integrated artificial pancreas in type 1 diabetes: modular closed-loop glucose control maintains near norglycemia. *Diabetes,* **61**: 2230-2237.

10. Campbell-Tofte, J.I.A., Mølgaard, P., Winther, K. 2012. Harnessing the potential clinical use of medicinal plants as anti-diabetic agents. *Botanics: Targets and Therapy,* **2**: 7-19.

11. Castellano, J.M., Guinda, A., Delgado, T., Rada, M., Cayuela, J.A. 2013. Biochemical basis of the anti-diabetic activity of oleanolic acid and related pentacyclic triterpenes. *Diabetes,* **62**: 1791-1799.

12. Chatzigeorgiou, A., Halapas, A., Kalafatakis, K., Kamper, E. 2009. The use of animal models in the study of diabetes mellitus. *In vivo,* **23**(2): 245-258.

13. Checker, R., Sandur, S.K., Sharma, D., Patwardhan, R.S. Jayakumar, S., Kohli, V., Sethi, G., Aggarwal, B.B., Sainis, K.B. 2012. Potent anti-inflammatory activity of ursolic acid, a triterpenoid antioxidant, is mediated through suppression of NF-kB, AP-1 and NF-AT. PLoS ONE, **7**(2): e31318. doi:10.1371/journal. pone.0031318.

14. Cooke, F.J. 2014. Infections in people with diabetes. *Medicine,* **43**: 41-43.

15. Eddouks, M., Chattopadhyay, D., Zeggwagh, N.A. 2012. Animal models as tools to investigate anti-diabetic and anti-inflammatory plants. *Evidence-Based Complementary and Alternative Medicine,* **2012**: 142087. doi:10.1155/2012/142087.

16. Fabricant, D.S., Farnsworth, N.R. 2001. The value of Plants used in Traditional Medicine for Drug Discovery. *Environmental Health Perspectives,* **109**: 69-75.

17. Fonseca, S.G., Gromada, J., Urano, F. 2011. Endoplasmic reticulum stress and pancreatic β-cell death. *Trends in Endocrinology and Metabolism,* **22** (7): 266-274.

18. Forbes, J.M., Cooper, M.E. 2013. Mechanisms of diabetic complications. *Physiological Reviews,* **93**: 137-188.

19. Ford, J.A., Lois, N., Royle, P., Clar, C., Shyangdan, D., Waugh, N. 2013. Current treatments in diabetic macular oedema: systematic review and meta-analysis. *BMJ Open*, **2013**: 3:e002269. doi:10.1136/bmjopen-2012-002269.

20. Greene, S. 2002. Diabetes in childhood and adolescents. *Medicine*, **30** (2): 60-65.

21. Hamada, S.H., El Debkry, H.M. 2014. Monitoring of motor function affection and postural sway in patients with type 2 diabetes mellitus. *Egyptian Journal of Ear, Nose, Throat and Allied Science*, **15**: 241-245.

22. Houghton, P.J., Howes, M.-J., Lee, C.C., Steventon, G. 2007. Uses and abuses of *in vitro* tests in ethnopharmacology: Visualizing an elephant. *Journal of Ethnopharmacology*, **110**: 391-400.

23. Hu, Y., Davies, G.E. 2009. Berberine increases expression of GATA-2 and GATA-3 during inhibition of adipocyte differentiation. *Phytomedicine*, **16**: 864–873.

24. Hu, Y., Davies, G.E. 2010. Berberine inhibits adipogenesis in high-fat diet-induced obesity mice. *Fitoterapia*, **81**: 358–366.

25. Hyman, M.A. 2010. Environmental toxins, obesity, and diabetes: an emerging risk factor. *Alternative Therapies in Health and Medicine*, **16** (2): 56-58.

26. IDF. 2014. *World diabetes atlas* 6th edition update, International Diabetes Federation.

27. www.idf.org (accessed 7 March 2015)

28. Kavishankar, G.B., Lakshmidevi, N., Mahadeva Murphy, S., Prakash, H.S., Niranjana, S.R. 2011. Diabetes and medicinal plants – a review. *International Journal of Biomedical Science*, **2** (3): 65-80.

29. Komaki, E., Yamaguchi, S., Maru, I., Kinoshita, M., Kakehi, K., Yasuhiro Ohata, Y., Tsukada, Y. 2003. Identification of anti-amylase components from olive leaf extracts. *Food Science and Technology Research*, **9**: 35–39.

30. Kumar, S., Singh, R., Vasudeva, N., Sharma, S. 2012. Acute and chronic animal models for the evaluation of anti-diabetic agents. *Cardiovascular Diabetology*, **11** (9): 1-13.

31. Lebovitz, H.E. 2011. Type 2 diabetes mellitus – current therapies and the emergence of surgical options. *Nature Reviews Endocrinology*, **7** (7): 408-419.

32. Lenzen, S. 2008. The mechanisms of alloxan-and streptozotocin-induced diabetes. *Diabetologia*, **51** (2): 216-226.

33. Li, A.P. 2005. Preclinical *in vitro* screening assays for drug-like properties. *Drug Discovery Today: Technologies*, **2**:179-185.

34. Lindsay, R.S. 2009. Gestational diabetes: causes and consequences. *British Journal of Diabetes and Vascular Disease*, **9**: 27-31.

35. Liu, M., Sun, J., Cui, J., Chen, W., Guo, H., Barbetti, F., Arvan, P. 2015. INS-gene mutations: From genetics and beta cell biology to clinical disease. *Molecular Aspects of Medicine*, In press, doi.org/10.1016/j.mam.2014.12.001

36. Longnecker, M.P., Daniels, J.L. 2001. Environmental contaminants as etiological factors for diabetes. *Environmental Health Perspectives*, 109: 871-876.

37. Martinez, R. 2013. Prevalence of Diabetes in the World, 2013. *Health Intelligence*.

38. http://healthintelligence.drupalgardens.com/content/prevalence-diabetes-world-2013 (accessed 7 March 2015).

39. Mastuda, H., Li, Y., Murakami, T., Matsumura, N., Yamahara, J., Yashikawa, M. 1998. Anti-diabetic principles of natural medicines III. Structure-related inhibitory activity and action mode of oleanolic acid glycosides on hypoglycemic activity. *Chemical and Pharmaceutical Bulletin*, 46: 1399-1403.

40. Nataraju, A., Saini, D., Ramachandran, S., Benshoff, N., Liu, W., Chapman, W., Mohanakumar T. 2009. Oleanolic acid, a plant triterpenoid, significantly improves survival and function of islet allograft. *Transplantation*, 88: 987–994.

41. Newman, D.J., Cragg, G.M. 2012. Natural products as sources of new drugs over the 30 years from 1981 to 2010. *Journal of Natural Products*, 75: 311-335.

42. Pescosolido, N., Campagna, O., Barbato, A. 2014. Diabetic retinopathy and pregnancy. *International Ophthalmology*, 34: 989-997.

43. Prabhakar, P.K., Kumar, A., Doble, M. 2014. Combination therapy: A new strategy to manage diabetes and its complications. *Phytomedicine*, 21: 123-130.

44. Rawal, L.B., Tapp, R.J., Wlliams, E.D., Chan, C., Yasin, S., Oldenburg, B. 2012. Prevention of type 2 diabetes and its complications in developing countries: A review. *International Journal of Behavioural Medicine*, 19:121-133.

45. Rees, D. A., Alcolado, J. C. 2005. Animal models of diabetes mellitus. *Diabetic Medicine*, 22 (4): 359-370.

46. Rolo, A.P., Palmeira, C.M. 2006. Diabetes and mitochondrial function: role of hyperglycaemia and oxidative stress. *Toxicology and Applied Pharmacology*, 212: 167-178.

47. Rupeshkumar, M., Kavitha, K., Haldar, P.K. 2014. Role of herbal plants in the diabetes mellitus therapy: an overview. *International Journal of Applied Pharmacology*, 6 (3): 1-3.

48. Sen, S., Chen, S., Feng, B., Iglarz, M., Chakrabarti, S. 2012. Renal, retinal and cardiac changes in type 2 diabetes are attenuated by macitentan, a dual enthelin receptor antagonist. *Life Sciences*, 91: 658-668.

49. Skovsø, S. 2014. Modeling type 2 diabetes in rats using high fat diet and streptozotocin. *Journal of Diabetes Investigation*, 5: 349-358.

50. Søfteland, E., Brock, C., Frøkjær, J.B., Brøgger, J., Madácsy, L., Gilja, O.H., Arendt-Nielsen, L., Simrén, M., Drewes, A.M., Dimcevski, G. 2014. Association between visceral, cardiac and sensory motor polyneuropathies in diabetes mellitus. *Journal of Diabetes Complications*, 28: 370-377.

51. Soria, N., Skoudy, A., Martín, F. 2001. From stem cells to beta cells: new strategies in cell therapy of diabetes mellitus. *Diabetologia*, 44: 407-415.

52. Suh, N., Honda, T., Finlay, H.J., Barchowsky, A., Williams, C., Benoit, N.E., Xie, Q.W., Nathan, C., Gribble, G.W., Sporn, M.B. 1998. Novel triterpenoids suppress inducible nitric oxide synthase (iNOS) and inducible cyclooxygenase (COX-2) in mouse macrophages. *Cancer Research*, **58**: 717–723.

53. Szkudelski, T. 2001. The mechanism of alloxan and streptozotocin action in B cells of the rat pancreas. *Physiological Research*, **50** (6): 537-546.

54. Teodoro, J.S., Gomes, A.P., Varela, A.T., Duarte, F.V., Rolo, A.P., Palmeira, C.M. 2013. Uncovering the beginning of diabetes: the cellular redox status and oxidative stress as starting players in hyperglycemic damage. *Molecular and Cellular Biochemistry*, **376**: 103-110.

55. Tiwari, A.K., Viswanadh, V., Gowri, P.M., Ali A.Z., Radhakrishnan, S.V.S., Agawane, S.B., Madhusudana, K., Rao, J.M. 2010. Oleanolic acid an α-glucosidase inhibitory and antihyperglycemic active compound from the fruits of *Sonneratia caseolaris*. *Open Access Journal of Medicinal and Aromatic Plants*, **1**: 19-23.

56. Wallia, A., Molitch, M.E. 2014. Insulin therapy for type 2 diabetes mellitus. *Journal of the American Medical Association*, **311** (22): 2315-2325.

57. Weiss, M.A. 2013. Diabetes mellitus due to the toxic misfolding of proinsulin variants. *FEBS Letters*, **587**: 1942-1950.

58. World Health Organization (WHO). 2005. Special Programme for Research and Training in Tropical Diseases (TDR): Operational guidance: Information needed to support clinical trials of herbal products.

59. http://www.who.int/tdr/publications/documents/operational-guidance-eng.pdf (accessed 7 March 2015).

60. Yin, J., Ye, J., Jia, W. 2012. Effects and mechanisms of berberine in diabetes treatment. *Acta Pharmaceutica Sinica* B, **2**: 327–334.

Chapter 7

Anti-Glycation and Glycation Reversing Potential of *Salacia reticulata* L. (Kothala Himbutu) Root, Stem, Leaf and Twig Extracts

G.A.S. Premakumara[1], W.K.S.M. Abeysekera[2] and P. Ranasinghe[3]

Industrial Technology Institute (ITI), 363, Bauddhaloka Mawatha, Colombo 07, Sri Lanka
E-mail: [1]gasp@iti.lk, [2]kancha@iti.lk. [3]pathmasiri@iti.lk

ABSTRACT

Glycation is a series of complex reactions between reducing sugars and proteins. This reaction ultimately produces multitude of detrimental advanced glycation end products (AGEs). Formation and accumulation of AGEs have been implicated in the development and progression of several diabetic complications, neurological diseases and aging. Thus, glycation inhibitors and glycation reversing agents offer a potential strategy as therapeutics for diverse diseases. *Salacia reticulata* L. is a scientifically well documented traditional anti-diabetic plant. However, anti-glycation and glycation reversing potential of this plant has not been studied. Present study reports anti-glycation and glycation reversing potential of *Salacia reticulata*.

Freeze dried hot water extracts of *Salacia reticulata* root, stem, leaf and twigs were used in this study. Different concentrations of root, stem, leaf and twig extracts were subjected to anti-glycation and glycation reversing assays in vitro. Rutin was used as the positive control.

Root, stem, leaf and twig extracts of *Salacia reticulata* showed significant (P < 0.05) anti-glycation activity in a dose dependent manner. IC_{50} values for anti-glycation activity of root, stem, leaf and twigs extracts were 13.06±0.69, 27.29±0.93, 144.53±1.12 and 171.90±0.88 μg/ml respectively. Root extract showed significantly high (P < 0.05) anti-glycation activity compared to other extracts and rutin (IC_{50}: 21.88±2.82 μg/ml). Glycation reversing potential of different parts of Salacia reticulata also showed significant (P < 0.05) and dose dependent relationship. EC_{50} values of root, stem, twig and leaf extracts were 101.60±11.57, 116.67±0.64, 180.53±7.41 and 264.40±9.30 μg/ml respectively. Potency of different parts of *Salacia reticulata* for anti-glycation and glycation reversing activities were root > stem > leaf > twig and root = stem > twig > leaf respectively.

It is concluded that all parts of *Salacia reticulata* possess both anti-glycation and glycation reversing activities. Further, this is the first study to report anti-glycation and glycation reversing potential of this plant.

Keywords: Salacia reticulata, Anti-glycation, Glycation reversing.

1. INTRODUCTION

Diabetes mellitus is a chronic metabolic disease increasing in epidemic proportions throughout the world (International Diabetes Federation, 2014; American Diabetes Association, 2010). It affected about 387 million people worldwide in 2014 and the number is projected to increase by another 205 million people by 2035 (International Diabetes Federation, 2014).

This chronic disease is characterized by hyperglycaemia due to defects in insulin secretion or insulin resistance (Wild *et al.*, 2004). Prolong hyperglycaemia results in the formation of advanced glycation end products in body tissues (Reddy and Beyaz, 2006; Wautier and Guillausseau, 2001). The complex AGEs formed can leads to protein cross linking and contribute to the development and progression of several diabetic complications such as peripheral neuropathy, cataracts, impaired wound healing, vascular damage, arterial wall stiffening, decreased myocardial compliances (Ahmed, 2005; Thomas *et al.*, 2005; Aronson, 2003; Wautier and Guillausseau, 2001), neurological diseases and aging (Reddy and Beyaz, 2006). Thus, glycation inhibitors and glycation reversing agents offer a potential strategy as therapeutics for diverse disease conditions.

Salacia reticulata L. is belongs to the genus *Salacia* and found in the submontane forests in Sri Lanka and India. In Sri Lankan and Indian traditional Ayurvedic medicine this plant is used in the treatment of diabetes, asthma, rheumatism, swelling, amenorrhea, dysmenorrhea, gonorrhoea, and skin diseases (Chandrasena, 1935; Jayaweera, 1981; Warrir, P.K. *et al.*, 1993; Ayurvedic pharmacopeia, 1994). Further, the root and stem decoctions of *Salacia reticulata* have been used for the management of diabetes (Karunanayake *et al.*, 1984; Ayurvedic pharmacopeia, 1994; Chandrasena, 1935). Traditional knowledge of anti-diabetic activity of roots and stems of this plant has been scientifically proven by many researches using *in vitro*, *in vivo* and clinical studies (Jayawardena *et al.*, 2005; Kajimoto *et al.*, 2000; Shimoda *et al.*, 1998; Yoshikawa *et al.*, 1998; Yoshikawa *et al.*, 1997; Serasinghe *et al.*, 1990). However, extremely limited studies have been showed that bark and root extracts can reduce HbA1c levels in diabetes patients (Jayawardena *et al.*, 2005; Kajimoto *et al.*, 2000). Further, there are no studies on anti-diabetic activity of leaves and twigs of *Salacia reticulata*. We have previously shown that root, stem, leaf and twigs extracts of *Salacia reticulata* have anti-oxidant activity (Ranasinghe *et al.*, 2007). As lot of oxidative reactions are known to participate in the process of AGEs formation (Nowotny *et al.*, 2015; Reddy and Beyaz, 2006) different parts of *Salacia reticulata* may have anti-glycation and glycation reversing activities. Present study reports anti-glycation and glycation reversing potential of root, stem, leaf and twigs extracts of *Salacia reticulata in vitro*.

2. MATERIALS AND METHODS

Materials

Leaves, stem, roots and twigs were collected from *Salacia reticulata* plantations of Eco-Tech Company Ltd at Nathandiya, Sri Lanka.

Chemicals and Reagents

Bovine serum albumin (BSA), D-glucose and trichloroacetic acid (TCA) were purchased from Sigma-Aldrich, USA. All the other chemicals used for the preparation of buffers and solvents were of analytical grade.

Preparation of Hot Water Extracts

Plant materials collected were air dried in an air-conditioned room (25±2 °C) for 6 days. Then, air dried samples were ground to fine powder using a laboratory grinder. Two grams from powdered root, stem, leaves and twigs samples were extracted in 50 ml of hot water for 20 min. Extracts were then filtered, centrifuged at 6000 rpm for 10 min and freeze-dried (Christ-Alpha 1-4 Freeze dryer, Biotech International, Germany). Freeze-dried extracts were used in anti-glycation and glycation reversing assays.

Anti-glycation Assay

The anti-glycation assay was performed according to the method of Matsuura *et al.* (2002) with some modifications. Freeze-dried extracts of *Salacia reticulata* root, stem, leaf and twigs (n=3 each) at 6 different concentrations (7.8, 15.6, 31.2, 62.5, 125.0 and 250.0 µg/ml) were used in the assay. Reaction mixtures containing 800 µg BSA, 400 mM glucose and different concentrations of *Salacia reticulata* extracts in a reaction volume of 1 ml in 50 mM phosphate buffer (pH 7.4) containing 0.02 per cent sodium aside (w/v) were incubated at 60 °C for 40 h. After cooling, aliquots of 600 µl were transferred to 1.5 ml eppendorf tubes and 60 µl of 100 per cent (w/v) TCA was added, stirred, centrifuged at 15,000 rpm at 4 °C for 4 min and supernatants were removed. The resulting AGEs-BSA precipitate was dissolved in 3 ml of phosphate buffer saline (pH 10) and fluorescence intensity was measured at an excitation wave length of 370 nm and emission wave length of 440 nm using a spectrofluorometer (Amino-Bowman®, Thermo Spectronic, USA). Rutin was used as the standard (positive control). Anti-glycation activity (inhibition per cent) of each *Salacia reticulata* extract and rutin was calculated using the following equation.

Inhibition (per cent) = $[(F_c-F_b)-(F_s-F_{sb})/(F_c-F_b)] * 100$ [1]

Where, F_c is the florescence of incubated BSA and glucose (control), F_b is the florescence of incubated BSA alone (blank), Fs is the florescence of the incubated BSA and glucose with *Salacia reticulata* extracts or the positive control (rutin) and F_{sb} is the florescence of incubated BSA with the *Salacia reticulata* extracts or the positive control.

Glycation Reversing Assay

Reaction mixture containing 800 µg BSA and 400 mM glucose in 1 ml of 50 mM phosphate buffer (pH 7.4) containing 0.02 per cent sodium aside (w/v) was incubated at 60 °C for 40 h. After cooling, aliquots of 600 µl were transferred to 1.5 ml eppendorf tubes and 60 µl of 100 per cent (w/v) TCA was added, stirred, centrifuged at 15,000 rpm at 4 °C for 4 min and supernatants were removed. The resulting AGEs-BSA precipitates were dissolved in 50 mM phosphate buffer (pH 7.4) and added with *Salacia reticulata* extracts (15.6, 31.2, 62.5, 125.0 and 250.0 µg/ml; n=3 each) in a final reaction volume of 1 ml for incubation at 60 °C for 40 h. After cooling, 60 µl of 100 per cent (w/v) TCA was added, stirred and centrifuged at 15,000 rpm at 4 °C for 4 min. The resulting precipitates were dissolved in 3 ml of phosphate buffer saline (pH 10) and fluorescence intensity was measured at an excitation wave length of 370 nm and emission wave length of 440 nm using a spectrofluorometer. Percentage glycation reversing was calculated using the following equation.

Glycation reversing (per cent) = $[(F_c-F_b)-(F_s-F_{sb})/(F_c-F_b)] * 100$ [2]

Where, F_c is the florescence of incubated BSA and glucose (control), F_b is the florescence of incubated BSA alone (blank), Fs is the florescence of the incubated BSA, glucose and *Salacia reticulata* extracts and F_{sb} is the florescence of incubated BSA with the *Salacia reticulata* extracts.

Statistical Analysis

Data represented as mean±SD (n=3). Data of each experiment were statistically analyzed using SAS version 6.12. One way analysis of variance (ANOVA) and the Duncan's Multiple Range Test (DMRT) were used to determine the differences among treatment means. $P < 0.05$ was regarded as significant.

3. RESULTS

Anti-glycation and glycation reversing activities of *Salacia reticulata* root, stem, leaf and twigs extracts are given in Tables 7.1 and 7.2 and Figures 7.1 and 7.2.

Root, stem, leaf and twigs extracts of *Salacia reticulata* showed significant ($P < 0.05$) anti-glycation activity in a dose dependent manner. IC_{50} values for anti-glycation activity of root, stem, leaf and twigs extracts were 13.06±0.69, 27.29±0.93, 144.53±1.12 and 171.90±0.88 µg/ml respectively. Root extract showed significantly high ($P < 0.05$) anti-glycation activity compared to other extracts and rutin (IC_{50}: 21.88±2.82 µg/ml). Potency of different parts of *Salacia reticulata* for anti-glycation activity was root > stem > leaf >twigs.

Table 7.1: Percentage Inhibition of Glycation by Water Extracts of *Salacia reticulata* Root, Stem, Leaf and Twigs

Salacia reticulata Extract	Concentration (µg/ml)						IC_{50} (µg/ml)
	7.8	15.6	31.2	62.5	125.0	250.0	
Root	32.09±3.47	59.55±5.42	68.96±0.18	77.92±1.17	83.79±0.57	93.81±0.74	13.06±0.69[d]
Stem	11.25±4.46	22.85±4.69	59.37±2.38	65.14±2.68	74.28±1.24	84.23±1.72	27.29±0.93[c]
Leaf	12.44±1.24	13.95±2.77	16.93±1.54	22.07±3.31	49.57±0.86	78.86±1.02	144.53±1.12[b]
Twigs	0.57±0.63	0.49±0.82	5.08±1.41	19.10±1.78	39.69±0.65	71.57±0.15	171.90±0.88[a]

Data represented as mean±SD. IC_{50} values superscripted by different letters are significantly different at $p < 0.05$.

Table 7.2: Percentage Glycation Reversing by Water Extracts of *Salacia reticulata* Root, Stem, Leaf and Twigs

Salacia reticulata Extract	Concentration (µg/ml)					EC_{50} (µg/ml)
	15.6	31.2	62.5	125.0	250.0	
Root	8.50±1.50	21.73±1.83	36.71±2.09	59.09±8.35	71.32±1.54	101.60±11.57[c]
Stem	5.71±1.81	18.15±2.84	28.72±4.55	52.62±1.94	72.98±4.38	116.67±0.64[c]
Leaf	2.56±1.74	8.05±2.74	21.00±2.43	32.10±3.12	45.96±0.25	180.53±7.41[b]
Twigs	8.50±1.50	15.23±3.41	32.14±4.87	45.40±1.73	60.22±1.99	264.40±9.30[a]

Data represented as mean±SD. EC_{50} values superscripted by different letters are significantly different at $p < 0.05$.

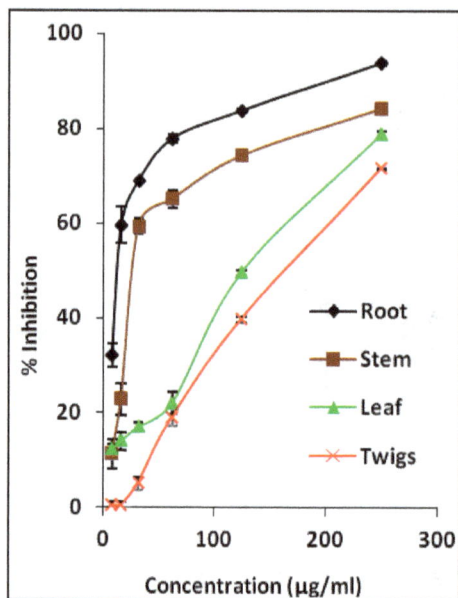

Figure 7.1: Percentage Inhibition of
Glycation by Water Extracts of
Salacia reticulata Root, Stem,
Leaf and Twigs.

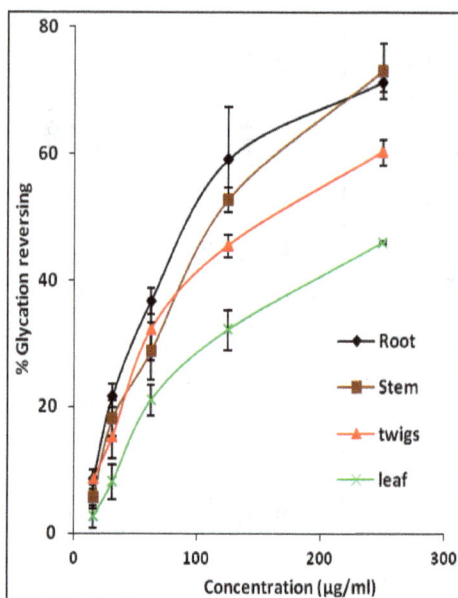

Figure 7.2: Percentage Glycation
Reversing Activity by Water Extracts of
Salacia reticulata Root, Stem,
Leaf and Twigs Extracts.

Glycation reversing potential of different parts of *Salacia reticulata* also showed significant (P<0.05) and dose dependent relationship. EC_{50} values of root, stem, twigs and leaf extracts were 101.60±11.57, 116.67±0.64, 180.53±7.41 and 264.40±9.30 µg/ml respectively. Potency of different parts of *Salacia reticulata* for glycation reversing activity was root = stem >twig> leaf.

4. DISCUSSION

Protein glycation (Maillard reaction) is a series of complex reactions between carbonyl groups of reducing-sugars with amino groups of proteins. Amino groups of proteins react initially with reducing sugars to form Schiff bases followed by their Amadori rearrangement products. These Amadori products undergo a rearrangement reaction giving a multitude of end products that are known as AGEs (Reddy and Beyaz, 2006). Some of the AGEs are intensely coloured compounds and have typical fluorescence characteristics (Reddy and Beyaz, 2006). Therefore, in this study anti-glycation and glycation reversing activities were measured in terms of inhibition of Maillard fluorescence formation.

Findings of this study clearly showed that all parts of *Salacia reticulata* possess anti-glycation activity. Interestingly, root extract had significantly high anti-glycation activity compared to the positive control, rutin. Therefore, especially root extracts can be used in the management of AGEs associated chronic diseases.

This plant is a well known anti-diabetic plant in Sri Lankan traditional knowledge and Indian system of Auyrveda (Chandrasena, 1935). Anti-diabetic activity of this plant has been explained through variety of mechanisms. However, anti-diabetic activity with respect to anti-glycation activity has been very poorly documented. Therefore, findings of this study showed a novel way of explaining the anti-diabetic mechanisms of this plant.

Glycation reversing is the reversing of already formed AGE cross links. It is an approach to attenuate AGE related complications. Such protein crosslink breakers might be useful as therapeutics for regulation of complications resulting from diabetes, neurological diseases and aging (Reddy and Beyaz, 2006). However, only few AGE crosslink breakers known to date and there are reports of their limited efficacy in *in vivo* studies (Nagai *et al.*, 2012; Reddy and Beyaz, 2006). Therefore, it is vital to explore compounds with AGEs reversing ability to manage AGE related complications. Findings of this study clearly showed that all parts of *Salacia reticulata* possess glycation reversing activity. These novel anti-diabetic properties further add values for this traditional medicinal plant as an anti-diabetic plant with multiple mechanisms. This is the first report of simultaneous presence of anti-glycation and glycation reversing activities of this plant.

Interesting and valuable findings of this study are that the presence of anti-glycation and glycation reversing activities in leaves and twigs. Therefore, leaves and twigs which can be repeatedly harvested in short cycles unlike root and stem can be used as a good natural source with anti-glycation and glycation reversing activities.

Different AGE inhibitors suppress AGE formation at different stages of glycation. Aspirin inhibits protein glycation at the early stage of glycation process by acetylating free amino groups of protein. Therefore, it causes to block the attachment of reducing sugars (Malik and Meek, 1994). The inhibitory activities of vitamin B1 and B6 derivatives such as pyridoxamine and thiamine pyrophosphate (Reddy and Beyaz, 2006; Booth *et al.*, 1996) have mainly been attributed to their abilities to scavenge reactive carbonyl compounds. We have previously shown that all parts of *Salacia reticulata* possess anti-oxidant activity (Ranasinghe *et al.*, 2007). Therefore, anti-glycation and glycation reversing activities may be due to the presence of anti-oxidative compounds. However, it is difficult to decide exactly at which stage of glycation process or in what way the intervention by *Salacia reticulata* extracts is exerted to reduce the glycation reaction. Further experiments are necessary to identify active compound/s, *in vivo* efficacy and mode of actions.

5. CONCLUSIONS

It is concluded that all parts of *Salacia reticulata* possess both anti-glycation and glycation reversing activities. Roots and stems were the most biologically active parts of the plant. Further, leaves and twigs, which can be repeatedly harvested in short cycles unlike roots and stems can be used as a good natural source with anti-glycation and glycation reversing activities. This is the first study to report anti-glycation and glycation reversing potential of this plant.

6. ACKNOWLEDGMENTS

The authors acknowledge Eco-Tech Company Ltd at Nathandiya, Sri Lanka for supplying samples for the study.

7. REFERENCES

1. Ahmed N. (2005). Advanced glycation end-products-role in pathology of diabetic complications. *Diabetes Res. Clin. Pract*. 67: 3-21.

2. American Diabetes Association. (2010). Diagnosis and Classification of Diabetes Mellitus. *Diabetes care*. 33: S62-S69.

3. Aronson D. (2003). Cross-linking of glycated collagen in the pathogenesis of arterial and myocardial stiffening of aging and diabetes. *J. Hypertens*. 21: 3-12.

4. Ayurvedic Pharmacopea (1994). Department of Ayurveda, Sri Lanka. Vol I (part II), 67.

5. Booth AA, Khalifah RG, Hudson BG. (1996). Thiamine pyrophosphate and pyridoxamine inhibit the formation of antigenic advanced glycation end-products: comparison with aminoguanidine. *Biochem. Biophys. Res. Commun*. 220(1): 113–119.

6. Chandrasena JPC. (1935). *The Chemistry and Pharmacology of Ceylon and Indian Medicinal Plants*, H and C Press, Colombo, Sri Lanka.

7. International Diabetes Federation. (2014). *Diabetes Atlas*, Sixth edition, 2014 update, Brussels, Belgium.

8. Jayawardena MH, de Alwis NM, Hettigoda V, Fernando DJ. (2005). A double blind randomised placebo controlled cross over study of a herbal preparation containing *Salacia reticulata* in the treatment of type 2 diabetes. *J Ethnopharmacol*. 97(2): 215-218.

9. Jayaweera, DMA. (1981). Medicinal Plants (Indigenous and Exotic) used in Ceylon, National Science Council of Sri Lanka. 76-77.

10. Kajimoto O, Kawamori S, Shimoda H, Kawahara Y, Hirata H, Takahashi T. (2000). Effects of a Diet Containing *Salacia reticulata* on Mild Type 2 Diabetes in Humans. A placebo-controlled, cross-over trial. *JJSNFS*. 53: 199-205.

11. Karunanayake, E.H., Welihinda, J., Sirimanne, S.R., Sinnadorai, G. (1984). Oral hypoglycaemic activity of some medicinal plants of Sri Lanka. *J. Ethnopharmacol*. 11, 223-231.

12. Malik NS, Meek KM. (1994). The inhibition of sugar-induced structural alterations in collagen by aspirin and other compounds," *Biochem. Biophys. Res. Commun*. 199: 683-686.

13. Matsuura N, Aradate T, Sasaki C, Kojima H, Ohara M, Hasegawa J. (2002). Screening system for the Maillard reaction inhibitor from natural product extracts. *J. Health Sci*. 48(6): 520-526.

14. Nagai R, Murray DB, Metz TO, Baynes JW. (2012). Chelation: a fundamental mechanism of action of AGE inhibitors, AGE breakers, and other inhibitors of diabetes complications. *Diabetes*. 61(3): 549-59.

15. Nowotny K, Jung T, Höhn A, Weber D, Grune T. (2015). Advanced glycation end products and oxidative stress in type 2 diabetes mellitus. *Biomolecules*. 5(1): 194-222.

16. Ranasinghe P, Premakumara GAS, Chanaka U. (2007). Free radical scavenging property of leaves of *Salacia reticulata* at different maturity stages. Proceedings of the Annual Scientific Sessions of the Institute of Biology, Sri Lanka. pp. 21.

17. Reddy VP, Beyaz A. (2006). Inhibitors of the Maillard reaction and AGE breakers as therapeutics for multiple diseases. *Drug Discov Today*. 11: 646-654.

18. Serasinghe S, Serasinghe P, Yamazaki H, Nishiguchi K, Hombhanje F, Nakanishi S. (1990). Oral hypoglycemic effects of *Salacia reticulata* in the Ssreptozotocin induced diabetic rat. *Phototherapy Research*. 4: 205-206.

19. Shimoda H, Kawamori S, Kawahara Y. (1998). Effects of aqueous extract of *Salacia reticulata*, a useful plant in Sri Lanka, on post-prandial hyperglycemia in rats and Humans. *JJSNFS*. 51: 279-287.

20. Thomas MC, Baynes JW, Thorpe SR, Cooper ME. (2005). The role of AGEs and AGE inhibitors in diabetic cardiovascular disease. *Curr. Drug Targets*. 6: 453-474.

21. Warrir, P.K., Nambir, V.P.K and Ramankutty, C. (1993). *Salacia reticulata* weight. In "Indian Medical Plant". Orient Longman, Chennai, pp. 47-48.

22. Wautier JL, Guillasseau PJ. (2001). Advanced glycation end products, their receptors and diabetic angiopathy. *Diabetes Metab*. 27: 535-542.

23. Wild S, Roglic G, Green A, Sicree R, King H. (2004). Global prevalence of diabetes estimates for the year 2000 and projections for 2030. *Diabetes Care*. 27: 1047-1053.

24. Yoshikawa M, Murakami T, Shimada H, Matsuda H, Yamahara J, Tanabe G, Muraoka O. (1997). Salacinol, potent anti-diabetic principle with unique thiosugar sulfonium sulfate structure from the Ayurvedic traditional medicine *Salacia reticulata* in Sri Lanka and India. *Tetrahedron Lett*. 38: 8367-8370.

25. Yoshikawa M, Murakami T, Yashiro K, Matsuda H. (1998). Kotalanol, a potent α-glucosidase inhibitor with thiosugar sulfonium sulfate structure, from anti-diabetic Ayurvedic medicine *Salacia reticulata*. *Chem. Pharm. Bull*. 46: 1339-1340.

II. Anti-Microbial Agents from Plants

Chapter 8

Potential Application of *Tylosema esculentum* Tuber Extractives in Treatment of Antibacterial Infections

Ngonye Keroletswe[1], R.R.T. Majinda[1], I.B Masesane[1]
and O. Mazimba[2]

[1]Chemistry Department, University of Botswana,
Private Bag 00704, Gaborone, Botswana
E-mail: ngosab4756@yahoo.com
[2]Nanomaterials Department,
Botswana Institute for Technology Research and Innovation,
Post Bag 0082, Gaborone, Botswana

ABSTRACT

Phytochemical analyses of *Tylosema esculentum* tuber yielded eight compounds; cyanogenic glycosides (lithorspermoside **1** and griffonin **2**,), flavan ((+)-catechin **3**), phenylacetic acid derivatives (2,4-dihydroxyphenyl acetic acid **4** and 2,4-dihydroxyphenyl acetic acid methyl ester **5**), triterpene (β-sitosterol **6**), benzofuran (griffonilide **7**), and a sugar (sucrose **8**). The structure of the isolated compounds were confirmed using spectroscopic techniques: NMR, MS and IR. The isolates were tested for antibacterial activities against *E. coli, S. aureus, B. subtilis, P. aeruginosa* and *C. albicans.* The 2,4-dihydroxyphenylacetic acid derivatives exhibited good antibacterial activities.

Keywords: *Tylosema esculentum, Lith ospermoside, 2,4-Dihydroxyphenylacetic acid, 2,4-Dihydroxyphenylacetic acid methyl ester, Cyanogenic glycosides, Benzofuran.*

1. INTRODUCTION

The current trend worldwide is integration of traditional medicine with primary health care systems (Fennell *et al.*, 2004). In this regard, medicinal plants have been and are still the major sources of medicine worldwide (Goyal *et al.*, 2011; Asari *et al.*, 2012). The unfortunate thing is that these herbal and traditional medicines are often not well researched and may contain adulterated products that may cause adverse effects to human life (Mills *et al.*, 2005). It is believed that traditional pharmacopoeias may provide leads to the discovery of novel compounds of pharmaceutical value (Wangchuk, 2008) and in clinical use (Harvey, 2008; Ntie-Kang *et al.*, 2013). Of all the 1135 new drugs that were approved over the time period, 1981 to 2010, 5 per cent were isolated natural product, 17 per cent as NP mimics produced by synthesis, 28 per cent as NP derivatives (semi-synthetic) and 50 per cent as synthetic drugs (Newman and Cragg, 2012). Although natural products are deemed safe and harmonious with biological systems (Tshibangu *et al.*, 2004), they are normally isolated in minute quantities. As such, they are mainly used as templates in drug discovery (Chin, 2006; Potterat and Hamburger, 2008; Harvey, 2008; Newman, 2008; Li and Vederas, 2009). Hence isolation, characterization and biological screening of both the pure secondary metabolites and plant crude extracts continue to be of utmost importance. A lot of work has been done on research tailored at finding secondary metabolites of plant origin that have therapeutic use even though much more needs to be done (Rout *et al.*, 2009). The driving forces in current research are the need to find cure for the continued outbreak of new human health ailments, resistance of pathogens to available drugs in clinical use, sensitization reactions and other undesirable side effects arising from synthetic drugs.

Botswana, just like many other African countries, India and China, houses a great wealth of medicinal plants. One of those plants that is the subject of this research project is *Tylosema esculentum* (Camel's foot, Morama). It is a creeping shrub found in southern Africa, particularly in Botswana, Namibia and South Africa. It belongs to the family *Fabaceae* (Leguminoseae). The plant is used by the native people of the Kalahari Desert as a source of food, medicine and for general health maintenance. The bean extract is used to treat diarrhea and improve blood pressure. Mixtures of morama tuber and devil's claw boost the immune system. The tuber extract and powdered *Acacia nigrescens* bean and *Acokanthera oppositifolia* tuber relieve diarrhoea, stomach cramps, headaches, and prevent hypertension (Chigwaru *et al.*, 2007; 2011). The decoction of morama root/tuber and flowers of *Tylosema fassoglence* are used to treat impotence (Dubois *et al.*, 1994). Previous investigation of the ethyl acetate *T. esculentum* tuber extract yielded griffonilide 7, 3,4-dihydrogriffonilide 9, morama epoxide 15 and behenic acid 17, Figure 8.1 (Mazimba, 2010). Compounds 7, 9 and 15 showed weak DPPH radical scavenging and antimicrobial activities (Mazimba *et al.*, 2011). The morama tuber ethyl acetate and methanol extracts exhibited high antioxidant and antimicrobial activities plus high total phenolic contents (Chingwaru *et al.*, 2011). Based on the above findings, it was obvious that there is need for a comprehensive investigation of the tuber to isolate secondary metabolites responsible for the observed activities or show that the observed activities were due to the combined effect of the isolated metabolites rather than the pure compounds. This is because the chemical complexity of an extract may contribute to the synergistic action of the extract. Thus, processes like purification and separation may result in partial or complete loss of specific activity due to the removal of chemically related substances contributing to the activity of the main components.

This project was part of an ongoing research program aimed at the isolation, characterization and testing secondary metabolites from local medicinal plants for pharmacological activities of therapeutic potential. Thus here we report the isolation and characterization of secondary metabolites from the tuber of *T. esculentum*, and their antimicrobial activities.

2. MATERIALS AND METHODS

General Experimental Procedures

The melting points were recorded on a Stuart Scientific Melting Point Apparatus and were uncorrected. 1D and 2D NMR spectra were recorded with a BRUKER DPX-300 and 600 MHz spectrometers using as solvent methanol-d_4, acetone-d_6 and D_2O, and the chemical shifts are reported in δ units (ppm) and coupling constants (J) in Hz. Sephadex LH-20 and Silica gel (E.M. Merck, 60-230 mesh) were used for column chromatography (CC), while aluminum plates impregnated with silica gel 60 F_{254} (E.M. Merck) were used for analytical (0.25 mm) TLC analyses. Spots on chromatograms were detected under UV light (254 and 366 nm) and by spraying the plates with 5 per cent H_2SO_4, followed by heating.

Plant Material

The *T. esculentum* tuber was collected in August 2011 from the Botswana College of Agriculture gardens, identified by Dr K. Mogotsi. A voucher specimen (voucher # MB4kg1990) is kept at the University of Botswana Herbarium in Gaborone, Botswana.

Extraction and Isolation

The tubers were skinned, cut into small pieces, air dried for one week and ground to powder. The dry tuber powder (289.41 g) was sequentially extracted at room temperature with EtOAc for 72 h followed by EtOAc/MeOH (7:3 v/v), then methanol, 70 per cent methanol and finally water. The ethyl acetate extract (TEf1=0.93 g) was subjected to CC using different combinations of *n*-hexane, acetone, ethyl acetate and methanol in increasing polarity. The CC work-up yielded 30 fractions which were combined into 5 sub-fractions using TLC profiles. Sub-fraction 2 gave compound **6** (17.2 mg) while sub-fraction 4 yielded compound **7** (3.9 mg) after prep TLC (E:M:W 9:1:1). Sub-fractions 1, 3 and 5 contained compounds **6** and **7** as major constituents. The EtOAc/MeOH (TEf2=28.89 g) and MeOH (TEf3=27.82 g) fractions were combined together after TLC profiling and compound **1** (9.6 mg) crystallized out of fraction TEf2 during the process of dissolution with methanol. CC on the combined extracts afforded 11 pre-combined sub-fractions. Prep TLC analysis (E:M:W 8:1:) on sub-fraction 3 (80.0 mg) furnished compounds **4** (3.9 mg) and **5** (32.0 mg). Sub-fraction 4 (16.0 mg) yielded compound **3** (5.2 mg) after CC,

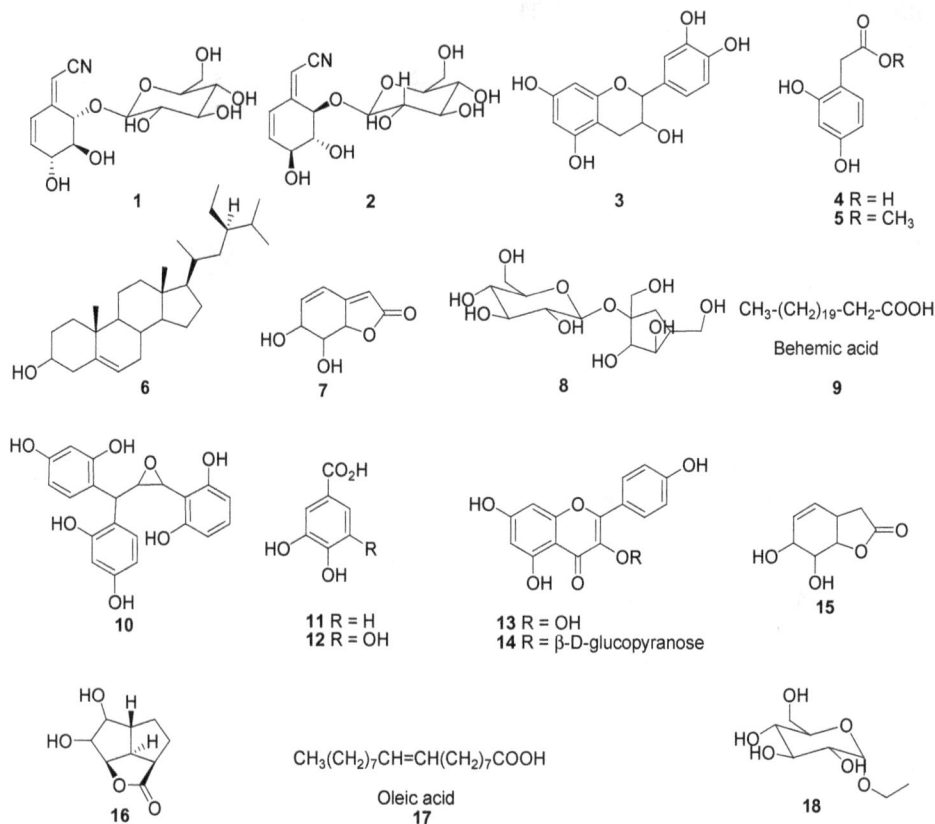

Figure 8.1: Compounds Isolated from *Tylosema esculentum*.

gradient elution. Compound **2** crystallized out of sub-fraction 10 whereas the major constituent of the filtrate was compounds **1** and **8**. Sub-fraction 11 yielded compound **8** (1.84 g combined mass) after prep TLC analysis (E:M:W 7:2:1). Fractions TEf4 and TEf5 comprised of compounds **2**, **3** and **8** as major constituents; as such they were not worked on (TLC profiling).

Antimicrobial Investigations

Test Organisms

The antibacterial activities were tested against *E. coli* (ATTC 11229), *S. aureus* (ATTC 9144), *B. subtilis* (ATTC 6633), *P. aeruginosa* (NCTC 10332) and *C. albicans* (ATTC 10231). The bacterial strain were obtained from the microbiology laboratory in the Department of Biological sciences, University of Botswana.

Preparation of Culture Media and Culture Suspensions

Nutrient broth (NB) was used for bacteria while Sabouraud dextrose broth (SDB) was prepared for fungi. Nutrient broth (13 g/L) and SDB (1g mycological peptone plus 4 g glucose) were each prepared by dissolving them in distilled water (100 mL). The media solutions were distributed into volumes of 50 mL in conical flasks (6 x 100 mL) covered with cotton wool plugs and aluminium foil. Nutrient agar (28 g) and sabouraud dextrose agar (39 g) were each dissolved in distilled water (1 L) by heating with continuous stirring to obtain clear solutions. The culture media were autoclaved at 121 °C for 15 minutes and kept in an oven set at 44 °C until they were used. The nutrient broth was left to cool to room temperature before use. The culture suspensions were prepared by aseptically transferring the microorganisms from the Petri-dishes to the nutrient broth using sterile nichrome wire. The microorganisms were incubated for 24 hrs in a Lab-Line Orbit Shaker/ incubator at 37 °C. The cloudy medium was observed confirming the growth of microorganisms in the cultures.

Agar well Diffusion Method

The Kirby Bauer Method (**Bauer** *et al.,* 1959; 1966) was adopted for the agar well diffusion method. The M001, HI media Nutrient agar (44°C) prepared as described above, poured into petri dishes (15 mL agar solution per a petri dish) and allowed to set. Five wells per a petri dish were cut with sterile cock borer into pre-labelled petri dishes. Stock solutions of 10 mg/L of *T. esculentum* EtOAc: MeOH (7:3 v/v) extract and compounds **4** and **5** in appropriate solvents were prepared and serially diluted to give final concentrations 5.0, 1.0, 0.1, 0.05, and 0.001 mg/mL. A 10 μg aliquot of the test sample/extract in appropriate solvent was put into each well corresponding to loading quantities of 50, 10, 0.5, 0.1 and 0.01 μg. The solvent was allowed to evaporate to dryness in the bio-hazard fume-hood before sterile cotton swabs were used to apply the overnight cultures. The inoculated petri dishes were incubated for 24 hours at 37°C. Zones of inhibition were determined to be the clear areas where there were no microbial growth, and were measured using a ruler (in mm). The larger the diameter, the more effective the isolate and or extract is.

Physical and Spectroscopic Data

Compound 1 (Lithospermoside)

White powder, mp 276-277 °C. $[\alpha]_{D}$ -37.3 ° (c = 0.02, H_2O) UV λ_{max} (H_2O) 288 nm; IR v_{max} cm^{-1} 3417, 2914, 2223, 1632, 1408, 1076, 1042 and 890. ^1H and ^{13}C NMR in D_2O (see Table 8.1). HR-TOF-EIMS m/z 329.1115 [M⁺] (15) (MF $C_{14}H_{19}O_8N$ calculated 329.1146), 311.1031 (20) [M-OH₂]⁺, 268.0809 (25), 222.0769 (20), 196.0614 (58), 191.0568 (25). ^1H and ^{13}C NMR (D_2O); 3.29 (2H, *m*; H-4′), 3.59 (1H, *dd*, *J* = 5.4; 12.6 Hz, H-6′), 3.76 (1H, *d*, *J* = 12.6 Hz, H-6′), 3.80 (1H, *d*, *J* = 3.9 Hz, H-8), 4.15 (1H, *dd*, *J* = 1.8; 12.6 Hz, H-6), 4.68 (1H, *br s*; H-7), 5.48 (1H, *br s*; H-2), 5.97 (1H, *dd*, *J* = 3.0; 10.2 Hz, H-5), 6.19 (1H, *d*, *J* = 10.2 Hz, H-4); 61.0 (C-6′), 69.8 (C-7), 70.1 (C-6), 72.9 (C-4′), 73.5 (C-5′), 76.2 (C-8), 97.2 (C-2), 102.7 (C-1′), 117.8 (C-1), 127.1 (C-4), 136.3 (C-5), 155.4 (C-3).

Compound 2 (Griffonin)

Brown powder, mp 260-262 °C. $[\alpha]_D$ +33.3 ° (c = 0.01, H_2O) UV λ_{max} (H_2O) 300 nm; IR v_{max} cm^{-1}. ^1H and ^{13}C NMR in D_2O (see Table 8.1). HR-EIMS m/z 329.1146 (22)

Table 8.1: ^1H (300 MHz) and ^{13}C NMR (75 MHz) Data for Compounds 1 and 2 in D_2O.

Position	Compound 1		Compound 2	
	δ_H	δ_C	δ_H	δ_C
1	–	117.8 (C)	–	117.6 (C)
2	5.48 (1H, br s)	97.2 (CH)	5.63 (1H, br s)	100.4 (CH)
3	–	155.4 (CH)	–	152.9 (CH)
4	6.19 (1H, d, J = 10.2 Hz)	127.1 (CH)	6.27 (1H, d, J = 10.2 Hz)	126.1 (CH)
5	5.97 (1H, dd, J = 10.2; 3.0 Hz)	136.3 (CH)	5.98 (1H, d, J = 10.2 Hz)	137.5 (CH)
6	4.15 (1H, dd, J = 12.6; 1.8 Hz)	70.1	4.23 (1H, t, J = 4.4 Hz)	68.8 (CH)
7	4.68 (1H, br s)	69.8 (CH)	4.66 (1H, d, J = 1.2 Hz)	65.7 (CH)
8	3.80 (1H, d, J = 3.9 Hz)	76.2 (CH)	4.80 (1H, d, J = 4.4 Hz)	77.2 (CH)
1´	4.74 (1H, d, J = 6.6 Hz)	102.7 (CH)	4.67 (1H, d, J = 1.2 Hz)	103.0 (CH)
2´, 3´	3.36 (2H, m)	75.9 (CH), 76.0 (CH)	3.40 (2H, m)	75.8 (CH), 76.2 (CH)
4´	3.29 (1H, m)	72.9 (CH)	3.14 (1H, t, J = 9.0 Hz)	73.1 (CH)
5´	(1H, t, J = 8.1 Hz)	73.5 (CH)	3.30 (1H, t, J = 9.0 Hz)	69.6 (CH)
6´	3.76 (1H, d, J = 12.6 Hz) 3.59 (1H, dd, J = 12.6; 5.4 Hz)	61.0 (CH₂)	3.82 (1H, d, J = 12.3 Hz) 3.65 (1H, dd, J = 12.3; 5.4 Hz)	60.7 (CH₂)

[M]⁺ (MF C₁₄H₁₉O₈N calculated 329.1146), 311.1032. ¹H and ¹³C NMR (D₂O); 3.14 (1H, *t*; *J* = 8.4 Hz; H-4´), 3.65 (1H, *dd*, *J* = 3.9 Hz, H-6´), 3.82 (1H, *d*, *J* = 12.0 Hz, H-6´), 4.80 (1H, *d*, *J* = 4.8 Hz, H-8), 4.23 (1H, *t*, *J* = 1.8; 12.6 Hz, H-6), 4.66 (1H, *d*, *J* = 1.2 Hz; H-7), 5.63 (1H, *br s*; H-2), 5.98 (1H, *d*, *J* = 10.2 Hz, H-5), 6.27 (1H, *d*, *J* = 10.2 Hz, H-4); 60.7 (C-6´), 65.7 (C-7), 68.8 (C-6), 73.1 (C-4´), 69.6 (C-5´), 75.8 (C-2´), 76.2 (C-3´), 77.2 (C-8), 100.4 (C-2), 103.0 (C-1´), 117.6 (C-1), 126.1 (C-4), 137.5 (C-5), 152.9 (C-3).

Compound 3 ((+)-catechin)

Brown solid, Melting point 173-174 (Lit. 175-177 RC (D-form)). [α] $_D$ = +10.6 (*c* = 0.001 g/mL, MeOH). UV λ$_{max}$ (MeOH) 277, 230 nm; IR v$_{max}$ cm⁻¹ 3325, 2900, 1622, 1470, 1389, 1021, 820. HR-EIMS *m/z* 290.0796 (50) [M⁺] (MF C₁₇H₁₄O₇, calculated 290.0790), 272.0686 (10) [M-OH]⁺, 152.0464 (50) [B]⁺, 139.0387 (100) [A + H•]. ¹H and ¹³C NMR (methanol-*d*₄); 2.76 (1H, *dd*; *J* = 2.4; 16.8 Hz; H-4ª), 2.89 (1H, *dd*, *J* = 4.5; 16.8 Hz, H-4ᵇ), 4.22 (1H, *d*, *J* = 2.4 Hz, H-2), 5.95 (1H, *d*, *J* = 2.1 Hz, H-8), 5.98 (1H, *d*, *J* = 2.1 Hz, H-6), 6.79 (1H, *dd*; *J* = 8.1 Hz; H-5´), 6.83 (1H, *dd*; *J* = 8.1; 2.1 Hz; H-6´); 27.9 (C-4), 66.1 (C-3), 78.5 (C-2), 94.6 (C-8), 95.1 (C-6), 98.7 (C-10), 114.0 (C-2´), 114.6 (C-5´), 118.1 (C-6´), 130.9 (C-1´), 144.4 (C-3´), 144.5 (C-4´), 156.0 (C-9), 156.2 (C-5), 156.6 (C-7),

Compound 4 (2,4-Dihydroxyphenylacetic acid)

Brown paste, UV λ$_{max}$ (MeOH) 292 nm. IR v$_{max}$ cm⁻¹ 3362, 2925, 1714, 1389 and 1017. HR-TOF-EIMS *m/z* 167.0338 [M-H]⁺ (30) (MF C₈H₈O₄; calculated 168.0422)., 150.0286 [M-OH₂] (30), 149.0225 (100), 111.1166 (30). ¹H and ¹³C NMR (CD₃OD); 3.40 (2H, *s*; H-2), 6.24 (1H, *dd*, *J* = 2.7; 8.1 Hz, H-5´), 6.30 (1H, *d*, *J* = 3.0 Hz; H-3´); 6.85 (1H, *d*, *J* = 7.8 Hz; H-6´); 41.6 (C-2), 103.8 (C-3´), 106.3 (C-5´), 115.4 (C-1´), 130.4 (C-6´), 156.9 (C-2´), 157.0 (C-4´), 180.2 (C-1) (see Table 8.3).

Compound 5 (Methyl 2-(2,4-dihydroxyphenyl)acetate)

Brown amorphous powder, melting point 102-104 RC. UV λ$_{max}$ (MeOH) 298, 448 nm; IR v$_{max}$ cm⁻¹ 3348, 3187, 1673, 1619, 1326 and 1104. HR-TOF-EIMS *m/z* 182.0572 (28) [M]⁺ (MF C₉H₁₀O₄ calculated 182.0579), 150.0304 (38) [M-OCH₃]⁺, 123.0425 (100) [M-CO₂CH₃]⁺, 122.0378 (41) [M-HCO₂CH₃]⁺, 95.0536 (25) [M-OCO₂CH₃]⁺. 94.0422 (30) [M-OHCO₂CH₃]⁺. ¹H and ¹³C NMR (CD3COCD₃); 3.52 (2H, *s*; H-2), 3.62 (3H, *s*, OCH₃), 6.31 (1H, *dd*, *J* = 2.4; 8.1 Hz, H-5´), 6.41 (1H, *d*, *J* = 2.4 Hz, H-3´); 6.94 (1H, *d*, *J* = 8.4 Hz; H-6´); 34.4 (C-2), 50.9 (OCH₃), 102.6 (C-3´), 106.5 (C-5´), 112.5 (C-1´), 131.3 (C-6´), 156.2 (C-2´), 157.6 (C-4´), 172.2 (C-1) (see Table 8.3).

3. RESULTS AND DISCUSSIONS

The phytochemical investigation of the tuber of *T. esculentum* afforded eight compounds. Compounds **1**, **3-5** were new to the species whereas compounds **2**, **6**, **7** and **8** were previously reported from *T. esculentum* bean husks, tubers, seed oil and bean cotyledons.

Compounds 1 and 2

Compounds **1** and **2** were isolated from the tuber ethyl acetate/methanol (7:3 v/v) extract of *T. esculentum*. The structure of compound **1** was proposed based on IR, MS and NMR spectra as well as comparison with that of griffonin **2**. The ¹H NMR

Table 8.2: ^1H (300 MHz) and ^{13}C NMR (75 MHz) Data for Compounds 4 and 5

Position	Compound 4 (ppm) (CD₃OD)		Compound 5 (ppm) (acetone d₆)	
	δ_H	δ_C	δ_H	δ_C
1	—	180.2 (C)	—	172.4 (C)
2	3.40 (2H, s)	41.6 (CH₂)	3.52 (2H, s)	34.4 (CH₂)
-OCH₃	—	—	3.62 (3H, s)	50.9 (CH₃)
1'	—	115.4 (CH₂)	—	112.5 (C)
2'	—	156.9 (CH₂)	—	156.2 (C)
3'	6.30 (1H, d, J = 2.9 Hz)	103.8 (CH)	6.41 (1H, d, J = 2.4 Hz)	102.6 (CH)
4'	—	157.0 (C)	—	157.6 (C)
5'	6.24 (1H, dd, J = 8.0; 2.9 Hz)	106.3 (CH)	6.31 (1H, dd, J = 8.2; 2.4 Hz)	106.5 (CH)
6'	6.85 (1H, d, J = 8.0 Hz)	130.4 (CH)	6.94 (1H, d, J = 8.2 Hz)	131.3 (CH)

spectrum of compound **1** showed characteristic glucose protons resonating within the range δ_H 3.38-3.95 and integrating for a total of six protons (6H). The anomeric proton resonated at δ_H 4.74 (1H, *d*, *J* = 6.6 Hz) directly attached to a carbon resonating at δ_C 102.7 (C-1) indicating the presence of a β-D-glucopyranoside (Jacobsen, 2007). Other protons were observed at δ_H 5.48 (1H, *s*), 6.19 (1H, *d*, *J* = 10.2 Hz), 5.97 (1H, *dd*, *J* = 10.2; 3.0 Hz), 3.80 (1H, *d*, *J* = 7.2 Hz) and 4.15 (1H, *br s*). The ^{13}C NMR spectrum (Table 8.1) showed fourteen carbon signals of which two were quaternary (δ_C 155.4 and 117.8), eleven methine and one methylene carbon (δ_C 61.0). Compound **1** was identified as lithospermoside, a 6,7,8-epimer of griffonin **2**. Lithospermoside has been previously isolated from the roots of *Thalictrum. rugosum* and *Thalictrum revolutum*, *Lithospermum purpureo-caerulum* and *L. officinale* (Boraginaceae). Griffonin **2** was previously isolated from the seed husk *n*-butanol extract of *T. esculentum* (Mazimba, 2010; 2011), from *Griffonia simplificola* (Kumar *et al.*, 2010) and from the Chinese traditional anticancer herb *Semiaquilegia adoxoides* (Niu *et al.*, 2006). It is worthy to mention that the major differences between the ^{13}C NMR spectra of the two compounds were observed at C-2 and C-3. In compound **2**, griffonin, C-2 (δ_C 100.4) lies in the de-shielding zone of the sugar moiety whereas it lies in the shielding zone in compound **1** (δ_C 97.2). C-3 (δ_C 152.9) lies in the shielding zone in compound **2** but the opposite was observed for compound **1** (δ_C 155.4). The other difference was observed in the ^{1}H-^{1}H COSY spectra of the two compounds. The ^{1}H-^{1}H COSY spectrum of compound **1** showed long range correlations between H-2 (δ_H 5.48) with H-8 (δ_H 3.80), H-7 (δ_H 4.68) and H-6 (δ_H 4.15). Such correlations are said to be observed in a case where a 'W-shape2 is present in a molecule (Siegler *et al.*, 2005). The same long range through bond correlations were not observed in the ^{1}H-^{1}H COSY spectrum of compound **2** where the 'W2 arrangement of atoms was absent. The IR spectra of compounds **1** and **2** showed absorption bands at 3417/3411cm^{-1} due to multiple hydroxyls, 2223/2225 cm^{-1} due to nitrile group and 1632 cm^{-1}due to C-O-C bonds (for compound **1/2**). The HR-TOF-EIMS spectrum of compound **1** showed the molecular ion peaks at *m/z* 329.1115 [M$^+$] (15) consistent with the molecular formula $C_{14}H_{19}NO_8$. Fragment ions were observed at *m/z* 311.1031 [M-OH$_2$]$^+$, 268.0809 and 167.0599 [M-Glu]$^+$.

Compound 3

Compound **3** was isolated as a brown powder from the tuber ethyl acetate/methanol (7:3 v/v) extract of *T. esculentum* with [α]$_D$ = + 10.6 R (c = 0.001, MeOH) and melting point of 173-175 RC (Lit. 175-177 RC (D-form) Hyde *et al.*, 2009). Its characteristic HR TOF EIMS data were [M]$^+$ at *m/z* 290.0796 (MF $C_{15}H_{14}O_6$), [M-OH]$^+$ at 272.0686 (10) and RDA fragments; [B]$^+$ at 152.0464 and [A + H$^•$]$^+$ 139.0387. Compound **3** was identified as 5,7,3,4-tetrahydroxyflavan-3-ol, commonly known as (+)-catechin. This is its first isolation from *T. esculentum* and from the genus Tylosema. It is said to be widespread in plants (Aggrawal *et al.*, 1989). It is present in cocoa, red wine and green tea. (+)-Catechin had been previously isolated from *Uncaria elliptica* (Law and Das, 1989) and from the extract of the red heartwood of *Acacia Catechu* tree (Hyde *et al.*, 2009). Compound **3** has shown good anti-inflammatory (Nakanishi *et al.*, 2010; Babu and Liu, 2008), anti-oxidant activities (Chung *et al.*, 2004) and anti-ulcer agent (Maity *et al.*, 2003 Srinivas *et al.*, 2013). It was also reported to exhibit hepatotronic (Gadjos *et al.*, 1972; Yu *et al.*, 2012).

Compound 4 and 5

Compounds **4** and **5** were isolated from the tuber ethyl acetate/methanol (7:3 v/v) extract of *T. esculentum*. These compounds were identified as benzenoids from their typical ABX proton spin systems observed at δ_H 6.30/6.41, 6.24/6.31 and 6.85/6.94 due to H-3´, H-5´ and H-5´, respectively and benzyl CH_2 protons resonated at δ_H 3.40/3.52 for compound **1/2** (Table 8.2). Compound **4** was identified as 2,4-dihydroxyphenylacetic acid whereas compound **5** is the methyl ester of compound **4**. The HR TOF EIMS data for compound **1**; 167.0338 [M-H]$^+$; 150.0286 [M-OH$_2$]$^+$; 149.0225 [M-OH$_3$]$^+$; 123.1164 [M-CO$_2$H]$^+$ and compound **2**; 182.0572 [M]$^+$; 150.0304 [M-OCH$_3$]$^+$ 123.0425[M-CO$_2$CH$_3$]$^+$ confirmed their identity. This is the first isolation of both compound **4** and **5** from *T. esculentum* and the genus Tylosema. Compound **4** was reported from the methanol extract of *Ouratea hexasperma* (Ochnaceae) branches (De Carvalho *et al.*, 2008), the *Nigella damascena* (Pamiculaceae) seeds (Figo *et al.*, 2000) and from the seeds of *Ilex aquifolium* (Aquifoliaceae) together with compound **5** (Nahar *et al.*, 2005). Compound **5** was also reported from *Mahuca pasquiery* (Sapotaceae) by Kamperdick and coworkers (Kamperdick *et al.*, 1997). Both compounds were reported to exhibit moderate levels of free radical scavenging activity compared to that of the quercetin (Nahar *et al.*, 2005).

The other compounds isolated from the tuber were **6-8**. Their structures were confirmed by comparing their spectroscopic data with literature (Mazimba, 2010; 2011). Compound **9-18** were previously reported from *T. esculentum* bean husks, bean cotyledons and seed oil (Mazimba, 2001, 2010, 2011).

Table 8.3: Antimicrobial Activities of Isolates and Extract Determined by Agar Well Diffusion Method

Extract/Compound	Test Organisms; MIQ µg				
	E. coli	B. subtilis	P. aeruginosa	S. aureus	C. albicans
EtOAc/MeOH (7:3)	0.01	50	10	0.01	0.01
4	50	na	5	5	na
5	0.01	na	50	0.5	0.5
Chloramphenicol	0.01	0.01	0.01	0.01	
Miconazole					0.01

Antimicrobial studies

The EtOAc/MeOH (7:3) as well as compounds **4** and **5** isolated from *T. esculentum* were screened for antimicrobial activities using the Agar well diffusion method. The results showed that compound **5** exhibits stronger anti-microbial activities against *E. coli*, *S. aureus* and *C. albicans* as compared to compound **4** (Table 8.3). The two compounds were inactive against *B. subtilis* whereas the extract exhibited weak activities. The observed differences in bioactivities between the two compounds could be explained in terms of enhancement of activity by the methyl group on compound **5**, which is absent in compound **4**. The EtOAc/MeOH (7:3) tuber extract exhibited better activity as compared to the two isolates. The loss in activity from the pure isolates might be due to lack of possible synergic action of different constituents present in an extract.

4. CONCLUSIONS

This study has revealed that the phytochemical constituents of the tuber of *T. esculentum* are lithospermoside **1**, griffonin **2**, (+)-catechin **3**, 2,4-dihydroxyphenyl acetic acid **4**, 2,4-dihydroxyphenyl acetic acid methyl ester **5**, β-sitosterol **6**, griffonilide **7** and sucrose **8**. Compounds **1-4** were new to *T. esculentum* while compounds **5-8** are known. Compounds **3-5** have shown moderate antibacterial and anti-oxidant activities. The 2,4-dihydroxyphenylacetic acid derivatives (**4, 5**) exhibited good antibacterial activities. Thus, validate the use of the *T. esculentum* tuber in treatment of anti-bacterial infections.

5. ACKNOWLEDGEMENTS

Special thanks to Dr K. Mogotsi for provision of Morama tubers and Mr I. Morobe for antimicrobial studies.

6. REFERENCES

1. Aggrawal P. K., 1989. Studies in Organic Chemistry 39: Carbon-13 NMR of flavonoids. Elsevier Science Publishing Company Inc., New York, United States of America, p. 13.

2. Ansari, S. H., Islam, F. and Samee, M., 2012. Influence of nanotechnology on herbal drugs: A Review. *J Adv Pharm Technol Res.*, 3(3), pp. 142-146.

3. Bauer, A. W., Kirby, W. M. M., Sherris, J. C., and Turck M., 1966. Antibiotic susceptibility testing by a standardized single disk method. *Am. J. Clin. Pathol.*, 36, pp. 493-496.

4. Bauer, A. W., Perry, D. M., and Kirby W. M. M., 1959. Single disc antibiotic sensitivity testing of *Staphylococci*. A.M.A. *Arch. Intern. Med.*, 104, pp. 208-216.

5. Babu, P. V., and Liu, D., 2008. Green tea catechins and cardiovascular health: an update. *Curr Med Chem.*, 15(18), pp. 1840-1850.

6. Chin, Y. W., Balunas, M. J., Chai, H. B., Kinghorn, A. D.; 2006. Drug discovery from natural sources. *AAPS J.*, 8(2), pp. E239-E253.

7. Chingwaru, W., Majinda, R. R. T., Yeboah, S. O., Jackson, J. C., Kapewangolo, P. T., Kandawa-Schulz, M and Cencic, A., 2011. *Tylosema esculentum* (Marama) tuber and bean extracts are strong antiviral agents against rotavirus infection. eCAM, Article ID 284795, doi:10.1155/2011/284795.

8. Chigwaru W., Faria M. I., Saravia C., and Cenciè A., 2007. Indegenous knowledge of health benefits of Morama plant among respondents in Gantsi and Jwaneng areas of Botswana. *AJFAND*, 7. Commentary sourced from http://www.ajfand.net/issue17commentary.htm.

9. Chung, J. E., Kurisawa, M., Kim, Y. J., Uyama, H. and Kobayashi, S., 2004. Amplification of antioxidant activity of catechin by polycondensation with acetaldehyde. *Biomacromolecules*, 5(1), pp. 113-118.

10. De Carvalho M. G., Suzart L. R., Cavatti L. C., and Kaplan M. A. C., 2008. New flavonoids and other constituents from *Ouratea hexasperma* (Ochnaceae). *J.B. Chem. Soc.*, 19(7), pp. 1423-1428.

11. Dubois, M., Lognay, G., Baudart, E., Marlier, M., Severin, M., Dardenne, G., and Malaisse, F., 1994. Chemical characterization of *Tylosema fassoglensis* (Kotschy) Torre and Hillc oil seed. *J. Sci. Food Agric.*, 67, pp. 163-167.

12. Fennell, C. W., Lindsey, K. L., McGaw, L.J. Sparg, S. G., Stafford, G. I, Elgorashi, E. E., Grace, O. M., and van Staden, J., 2004. Review: Assessing African medicinal plants for efficacy and safety: pharmacological screening and toxicology. *J. Ethnopharmacol.*, 94, 205-217.

13. Fico, G., Braca, A., Tome, F., and Morelli, I., 2000. Phenolic derivatives from *Nigella damascena* seeds. *Pharm Biol*, 38, pp. 371-373.

14. Gadjos, A., Gadjos-Torok, M. and Horn, R., 1972. The effect of (+)-catechin on the hepatic level of ATP and the lipid content of liver during experimentalsteatosis. *Biochem. Pharmacol.*, 21, pp. 594-600.

15. Goyal, M. and Mathur, R., 2011. Antimicrobial potential and phytochemical analysis of plant extracts of *Calotropis procera*. *Int. J. of Drug Discovery and Herbal Research*, 1(3), pp. 138-143.

16. Hamaishi, K., Kojima, R. and Ito, M., 2006. Anti-ulcer effect of tea catechin in rats. *Bio. Pharm. Bull.*, 29(11), pp. 2206-2213.

17. Harvey, A., 2008. Natural products in drug discovery. *Drug Discov. Today*, 13(19-20), pp. 894-901.

18. Hye, M. A., Taher, M. A., Ali, M.Y., Ali, M.U. and Zaman, S., 2009. Isolation of (+)-catechin from *Acacia Catechu* (Cutch tree) by a convenient method. *J. Sci. Res.*, 1, pp. 300-305.

19. Lee, J.-S. and Khitrin, A. K., 2006. Algorithmic subtraction of high peaks in NMR spectra. *Chem. Phys. Lett.*, 433(1), pp. 244-247.

20. Jacobsen, N. E., 2007. NMR Spectroscopy Explained: Simplified Theory, Applications and Examples for Organic and Structural Biology. John Wiley and Sons, New Jersey, pp. 16-20.

21. Kamperdick, C. Adam G., Van, N. H., and Sung T. V., 1997. Chemical Constituents of *Madhuca pasquiery*. Zeitschrift für Naturforschung C. *J. Biosci.*, 52(5-6), 295-300.

22. Kumar, P. S., Praveen, T., Jain, NishiPrakash and Jitendra, B., 2010. A review on *Griffonia simplicifollia* - an ideal herbal antidepressant. *IJPLS*, 1(3), 174-181.

23. Law, K.-H., and Das N. P., 1989. Production of (–)-epicatechin by *Uncaria elliptica* callus cultures. *Phytochemistry*, 28(4), pp. 1099-1100.

24. Li, J. W. H. and Vederas, J. C., 2009. Drug discovery and natural products: end of an era or an endless frontier? *Science*, 325(5937), pp. 161-165.

25. Maity, S., Chaudhuri, T., Vedasiromoni, J. R. and Gangul, D. K., 2003. Cytoprotection mediated anti-ulcer effect of tea root extract. *Indian Journal of Pharmacology*, 35(4), 213-219.

26. Mazimba, O., 2001. Phytochemical studies of Morama seeds: *Tylosema esculentum*. A research report for B.Ed Science Degree. University of Botswana, Botswana, p. 29.

27. Mazimba O., 2010. Phytochemical and bioactivity studies on *Tylosema esculutum* (Morama) and *Mundulea sericea*. Novel dimerization of styrene epoxides during attempts towards the chromene ring system. *Doctor of Philosophy Thesis*, University of Botswana, pp. 63-66.

28. Mazimba O., Majinda R. R. T., Modibedi C., Masesane I. B., Cenciè A and Chingwaru W., 2011. *Tylosema esculentum* extractives and their bioactivity. *Bioorg. Med. Chem.*, 19(17), pp. 5225-5230.

29. Mills, E., Cooper, C., Seely, D. and Kanfer, I., 2005. African herbal medicines in the treatment of HIV: Hypoxis and Sutherlandia. An overview of evidence and pharmacology. *Nutrition Journal*, 4, p. 19.

30. Nahar, L., Russell, W. R., Middleto, M., Shoeb, M. and Sarker, S. D., 2005. Antioxidant phenylacetic acid derivatives from the seeds of *Ilex aquifolium*. *Acta Pharmaceutica* 55(2), pp. 187-193.

31. Nakanishi, T., Mukai, K., Yumoto, H., Hirao, K., Hosokawa, Y. and Matsuo, T., 2010. Anti-inflammatory effect of catechin on cultured human dental pulp cells affected by bacteria-derived factors. *Eur. J. Oral Sciences*, 118(2), pp. 145-50.

32. Newman, D. J., 2008. Natural products as leads to potential drugs: an old process or the new hope for drug discovery. *J. Med. Chem.*, 51(9), pp. 2589-2599.

33. Newman, D. and Cragg, G. M., 2012. Natural Products as sources of new drugs over the 30 years from 1981 to 2010. *J. Nat. Prod.*, 75(3), pp. 311-335.

34. Niu, F., Chang, H. T., Jiang, Y., Cui, Z., Chen, F. K., Yuan, J. Z and Tu, P. F., 2006. New diterpenoids from *Semiaquilegia adoxoides*. *J. Asian Nat. Prod. Res.*, 8, pp. 87-91.

35. Ntie-Kang, F., Mbah, J. A., Mbaze, L. M., Lifongo, L. L., Scharfe, M., Hanna, J. N., Cho-Ngwa, F., Onguéné, P. A., Owono, L. C. O., Megnassan, E., Sippl, W. and Efange, S. M. N., 2013. CamMedNP: Building the Cameroonian 3D structural natural products database for virtual screening. *BMC Complement Altern. Med.*, 13, pp. 1-10.

36. Potterat, O., and Hamburger, M., 2008. In progress in drug research: natural compounds as drugs: Drug discovery and development with plant-derived compounds. Basel, Birhäusser Verlag AG: Edited by Petersen F, Amstutz R, pp. 45-118.

37. Rout, S. P., Choudary, K. A., Kar, D. M., Das, L. and Jain, A., 2009. Plants in traditional medicinal systems-Future source of new drugs. *Int. J. of Pharmacy and Pharmaceutical Sciences*, 1, pp. 1-23.

38. Siegler, D. S., Pauli, G.F., Frohlich, R., Wegelius, E., Nahrstedt, A., Glander, K. E. and Ebinger, J. E., 2005. Cyanogenic glycosides and menisdaurin from *Guazuma ulmifolia, Ostrya virginiana, Tiquilia plicata*, and *Tiquilia canescens. Phytochemistry*, 66, pp. 1567-1580.

39. Srinivas, T, L., Lakshmi, S. M., Shama, S. N., Reddy, G. K. and Prasanna, K. R., 2013. Medicinal plants as anti-ulcer agents. *J. Pharmacogn. Phytochem.*, 2, pp. 91-97.

40. Tshibangu, K. C., Worku, Z. B., De Jongh, M. A., van Wyk, A. E., Mokwena, S. O. and Peronovic, V., 2004. Assessment of effectiveness of traditional medicine in managing HIV/AIDS patients in South Africa. *East Afr. Med. J.*, 81, pp. 499-504.

41. Wangchuk, P., 2008.Health impacts of traditional medicines and bio-prospecting: A world scenario accentuating Bhutan's Perspective. *J. Bhutan Studies*, 18, pp. 117-134.

42. Yu, X., Sainz Jr, B., Petukhov, A. P. and Uprichard, S. L., 2012. Identification of hepatitis C virus inhibitors targeting different aspects of infection using a cell-based assay. *Antimicrob. Agents Chemother.*, 56, pp. 6109-6120.

Chapter 9

Detection of the Antimicrobial Activity of some Myanmar Medicinal Plants

Htet Htet Win, Khin Mar Mya,
Khin Htay Myint, Hla Myat Mon

Department of Biotechnology,
Mandalay Technological University,
Mandalay, Myanmar
E-mail: htethtetwin86@gmail.com

ABSTRACT

Medicinal plants constitute an important component of flora and are widely distributed in Myanmar. In this work, seven Myanmar medicinal plants; *Syzygium cumuni* Skells. (Black plum), *Terminalia bellerica* Roxb. (Belleric Myrobalan), *Murraya paniculata* Jack. (Orange jasmine), *Solanum xanthocarpum* Schard. and Wendl. (Indian Solanum), *Jacaranda mimosifolia* D.Don. (Jacaranda), *Capsicum minimum* Roxb. (Chilli) and *Vitis carnosa* Wall.; were screened for their antimicrobial activity against six pathogenic micro organisms, based on the ethno pharmacological literature. Preliminary phytochemical examination revealed the presence of different groups of chemical compounds in all plant samples. Then, the antimicrobial activity screening of the ethanolic crude plant extracts was carried out by Agar Well Diffusion method. Among the tested plant samples, *Terminalia bellerica* Roxb., was the most potent plant extract on all tested microorganisms while *Jacaranda mimosifolia* D.Don. showed the best activity against the pathogenic fungus, *Candida albicans*. The minimal inhibitory concentration

(MIC)values of *Terminalia bellerica* Roxb. were 6.654 mg/ml, 13.307 mg/ml, 3.327 mg/ml, 13.307 mg/ml, 3.327 mg/ml and 6.654 mg/ml for the standard strains of organisms: *Bacillus pumilus, Bacillus subtilis, Staphylococcus aureus, Pseudomonas aeruginosa, Escherichia coli, Candida albicans,* respectively.

Keywords: *Medicinal plants, Phytochemicals, Ethanolic extracts, Human pathogens, Antimicrobial activity, MIC.*

1. INTRODUCTION

The World Health Organization (WHO) estimated that 80 per cent of the population of the developing countries relies on traditional medicines, mostly plant-based drugs for their primary health care needs (Hussain; Majeed; Ismail; Sadikun; Ibrahim., 2009). Also, modern pharmacopoeia still contains at least 25 per cent drugs derived from plants while many others are synthetic analogues built on prototype compounds isolated from plants.

The demand for medicinal plants is steadily increasing in both developing and developed countries due to the growing recognition of drugs based on natural products. Being non-narcotic, having no side-effects and easy availability at affordable prices make these products sometimes the only source of health care available to the poor (Chauhan., 2006). Certainly, the great civilizations of the ancient Chinese, Indians and North Africans have provided written (OR documented) evidence of man's ingenuity in utilizing plants for the treatment of a variety of diseases.

A number of diseases are caused by pathogenic microorganisms. Infectious diseases are the leading cause of death worldwide. Unavoidable deaths from acute respiratory infections, diarrhoeal diseases, measles, AIDS, malaria and tuberculosis account for more than 85 per cent of the mortality from infection worldwide (WHO, 2001). Even when theretherapeutic agents are used to control several microbial diseases, in course of time the pathogen acquires resistance, making the therapeutic agent ineffective. The increasing failure of chemotherapeutics and antibiotic resistance exhibited by pathogenic microbial infectious agentshave lead to the screening of several medicinal plants for their potential antimicrobial activity (Parekh; Chanda., 2007).

The National Health Plan (NHP) has been aimed to solve the problem of six diseases as top priority in Myanmar. They include malaria, tuberculosis, diabetes, hypertension, diarrheaanddysentery. One of the main objectives of NHP is to effectively control and treat the diseases utilizing the locally available resources, including medicinal plants (The New Light of Myanmar, 2004). In Myanmar, there are 7,000 different known plants and most of them have been recognized as medicinal plants (MyoNgwe, 2004). There is a continuous and urgent need to discover new antimicrobial compounds with diverse chemical propertiesand novel mechanisms of action for new and re-emerging infectious diseases.

2. MATERIALS AND METHODS

Collection of Plant Samples

Seven Myanmar traditional medicinal plants were selected and collected from Mandalay region, Myanmar in 2006. The name of selected plants, used parts and collection sites are shown in Table 9.1. Identification of the plant samples was made by an authorized botanist from Pharmaceutical Research Department (PRD), Ministry of Science and Technology, Yangon.

Table 9.1: Selected Myanmar Medicinal Plants

Sl.No.	Botanical Names	Family Names	English Names	Myanmar Names	Used Parts
1.	*Jacaranda mimosifolia* D.Don.	Bignoniaceae	Jacaranda	Mye-bok-kayan	Barks
2.	*Syzygium cumuni* (Linn.) Skeels.	Myrtaceae	Black plum	Yu-za-na	Barks
3.	*Vitis carnosa* Wall.	Ampelidaceae	Nil	Mo-hmyaw-nga-yoke	Roots
4.	*Capsicum minimum* Roxb.	Solanaceae	Bird's eye, Chilli	Tha-bye	Fruits
5.	*Terminali abellerica* Roxb.	Combretaceae	Belleric Myrobalan	Thit-seint	Barks
6.	*Solanum xanthocarpum* Schard. and Wendl.	Solanaceae	Indian Solanum	Mann-thone-gwa	Whole plant
7.	*Murraya paniculata* Jack.	Rutaceae	Orange jasmine	Sein-ban-pya	Leaves

Microorganisms

The pure culture of two Gram-negative bacteria, three Gram-positive bacteria and one strain of fungus were obtained from Development Center for Pharmaceutical Technology (DCPT), Yangon.

Table 9.2: Strains of Tested Microorganisms and their Source

Sl.No.	Tested Microorganisms	Source
1.	*Staphylococcus aureus*	NCTC 5671
2.	*Bacillus subtilis*	NCIB 3610
3.	*Bacillus pumilus*	NCIB 9363
4.	*Escherichia coli*	NCIB 8134
5.	*Pseudomonas aeruginosa*	NCIB 8295
6.	*Candida albicans*	–

NCTC: National Collection of Type Culture, Central Public Health Laboratory, London, England.

NCIB: National Collection of Industrial Bacteria, Aberdeen, England.

Phytochemical Examination

Preliminary phytochemical examination of plant samples was carried out to screen the presence of 12 types of compounds; alkaloids, glycosides, reducing sugar, steroids and terpenoids, phenolic compound, carbohydrates, α-amino acids, flavonoids, tannin, saponin glycosides, acid or basic or neutral, and cyanogenic glycosides.

Preparation of Crude Plant Extracts

The selected plant samples were collected, cleaned, air-dried for about 5-7 days and powdered. The weighed sample were separately extracted with 95 per cent ethanol by using percolation method. Then, they were filtered and the filtrates were concentrated by using rotary evaporator. The concentrated plant extracts were stored at 4°C to test the antimicrobial activity.

Agar Well Diffusion Test

Preparation of Sample

Weighed quantity of each of the crude plant extracts was dissolved in 70 per cent ethanol to get the 500mg/ml concentration of stock solution.

Preparation of Media and Plates

Muller-Hinton agar (pH 7.2-7.4) was used as a standard medium for the antimicrobial activity testing.

Preparation of Plates for Inoculation and Incubation

Four or five similar colonies from the subculture of microorganisms were inoculated by streaking in five different directions over the entire sterile agar surface of a plate to obtain a confluent lawn of bacterial growth under sterile conditions. Using a 6 mm punch, nine equally spaced wells were made on the inoculated agar surface. Each antimicrobial agent 25mg/well (Stock Concentration: 500mg/ml) was introduced into each well. Tetracycline hydrochloride (30µg/well) was used as the positive control to compare the antimicrobial potential with that of crude extracts and 70 per cent EtOH was used as negative control. Then, the plates were placed in an incubator at 37°C for 18 hrs.

Reading and Interpretation

After incubation, the plates were examined and zone diameters of complete inhibition were measured and recorded by using a ruler. According to the zone diameters, all the extracts showed varying degrees of antimicrobial activity on the tested microorganisms.

Macrodilution Broth Test (Tube Dilution Test)

Preparation of Inoculums

To prepare the inoculums of the bacterial strains, four or five colonies from the subculture of the bacteria were inoculated into 5 ml Muller-Hinton broth and incubated on a water-bath shaker at 35°-37°C until a slightly visible turbidity appears (2-5 hrs). The inoculum was prepared to contain 10^5 to 10^6 CFU/ml by adjusting the turbidity of a broth culture by matching the turbidity standard and then further diluting it 1: 200 in broth. A turbidity standard was prepared by adding 0.5 ml of

0.048 M $BaCl_2$ (1.75 per cent [wt/vol] $BaCl_2$. $2H_2O$) to 99.5 ml of 0.36 NH_2SO_4 (1 per cent vol/vol). This turbidity is half the density of a Mcfarland 0.5 standard.

Preparation for Dilution and Inoculation

The antimicrobial solution (1.0 ml) was prepared by diluting the sample solution in Muller-Hinton broth by two-fold dilution method. After dilution, 1.0 ml of the adjusted inoculum was added toeach test tube. The final concentration of antimicrobial agent is half of the initial dilution series,because the addition of an equal concentration of inoculum in broth (NOT CLEAR). The final working dilution series for the plant extract were 53.23, 26.615, 13.307, 6.654, 3.327, 1.664 and 0.832 mg/ml while that of tetracycline hydrochloride were 32, 16, 8, 4, 2, 1 and 0.5μg/ml.Then, all the test tubes including one growth control (noantimicrobial agent) and one sterility control (no microbial inoculum) tubes were incubated at 35°-37°C for 16 to 20 hrs.

Reading and Interpretation

After 16-20 hrs incubation, MIC were determined. MIC was taken as the most diluted concentration of antimicrobial agent that remained sparkling clear and free of growth.

3. RESULTS AND DISCUSSION

Phytochemical Examination

The results of the phytochemical constituents of the plant samples are shown in Table 9.3. All selected plants could be potentially safe for further experiments since the presence of cyanogenic glycoside was not detected. In literature, alkaloids, flavonoids, phenolic compounds, saponins, tannins, etc. are the important plant compounds for antimicrobial activity. The results showed the presence of these compounds and hence all selected plants may have antimicrobial activity.

Antimicrobial Activity of Crude Extracts

The results of the antimicrobial screening assay of the crude extracts against the tested strains are shown in Table 9.4. Ethanolic extracts were carried out by percolation method because some of the chemical constituents can be changed or destroyed by heating hot extraction, Reflux method.

In studying the antimicrobial activity of the extracts of selected Myanmar medicinal plants, six microorganisms were used as test strains by agar-well diffusion method (Figure 9.1).

According to the results, *Terminalia bellerica* was the most effective plant to all test bacteria [three Gram-positive bacteria and two Gram-negative bacteria]. It showed the better activity against *Staphylococcus aureus* and *Bacillus pumilus*. It can be recorded as the broad spectrum agents are residing in that plant since it is effective to both Gram-positive and Gram-negative bacteria.

On screening the antimicrobial activity of *Jacaranda mimosifolia*, it had the strong antifungal activity because it showed the widest zone of inhibition (19mm)

Table 9.3: Preliminary Phytochemical Examination of Selected Myanmar medicinal plants

Plant Samples	Alkaloids	Glyco-sides	Reducing Sugar	Phenolic Compound	Flavo-noids	Saponin Glyco-sides	Cyano-genic Glycosides	α-amino Acids	Carbo-hydrates	Acid/Base/Neutral	Tannin	Steroids and Terpenoids
Jacaranda mimosifolia	+	+	+	+	+	+	–	–	+	Neutral	+	–
Syzygium cumuni	+	+	+	+	–	+	–	+	–	Base	+	+
Vitis carnosa	+	+	–	–	+	–	–	–	–	Base	+	+
Capsicum minimum	+	+	–	+	+	–	–	–	–	Base	+	–
Terminali abellerica	+	+	+	+	+	+	–	–	–	Neutral	+	+
Solanum xanthocarpum	+	+	–	+	+	+	–	+	–	Base	+	–
Murraya paniculata	+	+	–	+	+	+	–	–	+	Base	+	+

+: Presence; –: Absence.

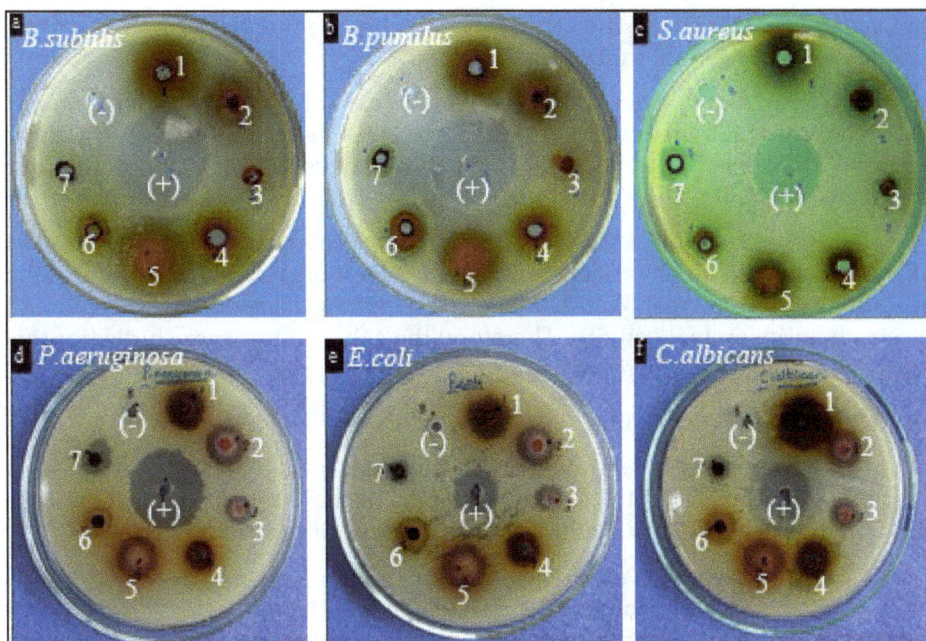

Figure 9.1: Culture Plates Showing the Inhibition Zones of the Crude Plant Extracts against Test Microorganisms.

1: *Jacaranda mimosifolia*; 2: *Syzygium cumuni*; 3: *Vitis carnosa*; 4: *Capsicum minimum*; 5: *Terminalia bellerica*; 6: *Solanum xanthocarpum*; 7: *Murraya paniculata*; (+) Drug control and (-) Solvent control.

Table 9.4: Antimicrobial Activity Crude Plant Extracts on Tested Microorganisms by Agar Well Diffusion Test

Plant Samples	Zone Diameter of Inhibition (mm)					
	B. pumilus	B. subtilis	S. aureus	P. aeruginosa	E. coli	C. albicans
Jacaranda mimosifolia	15	13	13	13	14	19
Syzygium cumuni	14	10	14	13	13	14
Vitis carnosa	10	7	9	9	8	12
Capsicum minimum	15	10	13	10	11	12
Terminalia bellerica	17	13	17	15	15	16
Solanum xanthocarpum	13	10	12	9	11	10
Murraya paniculata	13	7	10	13	6	8
Solvent Control	–	–	–	–	–	–
Drug Control	29	37	29	25	13	18

Well content of plant extracts' solution - 25mg.

Well content of Tetracycline hydrochloride - 30µg.

on *Candida albicans*. This is the second most potent plant with antifungal activity among the selected plant samples.

Next, *Syzygium cumuni* showed only moderate antimicrobial activity against test organisms while *Capsicum minimum* showed the effective antimicrobial activity against *Bacillus pumilus* although it had low antimicrobial activity on other test organisms.

The remaining plant extracts (*Vitis carnosa, Solanum xanthocarpum, Murraya paniculata*) showed no remarkable activity on the test organisms. According to the overall results from Agar Well Diffusion Test, *Terminalia bellerica* showed broad spectrum antimicrobial activity. Therefore, the MIC values were studiedfor this plant extract on tested microorganisms.

Macrodilution Broth Test

MIC is the lowest concentration of antimicrobial agent that inhibits the growth of a microorganism. The MIC values of the ethanolic extract of *Terminalia bellerica* was tested by tube dilution method and the results are shown in Table 9.5.

Table 9.5: MIC of *Terminalia bellerica* on Tested Strains of Microorganisms by Tube Dilution Method

Tested Microorganisms	Minimum Inhibitory Concentrations	
	Terminalia bellerica (mg/ml)	Tetracycline Hydrochloride (µg/ml)
Bacillus pumilus	6.654	2
Bacillus subtilis	13.307	2
Staphylococcus aureus	3.327	2
Pseudomonas aeruginosa	13.307	2
Escherichia coli	3.327	2
Candida albicans	6.654	0.5

4. CONCLUSIONS

Selected seven Myanmar medicinal plants were tested for their antimicrobial activity against six strains of human pathogens by agar-well diffusion method. Among these plant samples,*Terminalia bellerica* was the most effective plant extract ontested strains of microorganism. Therefore, this plant extract was selected to study its minimum inhibitory concentration.This preliminary testing pointed out that *Terminalia bellerica* is the potent antimicrobial plant and the antimicrobial agents are present in this extract. According to literature, *Terminalia bellerica* contains β-sitosterol, gallicacid, ellagicacid, ethylgallate, galloylglucose and chebulagic acid, which render the herb its therapeutic properties (Anonymous). Moreover, the gallic acid has a higher antimicrobial activity than some types of antibiotics (Khin, 2003). Therefore, further study should be done by isolation and identification of active constituents from that plant extract.

5. ACKNOWLEDGMENTS

We want to express our deepest gratitude to Department of Biotechnology, Mandalay Technological University, Mandalay and Myanmar Scientific and Technological Research Department, Yangon for the provision of the research facilities during this work.

6. REFERENCES

1. Anonymous, no date. Belliric Myrobalan. <http://www.himalayahealthcare.com/herbfinder/h-termebe.htm>

2. Ashok K. Chauhan., 2006. Microbes: Health and Environment; Microbiology Series (Volume 3); I.K. International Pvt Ltd.

3. Hussain K, Majeed MT, Ismail Z, Sadikun A, Ibrahim P.,2009. Complementary and alternative medicine: quality assessment strategies and safe usage. Southern Med Review.2; 1:19-23.

4. Khin Mar Mya, 2003. Study on Multiple Drug Resistant *S. aureus*:Antibiotic Incorporated Agar Plate Screening Test and Antibacterial Activity of Gallic Acid Isolated from *Lowsoniainermis* Linn., Department of Biotechnology, Yangon Technological University.

5. KyawSoeand Tin MyoNgwe, 2004. Medicinal Plants of Myanmar, Identification and Uses of Some 100 Commonly Used Species, SERIES (1), Myanmar.

6. Parekh J.,Chanda S., 2007. In Vitro Antimicrobial Activity and Phytochemical Analysis of Some Indian Medicinal Plants. Turk J Biol.31:53-58.

7. The New Light of Myanmar (article). Tuesday, 14th December, 2004.

8. World Health Organization. WHO global strategy for containment of antimicrobial resistance.2001.WHO/CDS/CSR/DRS/2001-2.

Chapter 10

Antimicrobial and Toxicity Studies of the Aqueous Extract of the Roots of *Zanthozylum zanthozyloides* (Lam. Zepern and Timler; Rutaceae) on Mice and Wistar Rats

E.O. Oshomoh[1] and M. Idu[2]

[1]Department of Science Laboratory Technology,
[2]Department of Plant Biology and Biotechnology,
Faculty of Life Sciences, University of Benin,
Benin City, Edo State Nigeria
E-mail: emmanuel.oshomoh@uniben.edu

ABSTRACT

During the last two decades, many plants have been screened for their biological activities (Tichy and Novak, 1998). There is no doubt that medicinal and aromatic plants have great potentiality in the amelioration of suffering in human and other living beings. Results showed significant zone of inhibition against selected bacteria and fungi associated with oral cavity, eight bacteria and six fungi were used, they are *Klebsiella pneumoniae, Streptococcus mutans, Pseudomonas aeruginosa, Staphylococcus epidermidis, Escherichia coli, Bacillus subtilis, Staphylococcus aureus, Proteus vulgaris, Aspergillus niger, Candida albican, Rhizopus oryzae, Aspergillus flavus, Penicilliumc hrysogenum* and *Saccharomyces cerevisiae*. *S. mutans* was the most sensitive bacterial organism tested with the aqueous extracts of the plant with inhibition zone of 14.33±2.08 mm in *Z. zanthozyloides*, while *A. flavus*, a fungus showed the highest sensitivity zone of 13.00±1.00 mm to *Z. zanthozyloides* extracts. The zone of inhibition is concentration dependent. The study also identified specific natural compounds present in the plant extracts. 10 natural product compounds were identified in *Z. zanthozyloides*. Based on the functional group analyzed by Fourier Transform Infra-Red (FTIR) and compounds identified by Gas Chromatography-Mass Spectrophotometer (GC-MS), the plant root extracts did not contain any known toxic compound. Some compounds detected or suspected from the chemical abstract library of the GC-MS were Octadec-9-enoic acid, Stearic acid (Octadecanoic acid), Hexadecanoic acid, Beta-sitosterol acetate, 1,3,5-tris(cyclohexyl) pent-1-ene and 9,12-Octadecadien-1-ol. The unsaturated fatty acids (6-octadecenoic acid, 9-octadecenoic acid-1, 2, 3-propanetriol ester) are important constituents of naturally occurring lipids. Result from short - term and long - term toxicological investigations of the plant extracts carried out with the use of wistar rats and mice confirmed the efficacy and general acceptability of the root of *Z. zanthozyloides* with zero side effect.

Keywords: *Phytopreparations, Medicinal plant, Chewing sticks, Intraperitoneal, Orogastric tube, Histopathological.*

1. INTRODUCTION

Quite a sizeable number of consumers are embracing the philosophy that natural products are better for their health and environment. In the developed countries such as the United states and Europe a lot of people are seeking products they perceive to be safer, healthier and without toxic chemical or synthetic ingredients (Adewunmi and Ojewale, 2004).

Natural products are "alternatives" to the conventional over-the-counter (OTC) cosmetic and therapeutic oral rinses that most dentists recommend and many individuals use, natural herbal mouth rinses (oral rinses or mouthwashes) are made of natural, plant-based ingredients that emphasize body/mind (holistic) health and wellness, and there has been the erroneous impression that herbal medicine have fewer adverse effects (Chan, 2009).

Development and assessment of herbal formulations for various beneficial health andfunctional effects in animal and human species are increasing daily. Despite well established scientific studies on phytopreparations of herbal ingredients for pharmacological properties, the safety details of herbal substances in compliance to internationally accepted guidelines seem to be inadequate (Joshua *et al.*, 2010; Mabeku *et al.*, 2007). In view of anticipated increase in use of herbal supplementation in future for various health needs, short-and long-term toxicological investigations are required for evaluation and classification of herbal preparations based on safety data.

In the last two decades, many plants have been screened for their biological activities (Tichy and Novak, 1998). There is no doubt that medicinal and aromatic plants have great potential in the amelioration of suffering in human and other living beings.Promotion, registration and encouragement on use according to WHO guidelines, by enlisting herbal medicines to be realistic in recognizing the role of traditional medicines in the health care delivery systems (W.H.O, 1977).

If a person's gums bleed when they brush their teeth, chances are they already have the mildest form of gum disease and this can increase their chances of getting diabetes or a heart problem. Gum disease can be sneaky. Many times the people who carry this disease do not even know it because it rarely shows any pain or irritation until permanent damage is done to the teeth (Atul*et al.*, 2011).

However there are only a few reports on the utility of medicinal plants in the treatment of specific disease. Sadangi *et al.* (2005) reported 10 species of medicinal plants used in the treatment of ear and mouth diseases by the tribal people of Kalahandi district. Jadhav, (2006) documented 15 species of medicinal plants used in different types of fever, while Kadel and Jain, (2008) reported that 34 plant species are used for the treatment of snakebite in Madhya Pradesh and Chhattisgarh states.

However, an anticariogenic organism was found to be resistant to many of the antibacterial agents viz, Penicillin, Chloramphenicol, Clindamycin, Ampicillin (Bhattacharya *et al.*, 2003). In addition they may lead to side effects including gastrointestinal problems (Crig, 1998). This drawback justifies further research and development of natural antimicrobial agents that are effective and safe for the host. The global need for alternative prevention and treatment options and products for oral diseases that are safe, effective and economical comes from the rise in disease incidence (particularly in developing countries), increased resistance by pathogenic bacteria to currently used antibiotics and chemotherapeutics, opportunistic infections and financial considerations in developing countries (Tichy and Novak, 1998; Badria and Zidan, 2004).

During the past two decades, reliability and usage of herbal product has become of increasing importance, due to the side effects and complications of many chemical and synthetic medicines. About 25 per cent of drugs are derived from plants and many others are formed from prototype compounds isolated from plant species (Kala *et al.*, 2006). Kanwar *et al.* (2006) reported that about two million traditional health practitioners use over 7500 medicinal plant species.

This research is aimed at studying the antimicrobial activities, phytochemical and toxicity of the tropical Nigerian chewing stick.

2. MATERIALS AND METHODS

Plant Material and Authentication

Fresh roots of *Z.zanthozyloides* (Lam, Zepern and Timler; Rutaceae) were collected and identified in Afeye- Okpameri in Akoko-Edo Local Government Area of Edo State, Nigeria.

Antimicrobial Investigation

Source of Microorganisms

Pure stock cultures of *Staphylococcus aureus, Staphylococcus epidermidis, Proteus mirabilis, Proteus vulgaris, Bacillus subtilis, Streptococcus mutans, Pseudomonas aeruginosa, Klebsiella pneumoniae, Candida albicans, Aspergillus flavus, Aspergillus niger, Rhizopus oryzae, Penicilium chrysogenum* and *Saccharomyces cerevasiae* isolated from patients with dental diseases were collected from the Department of Medical Microbiology, University of Benin Teaching Hospital (UBTH). Edo State, Nigeria. Keptin 4°C until when needed for further studies.

Microbial Inocula Preparation for Susceptibility Testing

The inocula of the bacteriaand fungi isolates were prepared for susceptibility testing and antimicrobial assay by methods of Ellof, 1998 and Oboh *et al.*, 2007.

Fourier Transform Infra-Red (FTIR)/Gas Chromatography-Mass Spectrophotometer (GC-MS)

Fine powder was mixed with KBr salt. Infrared spectra were recorded in KBr by a sophisticated computer-controlled FTIR Perkin Elmer spectrometer with He-Ne laser as reference, between 4,000.0-350.0 cm^{-1} while GC-MS was used to separate chemical mixture (the GC component) in the extracts and identified the various components at a molecular level (the MS component).

Experimental Animals: Acute and Sub Acute Toxicological Experiments

Healthy, mature male rats (*Rattus norvegicus*), weighing between 170–220g and females, weighing between 155–165g and mature male and female mice (*Mus musculus*) weighing between 20-35g were handled according to standard protocols for the use of laboratory animals (PHSP 2002). Approved by the Ethics Committee, University of Benin. The intraperitoneal, oral LD$_{50}$, acute and sub acute toxicity of the aqueous extracts of the roots of Z. *zanthoxyloides* were estimated by the Miller and Tainter method cited by Raymond, *et al.* (2010).

Haematological Assays

The haematological parameters were analyzed using auto haematology analyzer (Model BC-2800).

Biochemical parameters were assayed by Biuret method (Doumas *et al.*, 1981) for total protein and albumin; Alamine Aminotransferase (ALT) and Aspartate Aminotransferase (AST) were quantified by the method described by Schumann *et al.* (2002). Alkaline phosphate was assayed using the method described by Raymond-Habecker and Lott (1995). Bilirubin was assayed by Jendrassik Grof method (Doumas and Wu, 1991) and GGT (Gamma Glutamyl transferase) was analysed by method described by Szasz, 1969.Triglyceride, cholesterol were analysed using the methods of Rakesh and Sushma, (2009); Pande *et al.*, 2012).The isolated tissues of the liver, heart, kidney and spleen were properly stored, homogenized and analyzed using Corash Cells, Tissues and Organs Method of Study (1983).

Statistics:Results were expressed as Mean±SEM (Standard error of mean). Data were subjected to one way analysis of variance (ANOVA) and differences between samples were determined by newman-keuls multiple comparison test. All data were analyzed using Graph pad prism software (UK). $P < 0.05$ indicates statistically significant difference.

3. RESULTS AND DISCUSSION

The root extract of Z. *zanthozyloides* (Tables 10.1a and 10.b) indicate high sensitivity at highest concentration of 100 mg/ml against all oral microorganisms tested. Also the level of activity is concentration dependent as the zone of inhibition decreases as concentration reduces in all the tested microorganisms. Similar letters

Table 10.1a: Zone of Inhibition of Aqueous Extracts of *Z. zanthozyloides* on Selected Oral Microorganisms (Bacteria)

Organisms	100 mg/ml	75 mg/ml	50 mg/ml	25 mg/ml	12.5 mg/ml	6.25 mg/ml	3.125 mg/ml	S.D H$_2$O	Ciprofloxacin
S. epidermidis	12.33^a±2.08	11.33^a±1.53	9.00±0.00	7.67^b±0.58	8.00^b±1.73	8.00^b±1.00	7.33^b±1.15	0.00±0.00	11.12±0.23
E. coli	10.00^a±1.00	10.00^a±0.00	9.33^a±1.15	7.67^b±0.58	7.00^b±1.00	7.00^b±0.00	7.33^b±0.58	0.00±0.00	13.17±0.43
P. mirabilis	9.67^a±2.08	10.67^a±1.15	8.00±1.73	8.00^b±0.00	9.00±2.00	8.67^b±1.15	8.33^b±0.58	0.00±0.00	12.23±1.24
K.pneumoniae	8.33±0.58	9.00^a±1.73	9.00±0.00	8.67^b±0.58	9.33^b±1.53	10.67^b±2.08	7.67^b±1.15	0.00±0.00	11.13±0.33
B. subtilis	12.00^a±1.73	11.00^a±1.00	8.33^b±0.58	8.33^b±0.58	8.67^b±1.53	8.67^b±0.58	6.67^b±1.15	0.00±0.00	14.37±2.26
S. aureus	9.00^a±0.00	8.00^b±0.00	8.33^b±0.58	6.67^b±2.31	7.67^b±1.53	7.33^b±1.53	7.33^b±1.15	0.00±0.00	09.12±0.32
S. mutans	14.33^a±2.08	13.67^a±3.21	14.33^a±0.58	10.33^c±3.21	10.00^c±0.00	8.67^c±0.58	8.00^c±1.00	0.00±0.00	13.46±2.37
P. aeruginosa	11.00^a±1.00	10.00^a±2.00	9.67^a±0.58	9.00^a±0.00	9.00^a±1.73	9.00^a±1.00	8.00^a±1.00	0.00±0.00	10.35±0.24

Similar letter within a row indicate means that are not significantly different at P>0.05 using Duncan multiple range test (DMRT)

Key: Mean±SEM; S.D H$_2$O – Sterile distilled water; Ciprofloxacin- Positive control

Table 10.1b: Zone of Inhibition of Aqueous Extracts of *Z. zanthozyloides* on Selected Oral Microorganisms (Fungi)

Organisms	100 mg/ml	75 mg/ml	50 mg/ml	25 mg/ml	12.5 mg/ml	6.25 mg/ml	3.125 mg/ml	S.D H$_2$O	Ketoconanzone
A. niger	10.00^b±2.00	7.67^c±2.52	8.33±0.58	8.67^c±0.58	7.67±0.58	8.67^c±0.58	7.33^c±0.58	0.00±0.00	14.18±3.24
C. albicans	8.33^c±0.58	8.00^c±2.00	8.00±0.00	7.67^c±1.53	7.67±2.52	6.33^c±1.53	6.67^c±0.58	0.00±0.00	10.41±0.21
R. oryzae	11.33^a±2.31	10.67^a±1.15	9.33^c±2.31	8.00^c±1.00	8.00±1.00	8.00^c±1.00	6.00±0.00	0.00±0.00	13.47±2.25
A. flavus	13.00^a±1.00	8.00^b±2.00	8.67^b±0.58	10.00^b±1.00	9.33^b±1.53	9.33^b±1.15	7.67^b±2.08	0.00±0.00	14.11±0.32
P. chrysogenum	11.33^b±1.53	10.33^b±2.52	9.33±1.15	8.33^c±1.53	8.67±3.06	10.00^b±2.00	7.00±1.00	0.00±0.00	10.57±1.22
S. cerevisiae	11.00^b±1.00	11.00^b±1.73	10.00^b±0.00	10.33^b±2.08	9.67^b±1.53	9.00^b±1.00	10.33^b±0.58	0.00±0.00	13.07±4.23

Similar letter within a row indicate means that are not significantly different at P>0.05 using Duncan multiple range test (DMRT).

Key: Mean±SEM; S.D H$_2$O – Sterile distilled water; Ketoconanzone - Positive control.

within a row indicate means that are not significantly different from each other. Table 10.1a indicates various activities of the aqueous root extract of *Z. zanthozyloides* on test organisms, revealing the highest susceptibility when compared with other test organisms at all concentrations. *S. mutans* with inhibition zone of between 8.00 mm and 14.33 mm retained the highest sensitivity at 100, 75, 50, 25 and 12.5 mg/ml, where *K. pneumoniae* showed the highest zone of inhibition of 10.67 mm at the extract concentration of 6.25 mg/ml, and *S. cerevisiae* recorded the highest susceptibility to the plant extract at the low-est concentrations of 3.125 mg/ml with inhibition zone of 10.33mm closely followed by *P. mirabilis* with inhibition zone of 8.33mm. This is in agreement with the earlier report of Osho *et al.* (2011) who showed that the essential oils of *Fagara zanthozyloides* were able to inhibit the growth of *K. pneumoniae* and *P. vulgaris* even at the minimal concentration of 0.5 per cent and Oshomoh and Idu, (2012) also showed zone of inhibition of the aqueous and ethanol extract of the root of *Z. zanthozyloides* on selected oral pathogens.

The sensitivities of *C. albicans, B. subtilis, S. aureus* and *E. coli* is in agreement with earlier work carried out by Okiei *et al.* (2009) on the leaf of *E. coccinea*, Oshomoh and Idu, (2012) on the antimicrobial activities of ethanol and aqueous crude extracts of *Hymenocardia acida* stem against selected Dental Caries Pathogens.

A two way ANOVA test revealed that for tested chewing stick there was no significant difference (p>0.05) between the length of incubation, but a significant difference exists (p<0.05) between extract concentrations. This confirmed the report of Dinesh *et al.* (2013) that antimicrobial properties in plants vary.

The extracts were able to inhibit the growth of *S. epidermidis, A. niger, C. albicans* and *R. oryzae* even at the minimal concentration of 6.25 mg/ml. It is concluded in this study that the extracts of *Z. zanthozyloides* possess strong efficacies in the treatment of various dental infections as described by the capability of its extracts to inhibit the growth of *S. epidermidis, E. coli, S. mutans, K. pneumonia, B. subtilis, S. aureus, P. vulgaris, P. aeruginosa, A. niger, Candida albicans, R. oryzae, A. flavus, P. chrysogenum* and *S. cerevisiae* which are the organisms directly associated with the establishment of dental infections, therefore can be used as precursors for the synthesis of useful drugs (WHO, 1977).Thus, the presence of the outer-membrane in the gram-negative *E. coli* and the refractory nature of *C. albicans* cell wall to many antimicrobial agents could be possibly linked to their susceptibility profile in this study.

The effect of the commercial antibacterial drug (Ciprofloxacin) and antifungal drug Ketoconanzone) tested at the concentration of 200mg/ml against the test bacteria and fungi (Tables 10.1a and 10.1b) can be considered not better inactivity when compared with the extracts, particularly at the highest tested concentration of 100mg/ml which was two times lower in concentration than that of the fungal antibiotics. Ogundiya *et al.* (2006) implies that if the concentrations of the extracts were increased, it can lead to increased activity.

Results of FTIR spectroscopic studies have revealed the presence of various chemical constituents in aqueous extract of the plant as shown below in (Figure 10.1). The peak at 3412 and 2926 cm^{-1} are corresponded to Hydroxyl and CH stretching frequency respectively. A band at 1722 cm^{-1} is corresponded to carbonyl carbon.

IR

3406 ➜ OH

2931 ➜ C ‾ H stretch (CH$_2$, CH$_3$)

1663 ➜ $\diagup$C $=$ C$\diagdown$

1427 ➜ O ‾ H bending

1253 ↙ ➜ C ‾ O stretching

1042 ↖

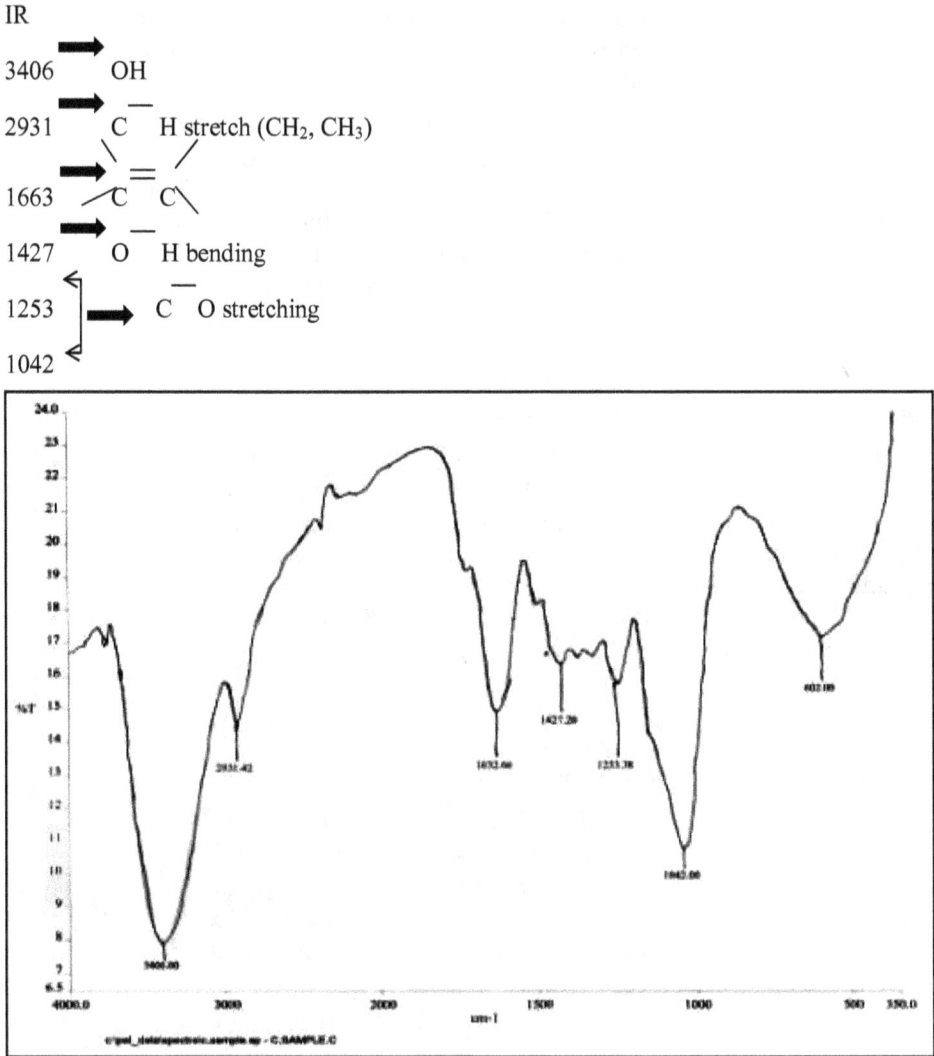

Figure 10.1: FTIR Spectrum Analysis of Aqueous Extract of *Z. zanthozyloides*.

The peak at 1635 cm^{-1} to assign C=C. the strong peak at 2862.36 cm-1 assigned to the CH3 stretching vibration and the peak at 2926.01 cm^{-1} assigned to C-H stretching which means that some alkane compound existed in these rare medicinal plant. The bands between 3000 and 2800 cm^{-1} represent C-H stretching vibration that are mainly generated by rapids (Wei *et al.*, 2009). The stretching assigned to the C-S linkage occurs in the region at 700-600 cm^{-1}. The weak absorption band of 601.79, 707.88 cm^{-1} indicates the presence of thiol, sulphite and sulphate group in the plant extract (Muruganantham *et al.*, 2009). The more intense band occurring at 1635.64, 1722.43, 1722.42, 1834.3, 3412.08, 1635.64 corresponding to C=O stretching indicate the presence of ketones, aldehydes, carboxylic acids and esters. The middle bands are primarily associated with the stretching motion of the C=O group. The C=O band is sensitive to the environments of the peptide linkage and also depends on the protein's overall secondary structure (Jagadeesan *et al.*, 2005).

Fourier Transform Infra Red (FTIR) Spectroscopy was useful for the compound identification; Figure 10.1 indicates that under IR region in the range of 350.0-4000.0 cm-1 there was a variation in the peaks in all the plant extracts (Thenmozhi *et al.*, 2011; Kalaiselvi *et al.*, 2012). FTIR allows detecting the whole range of infrared spectrum simultaneously providing speed and accuracy in measurements of biological specimens. FTIR is one of the most widely used methods to identify the chemical constituents and elucidate the compounds structures, and has been used as requisite method to identify medicines in pharmacopoeia of many countries (Liu *et al.*, 2006). Based on the functional group analyzed by FTIR, the plant extract does not contain any know toxic compounds.

The extract of the plant was subjected to GC-MS for further analysis of phytochemical components. Ten (10) phytochemical components were identified in the root extract of *Z. zanthozyloides* (Table 10.2) with the following compounds with high peak values of (a) Octadec-9-enoic acid, (b) Stearic acid (Octadecanoic acid), (c) Hexadecanoic acid, (d) Beta-sitosterol acetate, (e) 1,3,5-tris (cyclohexyl) pent-1-ene and (f) 9,12-Octadecadien-1-ol (Abdulrazak *et al.*, 2000). The unsaturated fatty acids (6-octadecenoic acid, 9-octadecenoic acid-1, 2, 3-propanetriol ester) are important constituents of naturally occurring lipids. They constitute among others the major storage lipid in plants and animals. The presence of these fatty acids (saturated and unsaturated) suggest that the plant under investigation is both medicinal and nutritional (Ndiokwere and Ukhun, 2001).

The detection of 9,12-octadecadienoic acid (Linolenic acid) from the chemical abstract service of the GC-MS spectra of sample of *Z. zanthozyloides* suggests that the plant sample constitute an essential fatty acid which act as precursor of the hormone-like prostaglandins (Ndiokwere and Ukhun, 2001). Most of the fatty acids detected can also act as emulsifiers in the food and drug industry.

Haematological indices in the rat after sub-acute oral treatment with 2 g kg^{-1} body weight of the extract are presented in Tables 10.3. White Blood Cell Count (WBC) and its differentials (lymphocytes and neutrophils), the heamatocrit (Packed Cell Volume or PCV), Haemoglobin (HB) and Platelet Count (PC) were all not significantly altered by the treatment. Haematologically, lack of effect on neutrophil levels indicates that the extracts did not induce inflammatory process since these cells are usually elevated in the course of inflammations (Zhang *et al.*, 2010; Idu *et al.*, 2006a). In the sub-acute toxicity study (Table 10.3); all the haemotological parameters (except 1000 mg/kg of lymphocyte in Table 10.3) of all the treated groups were not significantly different from the control group.

The use of much higher doses in toxicological tests gives an idea of the safety margin of the extracts (Ozolua *et al.*, 2009). Since there was no change in fluid intake by the animals given 2.5 g kg^{-1} body weight, indicating that the dose of the extracts may not have caused any renal damage (Tables 10.3–10.6). Organ- body weight

Table 10.2: Gas Chromatographic and Mass Spectral Data for Sample – CZZ Isolated from the Stem of *Z. zanthozyloides*

Sl.No.	Compound	Natural Product Class	Rt (mins)	Molecular Mass	Molecular Formula	Peak per cent	MS Fragments	Compound Structure
1	2,4-dimethyl heptanol	Alcohol	11.48	144	$C_9H_{20}O$	0.21	126, 113, 101, 84, 71, 57, 43, 41, 27	
2	Hexadecanoic acid, methyl ester	Ester	17.68	270	$C_{17}H_{34}O_2$	1.93	270, 239, 227, 171, 143, 129, 101, 87, 74, 69, 43, 41, 27	
3	Hexadecanoic acid	Saturated fatty acid	18.32	256	$C_{16}H_{32}O_2$	14.87	256, 227, 213, 185, 171, 129, 98, 73, 60, 43, 41, 27	
4	7-Octadecenoate	Unsaturated ester	19.38	296	$C_{19}H_{36}O_2$	7.74	264, 222, 180, 137, 123, 98, 84, 74, 69, 55, 41, 27	
5	Octadecanoic acid,	Saturated ester	19.58	298	$C_{19}H_{38}O_2$	1.12	298, 267, 255, 241, 213, 199, 143, 87, 74, 57, 43, 41, 27	
6	Octadec-9-enoic acid methyl ester (stearic acid ester)	Unsaturated fatty acid	19.99	282	$C_{18}H_{34}O_2$	27.32	282, 264, 222, 180, 151, 138, 125, 98, 97, 83, 69, 55, 41, 27	
7	Stearic acid (Octadecanoic acid)	Saturated fatty acid	20.15	284	$C_{18}H_{36}O_2$	17.03	284, 241, 199, 185, 143, 129, 115, 98, 85, 73, 60, 43, 41, 27	

Contd...

Table 10.2–Contd...

Sl.No.	Compound	Natural Product Class	Rt (mins)	Molecular Mass	Molecular Formula	Peak per cent	MS Fragments	Compound Structure
8	9,12-Octadecadien-1-ol	Alcohol	22.64	266	$C_{18}H_{34}O$	8.39	266, 149, 135, 121, 110, 95, 81, 67, 55, 41, 27	
9	Beta-sitosterol acetate	Lipids	23.58	456	$C_{31}H_{52}O_2$	10.72	396, 382, 288, 275, 255, 213, 159, 147, 133, 121, 105, 95, 81, 57, 43, 41	
10	1,3,5-tris(cyclohexyl) pent-1-ene	Alkene	24.68	316	$C_{23}H_{40}$	10.67	234, 220, 206, 151, 137, 125, 111, 97, 83, 67, 55, 41, 39	
						100 per cent		

Ten components were identified in 'CZZ' sample of *Z. zanthozyloides*. The major components of the plant extract fraction were Octadec-9-enoic acid (Unsaturated fatty acid) (27.32), Hexadecanoic acid (Saturated fatty acid) (14.87 per cent) and Stearic acid (Saturated fatty acid) (17.03 per cent), with unsaturated fatty acid being the most abundant (27.32 per cent).

Table 10.3: Effect of 28 Days Administration Extract of Aqueous Stem of Z. zanthozyloides on some Haematological Parameters

	Control	500 (mg/kg)	1000 (mg/kg)	2500 (mg/kg)
WBC(x10³/μL⁻¹)	13.73ᵃ±3.51	11.77ᵃ±3.68	12.61ᵃ±5.99	10.04ᵃ±1.94
LYP (per cent)	62.95ᵃ±6.85	61.48ᵃ±4.23	62.91ᵇ±15.95	71.61ᵃ±16.06
MO (per cent)	15.30ᵃ±2.30	12.12ᵃ±0.73	13.94ᵃ±5.88	11.37ᵃ±7.46
GR (per cent)	21.75ᵃ±5.08	26.40ᵃ±3.84	23.14ᵃ±10.15	13.84ᵇ±8.77
RBC (x10⁶ μL⁻¹)	7.12ᵃ±0.88	7.45ᵃ±0.61	12.53ᵇ±2.86	6.06ᵃ±0.75
Hgb (g/dL)	15.52ᵃ±1.66	14.62ᵃ±1.11	15.26ᵃ±1.60	13.78ᵃ±1.67
PCV (per cent)	42.98ᵃ±4.03	42.50ᵃ±3.06	43.83ᵃ±3.21	39.29ᵃ±4.31
PLT (x10³μL⁻¹)	603.00ᵃ±156.8	595.3ᵃ±193.9	542.4ᵃ±275.7	661.8ᵃ±168.9

Similar letter within a row indicate means that are not significantly different from each other.

Values are expressed as mean±S.E.M. P>0.05 – No significant Difference.

WBC: White Blood Cell count; LYP: Lymphocytes; PCV: Packed Cell Volume; HGB: Haemoglobin; PC: Platelet Count; RBC: Red Blood Cell; MO: Monocytes, GR: Granulocytes n=6 (control) or (treated).

Table 10.4: Effect of 28 Days Daily Administration of Aqueous Stem Extract of Z. zanthozyloides on some Biochemical Parameters

Parameters	Control	500 (mg/kg)	1000 (mg/kg)	2500 (mg/kg)
ALT(μ/L)	3.00 [a] ±1.53	5.29 [a] ±2.49	4.29 [a] ±3.15	3.29 [a] ±1.70
AST(μ/L)	21.14 [a] ±4.74	9.71 [b] ±5.35	14.57 [a] ±8.99	15.71 [a] ±3.30
ALP(μ/L)	148.10 [a] ±29.51	64.74 [b] ±41.89	149.20 ±213.40	186.3 ±115.4
DB(mg/dL)	0.73 [a] ±0.38	0.43 [a] ±0.31	1.38 [a] ±1.94	1.09 [a] ±2.03
TB(mg/dL)	0.32 [a] ±0.08	0.21 [b] ±0.07	0.10 [c] ±0.08	0.36 [a] ±0.17
TP(g/dL)	8.28 [a] ±2.90	8.87 [a] ±0.88	7.87 [a] ±1.28	6.72 [a] ±1.17
ALB (g/dL)	4.60 [a] ±0.55	4.71 [a] ±1.90	4.77 [a] ±1.43	7.12 [b] ±3.31
GGT (μ/L)	61.98 [b] ±29.08	88.61 [c] ±58.98	48.14 [a] ±42.11	48.19 [a] ±18.54
Cl$^-$(mmol/L)	85.29 [a] ±16.39	80.86 [a] ±13.04	77.71 [a] ±21.05	81.86 [a] ±15.08
HCO3$^-$(mmol/L)	11.29 [a] ±2.21	14.86 [a] ±1.68	14.57 [a] ±2.82	14.57 [a] ±3.64

Similar letter within a row indicate means that are not significantly different from each other.

Values are expressed as mean±S.E.M. P>0.05–No significant Difference.

GGT: Gamma Glutamyltransferase; AST: Aspartase Aminotransferase; ALT: Alanine Aminotransferase; ALP: Alkaline phosphatase; DB: Direct bilirubin; TB: Total Bilirubin; ALB: Albumin; Cl: Chloride ions; CHO_3^-: Bicarbonate ions; n=6 (control) or (treated).

Table 10.5: Effect of 28 Days Daily Administration of Aqueous Stem Extract of *Z. zanthoxyloides* on Renal Function of Rats

Dose	Urea	Creatinine	Na+	K+
Control	86.00 [a] ±70.64	0.02 [a] ±0.009	152.30 [a] ±7.25	8.30 [a] ±1.39
500 mg/kg	123.00 [a] ±43.32	0.06 [a] ±0.09	138.10 [a] ±7.34	4.93 [a] ±0.57
1000 mg/kg	145.10 [a] ±94.70	0.02 [a] ±0.007	134.40 [a] ±16.71	9.17 [a] ±2.49
2500 mg/kg	161.00 [a] ±74.99	0.02 [a] ±0.01	133.70 [a] ±15.30	6.79 [a] ±1.23

Similar letter within a row indicate means that are not significantly different from each other.

Values are expressed as mean±S.E.M. P>0.05 –No significant Difference.

*indicate P<0.05- Significant Difference.

Na+ - Sodium ions, K+ -Potassium ions.

Table 10.6: Effect of 28 Days Daily Administration of Aqueous Stem Extract of *Z. zanthozyloides* on the Lipid Profile of Rats

Dose	TCHOL	HDL	LDL	TRG
Control	188.30 [a] ±76.62	95.60 [a] ±27.00	84.07 [a] ±102.30	60.43 [c] ±70.93
500 mg/kg	166.10 [a] ±51.22	81.31 [a] ±38.95	65.45 [a] ±64.41	132.90 [a] ±63.56
1000 mg/kg	166.30 [b] ±84.14	82.44 [a] ±46.55	70.11 [a] ±62.02	229.30 [b] ±277.50
2500 mg/kg	237.10 [b] ±125.60	99.49 [a] ±29.84	89.90 [a] ±115.90	133.00 [a] ±178.60

Similar letter within a row indicate means that are not significantly different from each other.

Values are expressed as mean±S.E.M. P>0.05 – No significant Difference.

CHOL: Cholesterol, HDL: High density lipoprotein, LDL: Low density lipoprotein, TRI: Triglyceride.

ratios are indices which are often used in toxicological evaluations (Michael *et al.*, 2007), but do not necessary indicate the absence of lesion.

From the result of this study it is established that oral administration of the extracts of the plant is a safe route, as the number of death recorded immediately after administration was insignificant and independent of the dose administered to the animals. The animals in groups I, II and III have similar sign observed such as itching, writhing, calmness, sedation and pylori erection, excluding group IV and V having additional sign of irregular breathing and reduced motor activity and death. The number of deaths that were recorded in group IV (10,000 mg/kg) for mice and group V (20,000 mg/kg) for rats could be attributed to other physio-chemical or environmental factors but may not be toxicity of the extracts (Igbe *et al.*, 2013).

Table 10.4 indicates the serum biochemical parameters of the administration of the extracts at graded level on rats for 28 days. Serum alanine aminotransferase (ALT) and aspartate aminotransferase (AST) are useful indices for identifying inflammation and necrosis of the liver (Tilkian *et al.*, 1979). While the decrease in these animals was dose dependent, as the value obtained for the animals on the highest dose of the extract was significantly ($p<0.05$) lower than the value obtained from the control, the observed decrease in AST and ALT was not dose dependent. It may be infer, therefore, that the change may not be due to effect of the administered extract but rather due to some other intrinsic factors. The observed changes in AST was due to significant increase ($p<0.01$) in the value of the animal treated with 2500 mg kg^{-1} body weight compared to the control group. Those animals treated with higher dose however, did not show any significant difference ($p>0.05$) with the control. This is also an indication that these changes may not necessarily be related to the extract. Singh and Devkota, (2003) also reported that aqueous extract of piper methylsticum did not affect significantly the serum levels of ALT, AST and ALP activities, In a similar study, Ephraim *et al.* (2000) demonstrated that animals treated with aqueous extract of *Ocimum gratissimum*, used for the treatment of rheumatism and paralysis did not affect the activities of ALT, AST and ALP. Elevated levels of bilirubin, AST, ALT and ALP are often diagnostic of underlying cellular injuries (Karthikeyan *et al.*, 2006; Wittekind, 1995; Idu *et al.*, 2006b). These parameters are comparable between the treated groups and the control.

On the other hand, the result of the assessment of the serum triglyceride level (Table 10.6) depicts that there is no significant change as the level of triglyceride are largely regulated through synthesis in the liver. The plant extracts administration reduced the level of the triglyceride for the 28 days administration supporting the study of (Vasim *et al.*, 2012).This is an indication of milder risk factor for a coronary artery disease as a result of the increase in the level of triglyceride in the serum (Table 10.6). There was no elevation in the level of triglyceride except at the medium concentration of 1000 mg/kg which show an elevation of 229.30 mg/kg when compare with control of 60.43 mg/kg. This increased level may suggest stimulation of the synthesis or interference in feedback or in mobilization pathways associated with the organ (Mabeku *et al.*, 2007).

The significant increase in the total cholesterol level in 500 mg/kg (Table 10.6) depict that such animal is prone to have coronary artery disease as elevated serum

Table 10.7: Effect of 28 Days Daily Administration of Aqueous Stem Extract of Z. zanthozyloides on Body and Vital Organs Body Weight Change of Rats

Dose	Per cent Wt Change	Liver	Heart	Kidney	Spleen
Control	15.28[a]±5.43	34.38[a]±1.84	3.38[a]±0.20	2.82[a]±0.09	4.35[a]±0.73
500 mg/kg	31.57[a]±7.49	37.73[a]±1.70	3.63[a]±0.16	3.02[a]±0.17	4.73[a]±0.46
1000 mg/kg	44.49[a]±9.58	39.41[a]±2.14	3.87[a]±0.26	3.18[a]±0.10	4.98[a]±0.74
2500 mg/kg	60.05[b]±7.49	35.15[a]±1.79	3.12[a]±0.39	3.32[a]±0.12	4.03[a]±0.75

Similar letter within a row indicate means that are not significantly different from each other.

Values are expressed as mean±S.E.M. P>0.05 –No significant Difference

Table 10.8: Effect of Intraperitoneal Administration (7 days) of Aqueous Stem Extract of Z. zanthozyloides on Body and Vital Organs Body Weight Change of Rats

Parameters	Concentration (mg/kg)			
	500	1000	2500	Control
Body weight	6.64[b]±3.75	2.89[b]±3.42	14.59[a]±5.45	10.09[a]±0.43
Kidney	3.02[a]±0.09	3.32[a]±0.42	3.62[a]±0.24	3.11[a]±0.12
Lungs	6.70[a]±0.30	6.70[a]±1.31	7.43[a]±3.95	7.20[a]±1.90
Liver	36.60[a]±7.69	38.60[a]±5.11	32.53[a]±5.61	36.57[a]±1.05
Heart	3.53[a]±0.35	3.30[a]±0.26	3.17[a]±0.35	3.07[a]±0.06

Similar letter within a row indicate means that are not significantly different from each other.

P>0.05 –No significant Difference.

level of cholesterol are a major risk factor for coronary artery disease (Aloke *et al.*, 2012), but the decrease in 1000 mg/kg and 2500 mg/kg (Table 10.6) indicates the swift responses of the body system to such abnormal condition (Snigur *et al.*, 2008).

Daily fluid intake, weight change and organ- to-body weight ratio were not significantly different between the control and treated groups after 7 days and 28-days oral daily dosing with 0.5- 2.5g kg-1 body weight of the plant extracts on wistar rats. The administration of the extracts, *Z. zanthozyloides* (Table 10.7) caused no significant (p>0.05) increase in the body weight and organ body weight when compare with control, which support the literature published by (Nwozo *et al.*, 2011). The body weight increase for the 28 days administration was dose dependent and was highest in 2500 mg/kg from the result of this study. The results of the weight changes (Table 10.8) in the current study indicated no significant (p>0.05) effects in both acute and sub-chronic toxicity tests. This may be an indication that the drug does not affect the feed utilization ratio of the animals. Also the effect of intraperitoneal administration (28 days and 7 days) of aqueous stem extract of the plant extracts on wistar rats did not show significant different (p>0.05) on body and vital organs body weight change of Rats as shown on Tables 10.7 and 10.8.

Graded doses of the plant extracts (*Z. zanthozyloides*) were administered (500, 1000, and 2,500 mg/kg) to Wister rats and after six weeks, the heart, spleen, liver and kidneys were harvested and examines for toxicity changes. Bearing in mind the errors in tissue processing (ARTEFACTS), the following changes were observed. When treated with 500 mg/kg, the heart showed mild vascular congestion, interstitial oedema and infiltrates of chronic inflammatory cells in *B. micrantha* (Oshomoh *et al.*, 2014).

There were minimal changes in the heart at 500 mg/kg, 1000 mg/kg and 2,500 mg/kg, when treated with *Z. zanthozyloides* the extract induced mild perivascular inflammation without affecting the myocyte bundles. This effect represented across the graded doses by congestion, oedema and inflammatory cellular infiltrates (Oshomoh *et al.*, 2013). These are signs of chronic inflammation (myocarditis) (Figures 10.2a-d). Increasing the dosage to 1000 mg/kg and then to 2500 mg/kg produced little changes from the low dose apart from slight increase in the thickness of the wall of the blood vessels (Figures 10.2c-d).

When the liver was treated with *Z. zanthozyloides*, it showed mild periportal inflammation as well as activation of the sinusoidal Kupffer cells. This is same for all the doses (Figures 10.3a-d). In addition, it was observed across board increased prominence of the nucleoli. For *Z. zanthozyloides*, the spleen showed mild activation of the white pulp (lymphoid follicles), irrespective of the dose administered (Fogire 10.4a-d).

The kidneys showed mild interstitial vascular congestion, irrespective of the dose when treated with *Z. zanthozyloides* aqueous root extract (Figures 10.5a-d). Also *Z. zanthozyloides* induces a mild inflammation, as well as activating the local immune system in the target organs and there is no significant variation with the doses (Igbe *et al.*, 2013; Oshomoh *et al.*, 2013). However the parenchymal tissues are unaffected. There is no damage to the tissues.

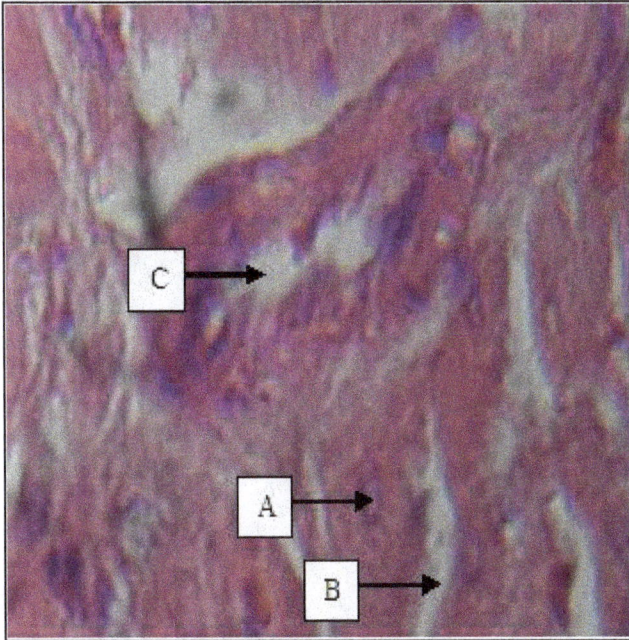

Figure 10.2(a): Rat Heart (Control) Showing Myocardial Bundles A, Separated by Interstitial Spaces B and Pierced by Coronary Blood Vessels C (H and E x 40).

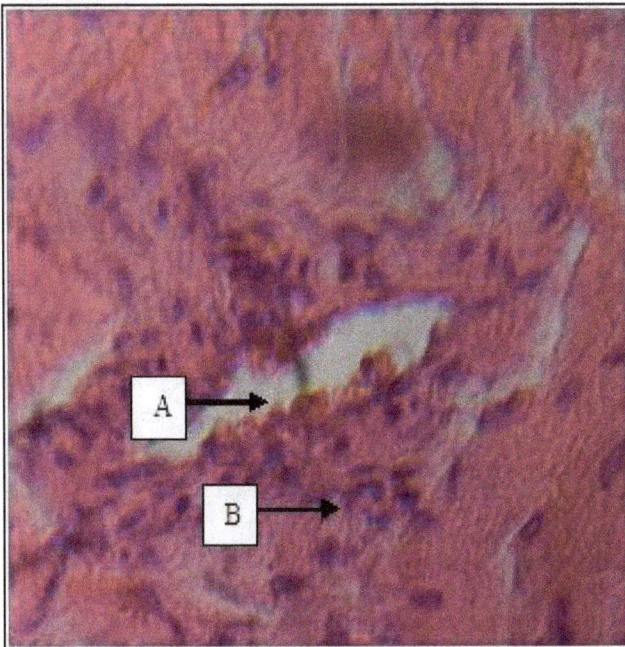

Figure 10.2(b): Rat Heart Treated with 500mg/kg *Z. zanthozyloides* for 28 Days Showing Mild Vascular Congestion A, and Mild Perivascular Infiltrates of Inflammatory Cells B (H and E x 40).

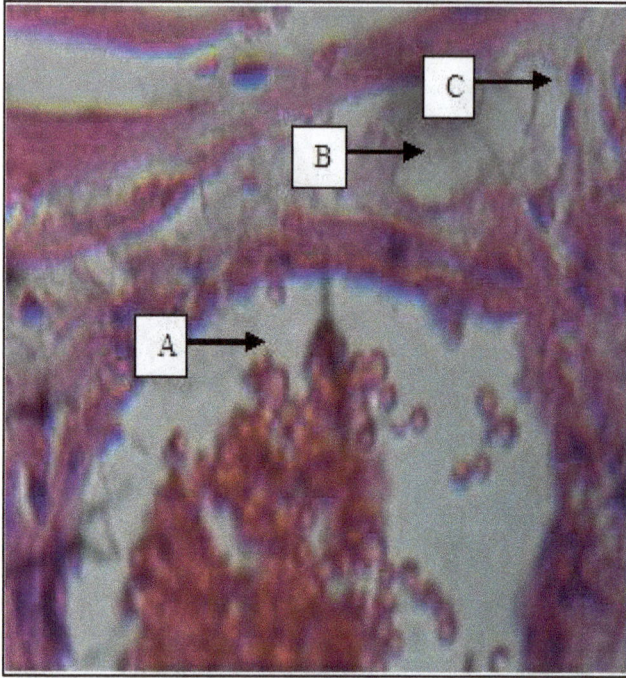

Figure 10.2(c): Rat Heart Treated with 1000mg/kg *Z. zanthozyloides* for 28 Days Showing Mild Vascular Congestion and Dilatation A, Interstitial Oedema B, and Mild Verevascular Infiltrates of Polymorphs C (H and E x 40).

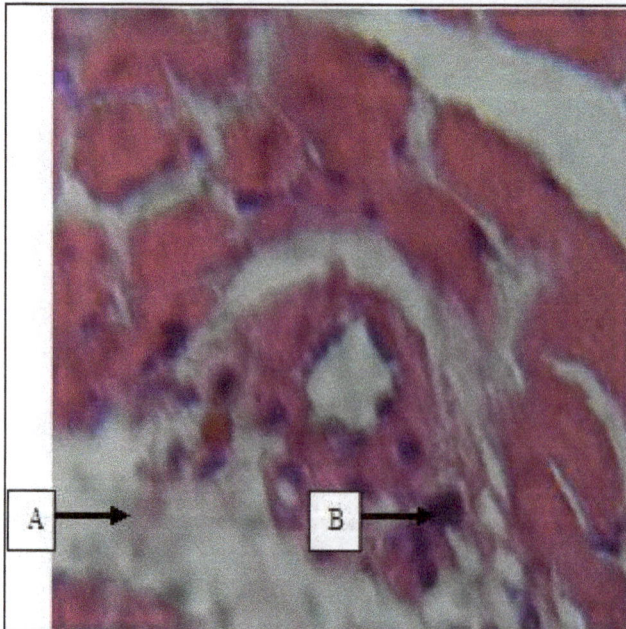

Figure 10.2(d): Rat Heart Treated with 2500mg/kg *Z. zanthozyloides* for 28 Days Showing Mild Interstitial Oedema A, and Mild Perivascular Infiltrates of Inflammatory Cells B (H and E x 40).

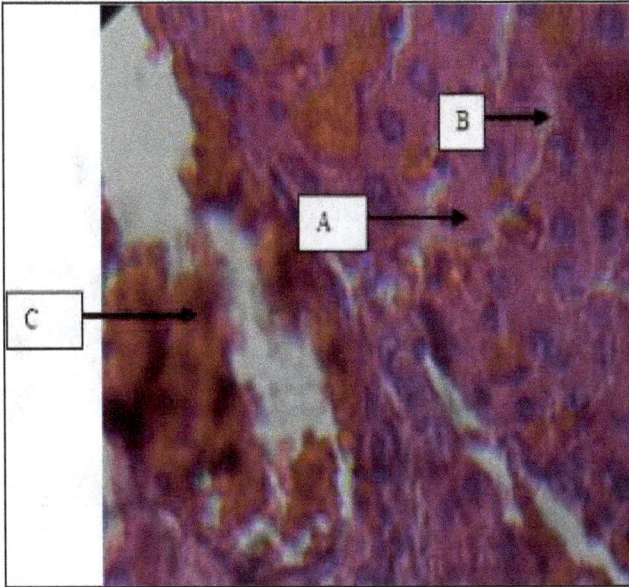

Figure 10.3(a): Rat Liver (Control) Showing Hepatocytes A, Sinusoids B and Portal Vein C (H and E x 40).

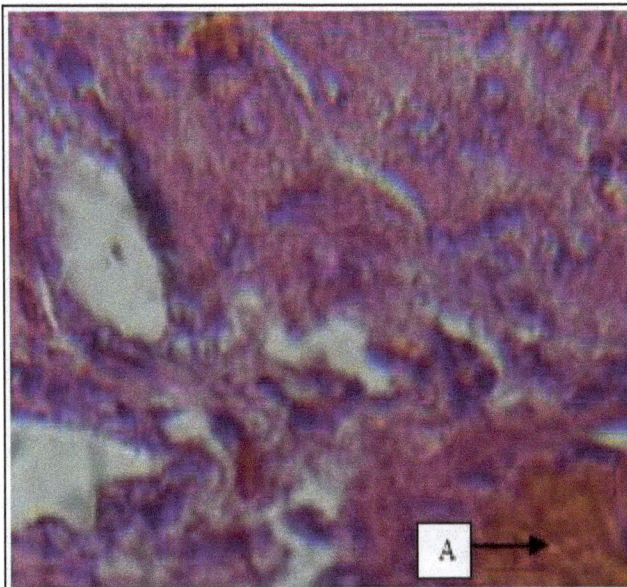

Figure 10.3(b): Rat Liver Treated with 500mg/kg of *Z. zanthozyloides* for 28 Days Showing Mild Vascular Congestion A (H and E x 40).

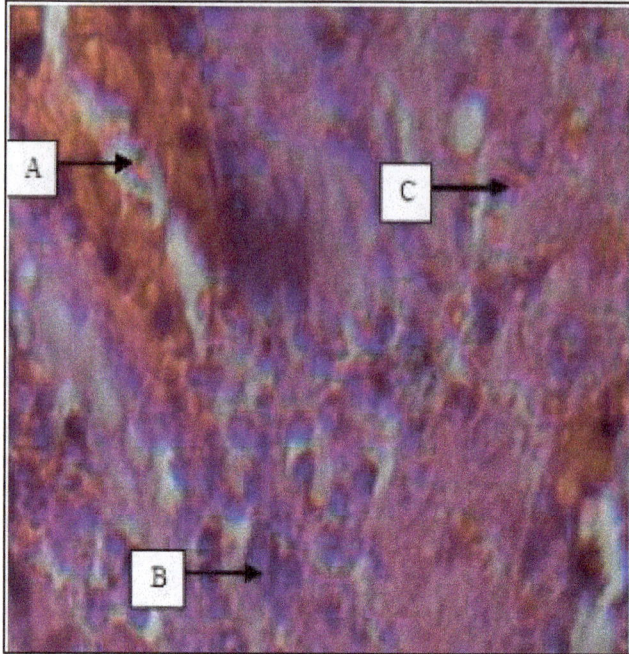

Figure 10.3(c): Rat Liver Treated with 1000mg/kg of *Z. zanthozyloides* for 28 Days Showing Mild Vascular Congestion A, Mild Periportal Infiltrates of Inflammatory Cells B and Kupffer Cell Hyperplasia C (H and E x 40).

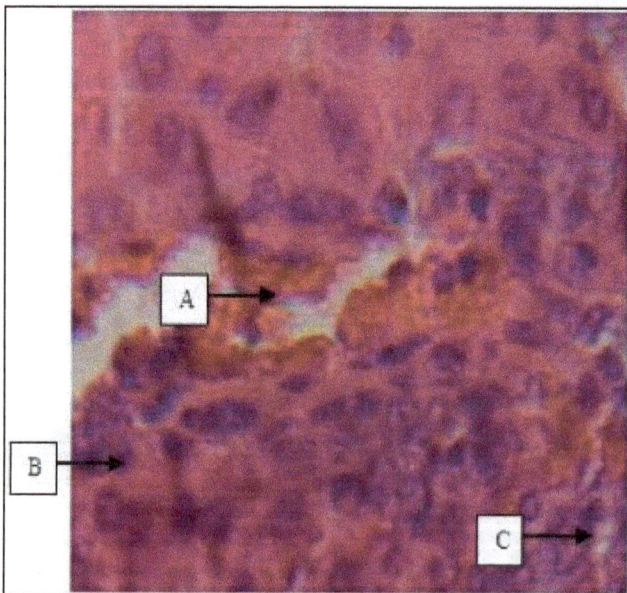

Figure 10.3(d): Rat Liver Treated with 2500mg/kg of *Z. zanthozyloides* for 28 Days Showing Mild Vascular Congestion A, Mild Periportal Infiltrates of Inflammatory Cells B and Kupffer Cell Hyperplasia C (H and E x 40).

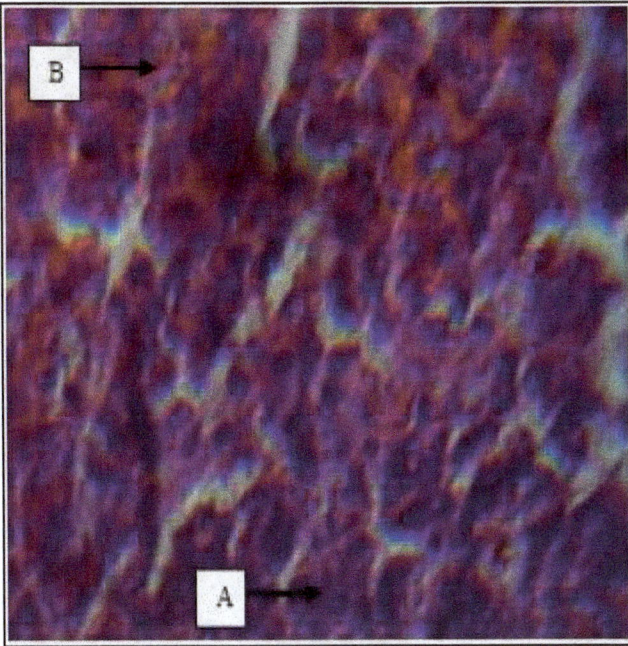

Figure 10.4(a): Rat Spleen (Control) Showing White Pulp A, Surrounded by Red Pulp B (H and E x 40), *Z. zanthozyloides*.

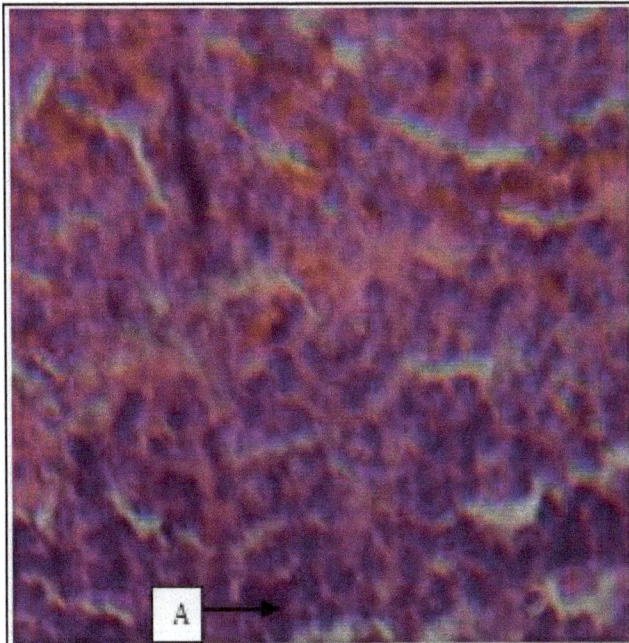

Figure 10.4(b): Rat Spleen Treated with 500mg/kg of *Z. zanthozyloides* for 28 Days Showing Mildly Activated Pulp A (H and E x 40).

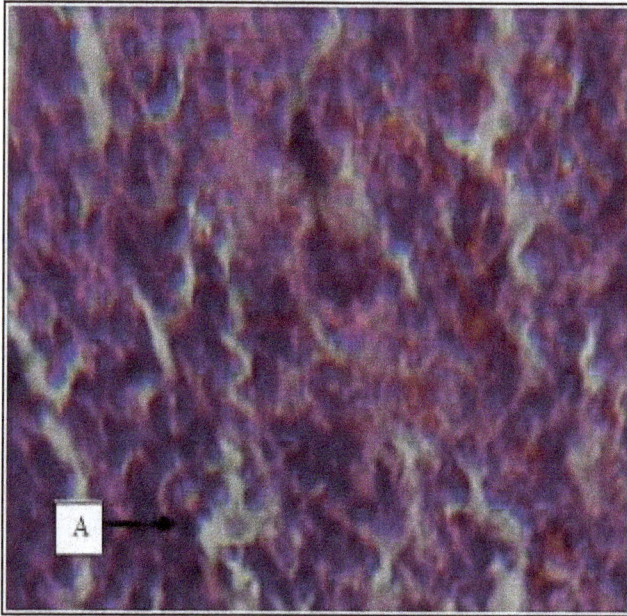

Figure 10.4(c): Rat Spleen Treated with 1000mg/kg of *Z. zanthozyloides* for 28 Days Showing Mildly Activated Pulp A (H and E x 40).

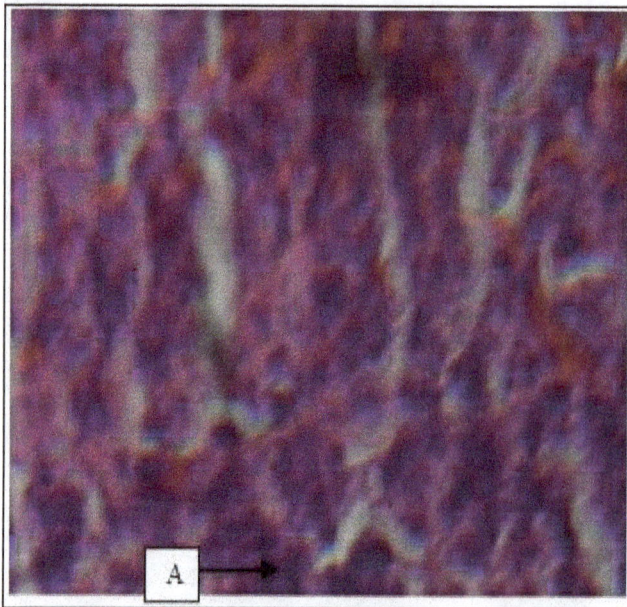

Figure 10.4(d): Rat Spleen Treated with 2500mg/kg of *Z. zanthozyloides* for 28 Days Showing Mildly Activated Pulp A (H and E x 40).

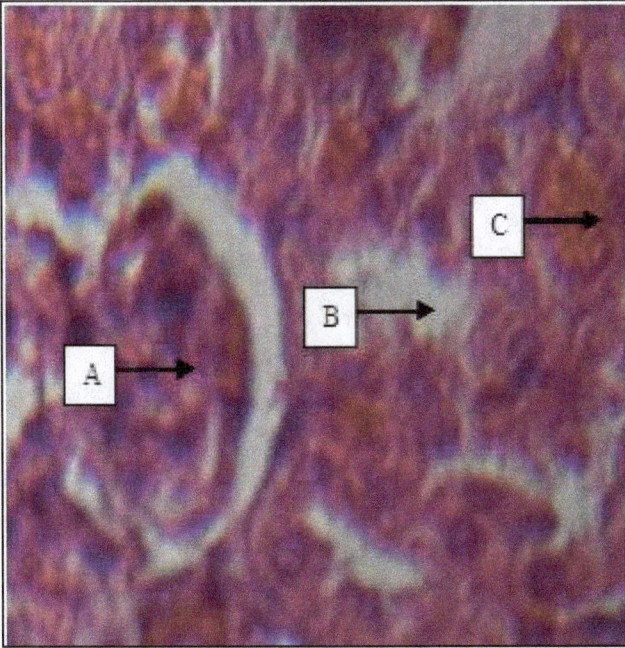

Figure 10.5(a): Rat Kidney (Control) Showing Cortical Glomeruli A, Tubules B Separated by Interstitial Space C (H and E x 40). *Z. zanthoxyloides.*

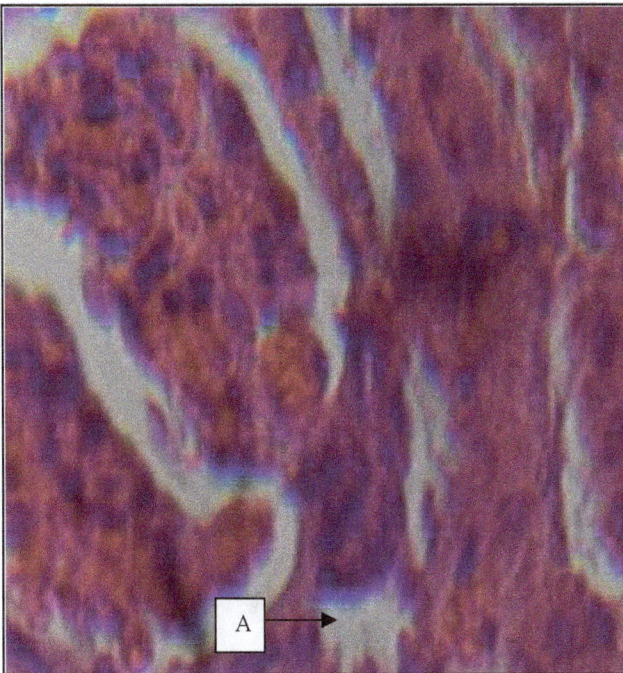

Figure 10.5(b): Rat Kidney Treated with 500mg/kg of *Z. zanthozyloides* for 28 Days Showing Mild Interstitial Vascular Congestion A (H and E x 40).

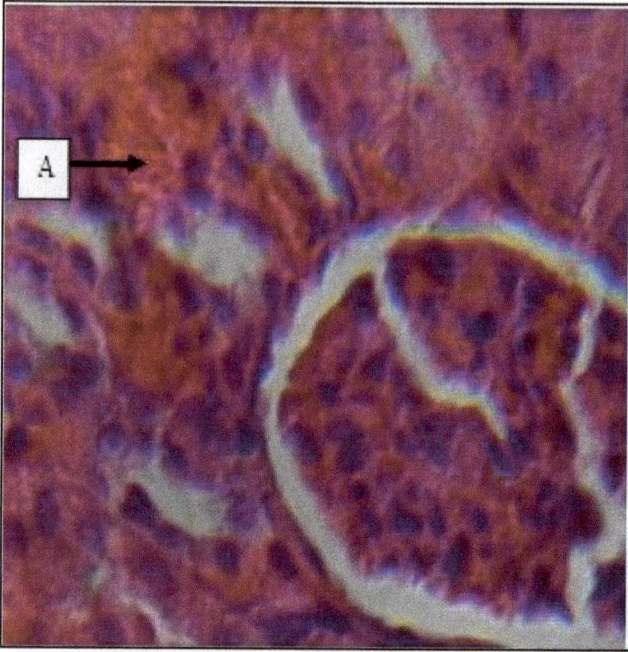

Figure 10.5(c): Rat Kidney Treated with 1000mg/kg of *Z. zanthozyloides* for 28 Days Showing Mild Interstitial Vascular Congestion A (H and E x 40).

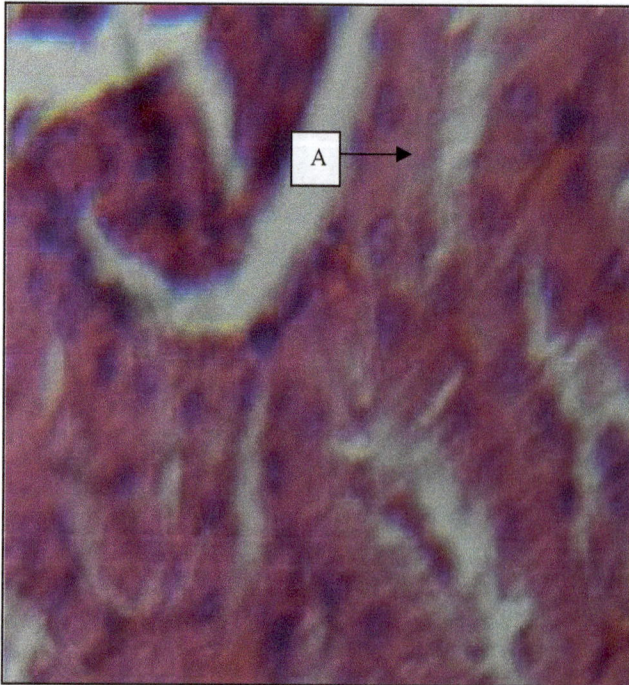

Figure 10.5(d): Rat Kidney Treated with 2500mg/kg of *Z. zanthozyloides* for 28 Days Showing Mild Interstitial Vascular Congestion A (H and E x 40).

4. CONCLUSIONS

It is shown in this study that the extracts of the plant, *Zanthozylumzanthozyloides* possess strong efficacies in the treatment of various dental infections as described by the capability of their extracts to inhibit the growth of *S. epidermidis, E. coli, S. mutans, K. pneumonia, B. subtilis, S. aureus, P. vulgaris, P. aeruginosa, A. niger, Candida albicans, R. oryzae, A. flavus, P. chrysogenum* and *S. cerevisiae* which are the organisms directly associated with the establishment of dental infections. The use of much higher doses in toxicological tests gives idea of the safety of the extracts, since there was no change in fluid intake by the animals administered with higher dose of 2.5 g kg^{-1} body weight, indicating that the dose of the extracts may not have caused any significant renal damage.

5. REFERENCES

1. Abdulrazak, S. A., Fujihara, T., Ondiek, J. O., Orskov, E. R., 2000. Nutritive evaluation of some Acacia tree leaves from Kenya. *Animal Feed Science and Technology,* **85:** 89-98.

2. Adewunmi, C. O., Ojewole, J. A. O., 2004. Safety of traditional medicines, complementery and alternative medicines in Africa. *African Journal of Traditional Complementary and alternative Medicines* 1: 1-3.

3. Atul, A. P., Rajendra, D. P., Upendra, C. G., Sadanand, Y. P., Aditya, S. D., Arvind, K. P., Rebecca, T., 2011. Antimicrobial activity of a polyherbal extract against dental micro flora. *Reseach Journal of Pharmaceutical, Biological and Chemical Sciences,* **2(2):** 533-539

4. Badria, F.A., Zidan, O.A., 2004. Natural products for dental cariesprevention. *Journal of Medicinal Food.***7(3):** 381-384.

5. Bhattacharya, S., Virani, S., Zavro, M., Hass, G.J., 2003.Inhibition of *Streptococcus mutans* and other oral Streptococci by (*Humulu slupulus* L.) constituents. *Economic Botany,* **57:** 118-125.

6. Chan, T. Y. K., 2009. Potential risk associated with the use of herbal anti-obesity products. *Drug Safe,* **32:** 453-456.

7. Crig, A., 1998. Antimicrobial resistance, danger signs all around. *Tennessee Medicine,* **91:** 433-455.

8. Dinesh, M. D., Uma, M. S., Anjali, V. M., Meenatchisundaram, S., Shanmugan, V., 2013. Inhibitory properties of aqueous extracts of selected indigenous medicinal plants against dental caries causing *Streptococcus mutans* and *Streptococcus mitis.African Journal of Basic and Applied Sciences,* **5(1):** 08-11.

9. Doumas, B. T., Bayse, D. D., Carter, R. J., Peters Jr, T., Schaffer, R., 1981. A candidate reference method for determination of total protein in serum I development and the validation. *Clinical Chemistry,* **27:** 1642 – 1650.

10. Doumas, B. T., Wu, T. W. 1991. The measurement of bilirubin fractions in serum. Crit. *Rev. Clin. Lab. Sci.* **28:** 415 – 445.

11. Eloff, J. N., 1998. A sensitive and quick microplate method to determine the minimal inhibitory concentration of plant extracts for bacteria. *PlantaMedica,* **64:** 711-713.

12. Ephraim, K. D., Salami, H. A., Osewa, T. S., 2000. The effect of aqueous leaf extract of *Ocimum gratissimum* on haematological and biochemical paramaters in rabbits. *African Journal of Biomedical Research,* **3:** 175-179.

13. Idu, M., Omogbai, E. K. I., Amaechina, F., Ataman, J. E., 2006a. Some cardiovascular effects of the aqueous extract of the leaves of *Starchytarpheta jamaicensis* (L.)Vahl. *International Journal of Pharmacology,* **2:** 163-165.

14. Idu, M., Ataman, J. E., Akhigbe, A. O., Omogbai, E. K. I., Amaechina, F., Odia, E. A., 2006b. Effects of *Stachytarpheta jamaicensis* (L.)Vahl.onWistar rats: serum biochemistry and ultrasonography. *Journal of Medical Sciences.***6(4):** 646-649.

15. Igbe, I., Eze, G. I., Ojameruaye, O., 2013. Sub-acute toxicity of aqueous fruit pulp extract of hunteria umbellate in Albino wistar rats. *Niger Journal of Physiological Science,* 077-082.

16. Jadhav, D., 2006. Plant sources used for the treatment of different types of fevers by Bhil Tribe of Ratlam district, Madhya Pradesh, India. *Journal of Economic and Taxonomic Bototany,* **30:** 909-911.

17. Jagadeesan, G., Kavitha, A.V., Subashini, J., 2005. FT-IR study of the influence of *Tribulus terrestris* on Mercury intoxicated mice, *Musmus culus* liver. *Tropical Biomedical,* **22:** 15-22.

18. Joshua, A.J., Goudar, K.S., Sameera, N., 2010. Safety assessment of herbal formulations, Rumbion and Tyrel in Albino Wistar rats. *American Journal of Pharmacology and Toxicology,* **5(1):** 42-47.

19. Kadel, C., Jain, A.K., 2008. Folklore claims on tribal communities of Central India. *Indian Journal of Traditional Knowledge.* **7:** 296-299.

20. Kala, C.P., Dhyani, P.P., Sajwan, B.S., 2006. Developing the medicinal plants sector in northern India: challenges and opportunities.*Journal of Ethnobiology and Ethnomedicine,* **12(4):** 356-364.

21. Kalaiselvi, M., Gomathi, D., Vidya, B., Uma, C., 2012. Evaluation of antioxidant potential and Fourier transform infrared spectroscopy analysis of *Ananas comosus* (L.) Merr peel. *International Research Journal of Pharmacy,* **3:** 237-242.

22. Kanwar, P., Sharma, N., Rekha, A., 2006. Medicinal plants use in traditional healthcare systems prevalent in Western Himalayas. *Indian Journal of Traditional Knowledge,* **5:** 300-309.

23. Karthikeyan, S., Gobianand, K., Pradeep, K., Mohan, C.V., Balasubramanian, M. P., 2006. Biochemical changes in serum, lung, heart and spleen tissues of mice exposed to sub-acute toxic inhalation of mosquito repellent mat vapor. *Journal of Environmental Biology,* **27:** 355-358.

24. Liu, H.X., Sun, S.Q., Lv, G.H., Chan, K. K., 2006. Study on Angelica and its different extracts by Fourier transform infrared spectroscopy and two-dimensional correlation IR spectroscopy. *Mol Biomol Spec,* **64:** 321-326.

25. Mabeku, L. B. K., Beng, V. P., Kouam, J., Essame, O., Etoa, F. X., 2007. Toxicological evaluation of ethyl acetate extract of *Cyclicodiscus gabonensis* stem bark (Mimosaceae). *Journal of Ethnopharmacology,* 111(3): 598-606.

26. Michael, B., Yano, B., Sellers, R. S., Perry, R., Morton, D. (2007). Evaluation of organ weights for rodent and non-rodent toxicity studies: A review of regulatory guidelines and a survey of current practices. *Toxicology Pathology,* 35: 742-750.

27. Muruganantham, S., Anbalagan, G., Ramamurthy, N., 2009. FT-IR and sem-eds comparative analysis of medicinal plants, *Ecliptaalba* HASSK and *Eclipta prostrata* Linn. *Romanian Journal of Biophysics,* 19: 285-294.

28. Ndiokwere, C. L., Ukhun, M. E., 2001. University Organic Chemistry, Sylva Publ. Inc., Ondo State. 247p.

29. Ogundiya, M. O., Okunade, M. B., Kolapo, A. I., 2006. Antimicrobial activities of some Nigerian,chewing sticks. *Ethnobotanical Leaflets,* 10; 265-271.

30. Oboh, I.E., Akerele, J.O., Obasuyi, O., 2007. Antimicrobial activity of the ethanol extract of the aerial parts of *Sida acuta* Burm f. (Malvaceae). *Tropical Journal of Pharmaceutical Research,* 6: 809-813.

31. Okiei, W., Ogunlesi, M., Ademoye, M. A., 2009.An assessment of the antimicrobialproperties of extracts of various polarities from *Chasmanthera dependens, Emilia coccinea* and *Cuscuta australis,* herbal medications, for eye diseases. *Journal of Applied Science.* 9: 4076 – 4080.

32. Osho, A., Bello, O. O., Fayemi, S. O., Adetunji, T. (2011). *In vitro* screening of some selected Nigerian medicinal plants (*Fagara zanthoxyloides, Vernonia amygdalina, Prosopis Africana,* and *Azadirachta indica*) for antibacterial activity. *Advances in Bioresearch,* 2(1): 190-195.

33. Oshomoh, E. O., Idu, M., 2012.Phytochemical screening and antimicrobial activities of ethanolic and aqueous root extracts of *Zanthoxyllum zanthoxylloides* (Lam.) Waterm. on selected dental caries causing microbes. *International Journal of Medicinal and Aromatic Plants,* 2(3): 411-419.

34. Oshomoh, E. O., Idu, M., 2012. Antimicrobial activities of ethanol and aqueous crude extracts of *Hymenocardia acida* stem against selected Dental Caries Pathogens. *Pharmacognosy Journal,* 4(29): 55-60.

35. Oshomoh, E.O., Idu, M., Uwaya, D., 2013.Acute and sub-acute toxicity of the Aqueous Extract of the stem of *Masularia acuminata* (G. Don) Bullock ex Hoyle on Albino rats and mice. *Pharmacopia,* 4(11): 606-616.

36. Oshomoh, E. O., Idu, M., Uwaya, D., 2014. Toxicity studies of aqueous extract of *Bridelia micrantha* stem administered on mice and rats. *International Journal of Pharmacuetical Sciences and Research,* 5(7): 2856-2865.

37. Ozolua, R. I., Anaka, O. N., Okpo, S. O., Idogun, S. E., 2009. Acute and sub-acute toxicological assessment of the aqueous seed extract of *Persea americana* Mill (Lauraceae) in Sprague-Dawley rats. *African Journal of Traditional, Complementary and Alternative Medicine.* 6: 573-578.

38. Pande, S., Platel, K., Srinivasan, K., 2012.Antihypercholesterolaemic influence of dietary tender cluster beans (*Cyamopistetra gonoloba*) in cholesterol fed rats. *Indian Journal of Medicinal Research*, **135**: 401-406.

39. Public Health Service Policy (PHSP)., 2002. On humane care and use of Laboratory Animals, National Institute of Health USA.

40. Rakesh, K., Sushman, S., 2009. Lipid profile changes in mouse gastroenemius muscle after denervation and beta-adrenoceptor stimulation. *Indian Journal of Experimental Biology*, **47**: 314-319.

41. Raymond-Habecker, J., Lott, J.A., 1995.Principle of Analysis of Alkaline Phosphatase. In: Clinical Chemistry, Theory, Analysis and Correlation, Kaplan, L.A., A. Pesce and S. Kazmierczak (Eds.). Mosby, St. Louis, 632p.

42. Raymond, I. O., Sylvester, E., Idogun, G., Eshiagiamhe, T., 2010. Acute and sub-acute Toxicological Assessment of aqueous leaf extract of *Bryophyllum pinnatum* (Lam.) in Sprague –Dawley rats. *American Journal of Pharmacology and Toxicology*, **5(3):** 145 – 157.

43. Raymond, I. O., Edward, O., Salami, S. E. I., Dickson, O. U., 2011. Some toxicological parameters in rats treated sub-chronically with aqueous Extract of *Bryophyllum pinnatum*. *Annals of Tropical Pathology*, **2(2):** 135-146.

44. Reyes, B. A. S., Bautista, N. D., Tanquilat, N. C., Anunciado, R. V., Leung, A. B., Sanchez, G. C., Magtoto, R. L., Castronuevo, P., Isukamura, H., Maeda, K. I., 2006. Anti-diabetic potentials of *Momordica charantia* and *Andrographis paniculata* and their their effects estrouscyclicity of alloxan-induced diabetic rats. *Journal of Ethnopharmacology* 9: 196-200.

45. Singh, D., Dhakre, J.S., 1989. Some medicinal plants of Mathura district (U.P.). *Mendel*, **6:** 60-66.

46. Schumann, G. R., Bonora, F., Ceriotti, G. F., Ferrero, C. A., 2002.IFCC Primary reference procedures for the measurement of catalytic activity concentrations of enzymes at 37 degrees C. International Federation of Clinical Chemistry and Laboratory Medicine. Part 5. Reference procedure for the measurement of catalytic concentration of aspartate aminotransferase.*Clinical Chemistry and Laboratory Medicine*, **40**: 725 – 733.

47. Snigur, G. I., Samokhina, M. P., Pisarev, V. B., Spasov,A. A., Bulanov, A. E., 2008. Structural alteration in pancreatic islets in streptozotocin-induced diabetic rats treated with bioactive additive on the basis of *Gymnema sylvestre*. *Morfologiia*, **133(1):** 60-64.

48. Szasz, G., 1969. A kinetic photometric method for serum gamma-glutamyl transpeptidase. *Clin. Chemistry*, **22:** 124–136.

49. Thenmozhi, M., Bhavya, P.K., Rajeshwari, S., 2011. Compound identification using HPLC and FTIR in *Eclipta alba* and *Emilia sonchifolia*. *International Journal of Engineering Science Technology*, **3:** 292-298.

50. Tichy, J., Novak, J., 1998. Extraction assay, and analysis of antimicrobials from plants with activity against dental pathogens (*Streptococcus* sp). *Journal of Alternative and Complementary Medicine*.**4(1):** 39-45.

51. Tilkian, S. M., Conover, M. B., Tilkian, A. G., 1979. Clinical implications of laboratory test. C. V. Mosby company, St Louis, Toronto, London. 352p.

52. Vasim, K., Abul, K. N., Mohd, A., Mohd, A., Mohd, M., Pillai, K. K., 2012.A pharmacological appraisal of medicinal plants with anti-diabetic potential. *Journal of Pharmacy and Bioallied Sciences,* **4(1):** 27-42.

53. Wei, Z. L., Dong, L., Tian, Z. H., 2009. Fourier transform infrared spectrometry study on early stage of cadmium stress in clover leaves. *Pakistan Journal of Botany,* **41:** 1743-1750.

54. WHO., 1997. Resolution- promotion and development of training and research in traditional medicine, W.H.O., 30: 49.

55. Zhang, W., Fievez, L., Cheu, E., Bureau, F., Rong, W., 2010. Anti-inflammatory effects of formoterol and ipratropium bromide against acute cadmium-induced pulmonary inflammation in rats. *European Journal of Pharmacology,* **628:** 171-178.

Chapter 11

Chemical Screening and *In vitro* Antibacterial Activity of Plants Used by Rwandan Traditional Healers to Treat HIV/Aids Opportunistic Diseases

Jean Pierre Nkurunziza[1], R. Muganga[1], J.B. Nkuranga[1], J.L. Seburanga[1], H. Rudakemwa[1], S. Were[2], P. Aduma[2], J. Tabuti[3] and S. Suguru[4]

[1]*University of Rwanda, Rwanda*
[2]*Maseno University, Kenya*
[3]*Makerere University, Uganda*
[4]*Great Lakes University, Kenya*

ABSTRACT

Traditional medicinal plants have been used for a long time mostly by local people of the East-African region to treat many infectious diseases. Even with the emergence of new modern drugs capable of alleviating HIV/AIDS opportunistic disease, early surveys showed that a large number of people still rely on these prescriptions from traditional healers. This study has focused at determining the phytochemical compositions and antibacterial activities of some species that traditional healers mostly rely upon when it comes to alleviating dysentery as an HIV/AIDS opportunistic disease.

The survey was carried on the following thirteen medicinal plants: *Eucalyptus globulus, Ensente ventricosum, Girardinia diversifolia, Acacia hockii, Cardiospernum halicacabum, Ipomea involcrata, Cyphostemma adenocaula, Bidens pilosa, Rumex abyssinicus, Musa acuminate,*

Synadenium grantii, Psidium guajava and *Hibiscus fuscus*. The bacterial strains that were used for antibacterial tests are *Staphylococcus aureus, Eschericia coli, Shigella sonnei* and *Pseudomonas aeruginosa*. Even though the phytochemical screening was conducted on all the above plants and only seven of them have been used for the antibacterial tests.

It was found that the predominant secondary metabolites are alkaloids, tannins, flavonoids and phenolics. Amongst all the tested species, flavonoids tend to be continuously present. It was also noticed that many of the plants used by the traditional healers display considerable antibacterial activities. This must be due to the many bioactive compounds as displayed in their phytochemical compositions. Particularly, it was experimentally noticed that *Acacia hockii* and *Psidium guajava* crude extracts are the most active ones against the tested microorganisms.

Keywords: *Chemical screening, In vitro antibacterial activity, Medicinal plants, Rwanda, Traditional healers, Secondary metabolites, HIV-Aids.*

1. INTRODUCTION

Medicinal plants play a vital role for human health care. In fact, they are used as remedies for many infectious diseases (Ahmed *et al.,* 1998). Plants have also been used as healers and health rejuvenators since immemorial time (Furnsworth and Loub, 1983). Medicinal plants research has also become the focus for discovering new drugs in the pharmaceutical industry (Sergio and Gloria, 2003; Paul and John, 2002). The search for new antimicrobials from the medicinal plants is still increasing and is of great interest with the emergence of antibiotic resistance development in the pathogens (Fyrquist, 2002).

HIV/AIDS opportunistic diseases, which have been defined as infections that are more frequent and more severe because of immunosuppression in HIV-infected persons resulting with illness and mortality in many countries. As dysentery is one of the opportunistic diseases, people living with HIV/AIDS are also highly vulnerable to this disease (WHO, 2005). The disease kills a large number of both young and adult and it shows a large persistence to antibiotic. Because of its high mortality and the increasing cost of antibiotic, poor people sometimes prefer to consult traditional healers for their primary health care.

However, the rate of microbial resistance to existing therapeutic agents is an increasing serious medical problem and it is believed to be common cause of treatment failures in bacterial infectious diseases. Moreover, the management of bacterial infection in patients infected with HIV disrupts the body's own immune system (Siegel *et al.,* 2006; Domin, 1998)

The present study intends to value a group of medicinal plants for their susceptibility to antibacterial potential. In fact, some phytochemicals have antiviral and antimicrobial activities (Bruneton, 1999; Newman, 2007),different interviews conducted on traditional healers across Rwanda have reported that these plants are used in the treatment of opportunistic diseases of HIV/AIDS. We are scientifically investigating whether certain of these plants can effectively treat the opportunistic diseases. Thus, the main objective is to evaluate the antibacterial activity of these medicinal plants against HIV/AIDS and dysentery.

2. MATERIALS AND METHODS

Methods

As previously mentioned, the main objective of the research was the determination of a number of active compounds in studied plant's leaves. In so doing, the powder obtained from the plant leaves were subjected to analytical phytochemical screening (see details in Table 11.1) using various standard methods and reagents. Then a study was undertaken to examine their antibacterial activity. Therefore, crude extracts from each of these plants were tested against three bacterial strains using agar diffusion assay.

Table 11.1: Description of the Methods Used

Criteria Secondary Metabolite	Reagent or Procedure Used	Expected Theoretical Observations Case the Metabolite Presence
Alkaloids	Dragendorf Mayer Wagner	Precipitation
Anthraquinones	Ether 20 per cent, HCl 10 per cent, NaOH 10 per cent	Red Coloring
Flavonoids	HCl, Methanol, Water	Red cerise color
	Ethylacetate, diluted ammonia	Yellow coloration
	Methanol, HCl	Pink-tomato red
Glycosides	Borntrager	Pink colour
	Acetic anhydride, H_2SO_4, Chloroform	Red, then to blue and finally green
Phlobatannins	Mix powder with distilled water and filter. Mix filtrate with 2 per cent HCl solution and boil.	Red precipitate
Anthocyanins	Ether, CH_3OH, H_2O	Formation of precipitate
Saponins	Mix powder with distilled water and shake vigorously.	Formation of persistent Foam
Steroids	Acetic anhydride, Concentrated H_2SO_4	Change from violet to blue or green
Terpenoids	$CHCl_3$, Concentrated H_2SO_4	Reddish brown coloration at the interface
Phenolics	Ferric Chloride	Dark green colour
	Gelatin test	White precipitate
	Lead acetate	Bulky white precipitate
Tannins	Salted gelatin	White precipitate
	$CuSO_4$ 2 per cent	Brownish precipitate
	$FeCl_3$ 5 per cent	White precipitate

As described above, we chose to start by a phytochemical screening because we needed a sound knowledge of the nature of the secondary metabolites from each plant, which could act as active principles. In fact, these secondary metabolites of the plant constitute the medicinal value of a drug plant, which produces definite physiological action on human body (Sharma *et al.*, 2007).

Test of Antibacterial Activity

Plant materials were collected and washed with distilled water then dried. The powder was obtained and macerated with a set of solvents namely petroleum ether, ethyl acetate, aqueous methanolic or methanol. The different dilutions were used for the preparation of plant extracts in different concentrations. The extracts were dissolved with 0.1 N DMSO (1 ml of concentrate DMSO then complete to 10 ml with distilled water) and used Mueller Hinton Agar as the liquid culture medium. For testing antibacterial activities three isolated species were used: *Escherichia coli*, *Salmonella typhi* and *Shigella dysenteriae*. To evaluate their sensitivity, the strains were directly cultivated. The antimicrobial activity testing as well as the assessment of the minimum inhibitor concentration (MIC) per formed using the diffusion method.

Description of the Plant Material

1. *Eucalyptus globulus*

Eucalyptus globulus commonly known as Tasmanian blue Gum, is a tree of Myrtaceae's family (Orwa *et al.*, 2009). It has alternate leaves drooping on flattened, yellowish leafstalks. Its flowers have with stamens and odor camphor. The fruits or seed capsules single at leaf base are broadly top-shaped or rounded (Orwa *et al.*, 2009).

2. *Ensente ventricosum*

Ensente vatricosum is commonly known as Ethiopian banana, Abbyssinian banana and false banana due to its physical resemblance to a *Banana plantor ensete*. It is a flowering species of flowering plants in the genus of ensete of the banana family Musaceaea. This plant only flowers and bears fruits once and then it dies (Randy *et al.*, 2007).

3. *Girardinia diversifolia*

Commonly known as Nilgiri nettle or Himalayan Giant nettle, *G. diversifolia* is a species of flowering plants in Urticaceae family, found abundantly in open forest land and river side of moist habitat in Nepal. It is an erect annual or short-lived perennial herb up to 1.5 m tall (Njogu *et al.*, 2007)

4. *Acacia hockii*

Acacia hokii belongs to the family of *Fabaceae* and the sub family of *Mimosaceae*. Species of this family are well adapted to arid environments with high temperature and moisture stress. Umbrella shaped canopy when adult, this plant has thin usually scarce foliage and its leaves are dark green with 4-12 pairs of pinnate having each 10-22 pairs of leaflets (Epiphanie, 2012). They are abundant on tropics soils of Australia, Africa, India and America (Arias *et al.*, 2004). They are reported to treat tuberculosis, dysentery, diarrhea and jaundice (Mukazayire *et al.*, 2010).

5. *Cardiospernum halicacabum* (Ubuzibira)

C. halicacabum is a climbing plant that belongs to the family of Spindaceace. It is widely distributed in tropical and sub-tropical regions. It is originated all through the plains of Africa, America, Bangladesh, India, Malacca and Pakistan. It is also known as ballon wine, heart vine, heart pea, love-in-a-puff, and heart seed (Huma *et al.*, 2012; Mattew and Dempsy, 2011; Shekhawat, 2012).

6. *Ipomea involcrata* (Bankura)

Ipomea is a shrub of some decimeters of height. It is just a flowering plant of the Convolvulaceae family. It has a broadly range of medicinal folklore uses. It reported to be of the list of plants eaten by Gorillas in Bwindi. The flower colour is pink, a complete flower (pistils and stamens present), filaments and anthers are each five in number, one style is present per flower (Adeyini and Olatunde, 2012).

7. *Bidens pilosa*

Bidens pilosa is an erect, branching, annual herb growing up to 1.5 m. Its stems are four angled in cross section and its leaves opposite and sparsely hairy. The abundant yellow flowers are borne in heads on long stalks and produce black, barbed seeds characteristically radiating in all directions from a common base (Dennis, 2009).

8. *Rumex abyssinicus*

This plant belongs to a genus of about 200 species of annual, biennial and perennial herbs in the buckwheat family of Polygonaceae. Members of this family are very common perennial herbs growing mainly in the Northern hemisphere, but various species have been introduced almost everywhere (Australia, 2009).

9. *Musa acuminata*

The banana family is Musaceae, and is related to the heliconias and travelers palm. Banana is not a "tree" but an herbaceous perennial. As the plant grows, the older leaves will yellow as new leaves appear. The fruit is consumed as food (Randy *et al.*, 2007).

10. *Synadenium grantii*

Synadenium grantii is a medicinal plant native to East central Africa which belongs to the Euphorbiaceae family. In nature, the plants reach up to 12 feet in height with an equal spreadlso called the African milk bush. The sap of the plant can be harmful if touched (Orwa *et al.*, 2009).

11. *Psidium guajava*

Psidium guajava is native of tropical America and belongs to the Myrtaceae family. It is a low evergreen tree or shrub 6 to 25 feet high, with wide spreading branches and square, downy twigs. Guajava is tropical and semitropical plant which is well known for its edible fruits (Kokwaro, 2009).

12. *Hibiscus fuscus*

Hibiscus fuscus garcke is a plant with alternate leaves, ovate to lanceolate, often with a toothed or lobed margin. It is a medicinal plant native from Ethiopia and South Africa which belongs to the Malvaceae family (Giday *et al.*, 2010).

Table 11.2: Results of the Phytochemical Analysis

	Alcaloids	Tannins	Phloba-tanins	Flavo-noids	Saponins	Glyc-osides	Steroids	Phenolics	Anthra-quinones	Antho-cyanins	Terpenoids
Acacia hockii	+	++	NT	+	+	-	-	++	NT	NT	NT
Bidens pilosa	+	+	+	++	+	++	++	+	++	NT	++
Cardiospernum halicacabum	+	+	NT	+	++	-	-	++	NT	NT	NT
Cyphostemma adenocaula	-	++	NT	+	+	-	-	+	NT	NT	NT
Ensente ventricosum	-	+	NT	+	-	-	-	+	NT	NT	NT
Eucalyptus globulus	++	++	NA	++	-	NA	-	++	NA	NA	NA
Girardinia diversifolia	++	+	NT	++	++	-	-	+	NT	NT	NT
Hibiscus fuscus	+	+	NT	++	+	NA	++	+	NT	-	+
Ipomea involcrata	+	-	NT	+	+	+	-	++	NT	NT	NT
Musa acuminate	++	++	-	+	++	+	-	+	+	NT	+
Psidium guajava	+	++	NT	+	N/A	++	++	++	NT	+	-
Rumex abyssinicus	+	++	±	++	+	±	-	-	++	NT	+
Synadenium grantii	+	+	NT	+	++	-	+	+	NT	+	+

++: Positive (present) +: Partially positive (slightly present) - : Negative (absent) ±: undecided.

N/T: Not tested N/A: Not applicable.

13. *Cyphostemma adenocaula*

This plant belongs to the family of Vitaceae and is likely to East and West tropical Africa habitat. It is a perennial herb climbing or scrambling with leaf-opposed, branched tendrils; roots tuberous; slender stem of about 3 to 6 cm of height. Leaves and fruits do contain oxalic acid which is even responsible for the acid taste of *Cyphostemma adenocaula* (Yang, 2005).

Description of the Bacteria

1. *Shigella sonnei*

Shigella sonnei is non-motile, nonspore-forming, facultative anaerobic gram-negative bacterium. Its non-motile characteristic without any flagella to facilitate its movement like many other human enterobacteria. *Shigella sonnei* is a rod-shaped, lactose- fermenting bacterium causing dysentery. Its natural habitat is in low pH environment such as the human gastrointestinal tract (Harris *et al.*, 2002)

2. *Staphylococcus aureus*

Staphylococci belong to gram-positive bacteriawith diameters of 0.5–1.5 µm. They are characterised by individual cocci which divide in more than one plane to form grape-like clusters. *Staphylococcus aureus* is a major pathogen of increasing importance due to the rise in antibiotic resistance. The species named *aureus* which refers to the fact that colonies (often) have a golden colour when grown on solid media (O'Sullivan *et al.*, 2006).

3. *Escherichia coli*

Escherichia coli is predominant facultative organism in the human gastrointestinal tract. *Escherichia coli* is gram-negative straight rod which either uses peritrichous flagella for mobility or is nonmotile. Pathogenic forms of *E. coli* can cause a variety of diarrhoeal diseases in hosts due to the presence of specific colonisation factors, virulence factors and pathogenicity associated genes. These properties are generally not present in all *E. coli* Species. Of the strains that cause diarrhoeal diseases, six pathotypes are now recognised (Chhetri *et al.*, 2008).

3. RESULTS

The repartition of secondary metabolites is the first set of results that were investigated. The yields of various extraction as well as their antibacterial activities using agar well diffusion have also been compiled. All these results are displayed in the tables below.

4. DISCUSSION

It was generally found that, the secondary metabolites are not uniformly distributed within the studied plants. However, a closer look at Table 11.2 shows that the predominant secondary metabolites are alkaloids, tannins, flavonoids and phenolics. Flavonoids tend to be present amongst all tested species.

Table 11.3: Yields of the different Extractions (in per cent)

Exact	Petroleum Ether	Ethyl Acetate	Aqueous methanolic	Methanol Plant
Acacia hockii	4.28	Np	Np	17.49
Cardiospernum halicacabum	2.485	Np	Np	5.43
Cyphostemma adenocaula	9.30	Np	Np	3
Hibiscus fuscus garcke	Np	0.67	0.83	0.61
Ipomea involcrata	2.01	Np	Np	2.57
Psidium guajava	Np	6.315	19.82	0.99
Synadenium grantii	Np	0.72	0.82	0.77

Np : Extraction was not performed.

Many of the plants used by these traditional healers display some antibacterial activities. This must be due to the many bioactive compounds as displayed in their phytochemical composition. *Acacia hockii* and *Psidium guajava* extracts have been demonstrated to be the most active against the tested microorganisms. They present higher antibacterial activity comparatively to all subjected plants as shown in Tables 11.4 and 11.5.

We could attribute these antibacterial activities to a combination of flavonoids and other bioactive secondary metabolites which may act by synergism. In fact, *In vitro* studies showed that flavonoids have anti-allergic, anti-inflammatory, anti-microbial, anti-cancer, and anti-diarrheal activities and may play a major role in the treatment of diarrheic diseases (Gnan and Demello, 1999).

In both cases, extracts from polar solvents (methanol or aqueous methanol) showed strong inhibition of bacteria than extracts from non-polar solvents (petroleum ether or ethyl acetate). This may due to a greater concentration of bioactive compounds in polar solvents than in non-polar solvents.

The studies also showed that many extracts are more active against Gram positive bacteria (*Sptahylococcus aureus*) than gram negative bacteria (*Escherichia coli* and *Shigella sonei*). These results could be due to the presence of a double membrane surrounding each bacterial cell of *E. coli* and *S. sonnei* which excludes certain drugs for penetrating the cell, partially accounting for the greater resistance of gram-negative bacteria to antibiotics (Fleischer, 2003; Calixto 2000).

The resistance of some bacterial strains against our extracts may be due to the climate change which affects the synthesis of secondary metabolite in plants. But for other plants (*Synadenium grantii* and *Hibiscus furkus*), the lack of antibacterial properties may also be attributable to the age of the plant used, freshness of plant materials, physical factors (temperature, light and water), contamination by field microbes, adulteration and substitution of plants, incorrect preparation an dosage (Calixto, 2000; Okigo and Igwe, 2007; Okigo and Omodamino, 2006). Because also these plants are traditionally used in association, the inactivity observed might due to the lack of synergy they have while associated.

Table 11.4: Inhibition Zone (mm) for the Extracts from Acacia hockii, Ipomea involcrata, Cardiospermum halicacabum and Cyphostemma adenoacaule

Plant	Petroleum Ether (mg/l)				Methanol (mg/l)				Bacteria
	0.4	0.6	0.8	1	0.4	0.6	0.8	1	
Acacia hockii	–	6	8	12	6	10	13	15	Staphylococcus aureus
	–	–	6	10	10	13	15	16	Escherichia coli
	–	7	8	10	6	10	13	15	Pseudomonas aureuginosa
Ipomea involcrata	–	6	6	8	6	6	8	8	Staphylococcus aureus
	–	–	–	6	–	–	6	8	Escherichia coli
	–	–	–	–	–	–	–	6	Pseudomonas aeruginosa
Cardiospermum halicacabum	–	–	6	7	6	7	8	10	Staphylococcus aureus
	–	6	6	8	6	8	11	11	Escherichia coli
	–	–	–	–	–	–	6	8	Pseudomonas aeruginosa
Cyphostemma adenocaule	–	6	6	8	–	6	8	8	Staphylococcus aureus
	–	–	–	6	–	6	8	8	Escherichia coli
	–	–	–	–	–	–	–	6	Pseudomonas aeruginosa

Table 11.5: Inhibition Zones (mm) for the Extracts from for *Hibiscus fuscus*, *Psidium guajava* and *Synandenium grantii*

Extract	Ethyl Acetate (mg/ml)			MeOH-H$_2$O (mg/ml)			Methanol (mg/ml)			Extract
Plant	1.5	2	2.5	1.5	2	2.5	1.5	2	2.5	Bacteria
Hibiscus fuscus	0	0	0	0	0	0	0	0	0	Staphylococcus aureus
	0	0	0	0	0	0	0	0	0	Escherichia coli
	0	0	0	0	0	0	0	0	0	Shigella sonnei
Psidium guajava	10	13	15	15	17	19	Np	Np	Np	Staphylococcus aureus
	0	0	0	2	4	7	2	5	8	Escherichia coli
	0	0	0	Np	Np	3	Np	2	4	Shigella sonnei
Synandenium grantii	0	0	0	0	0	0	0	0	0	Staphylococcus aureus
	0	0	0	0	0	0	0	0	0	Escherichia coli
	0	0	0	0	0	0	0	0	0	Shigella sonnei

Np: Data not provided.

5. CONCLUSIONS AND RECOMMENDATIONS

The present study aimed to evaluate the antibacterial activity of the extracts from fourteen plants which have been used by traditional healers to treat dysentery which is an opportunistic disease in people living with HIV/AIDS. The results showed that the extracts from *Acacia hockii* and *Psidium guajava* leaf tend to be more active against *Staphylococcus aureus, Eschericia coli, Shigella sonnei* and *Pseudomonas aeruginosa*. This means that they are potentially active against the bacteria which cause dysentery.

We recommend that further researches should be performed on the viral activities of the plants extracts. We also recommend further researches on those inactive extracts which might not undergo further transformations once they are in intestinal tract. Maybe some of the drugs we thought to be inactive might be displaying their antibacterial activities after undergoing some metabolic changes in the body.

As generally, the drugs from natural resources are better accepted by the body than substances made in the laboratory, we also recommend continuous further researches on other plant species that might potentially fight against opportunistic diseases of HIV/AIDS. These surveys would lead to a promotion of herbal medicine that may be benefic for the population as it is part of the culture and can be obtained at low cost.

6. ACKNOWLEDGEMENTS:

☆ *SIDA and Inter-University Council for East Africa, IUCEA: for the financial support*

☆ *Lac Victoria Basin research Initiative, VicRes: for the project coordination*

☆ *National University of Rwanda, NUR: for facilities provided*

☆ *National Association of Traditional Healers, Rwanda : for information and contact with traditional healers*

☆ *All students at NUR in Chemistry and Bitotechnology department: for field survey, data collection and laboratories.*

7. REFERENCES

1. Ahmed, I., Mehmod, Z., Mohammad, F., 1998. Screening of some Indian medicinal plants for their antimicrobial properties. *J. Ethnopharmacol.* 62. 183-193. Publisher

2. Farnsworth, N.R. and Loub, W.D., 1983. Information gathering data and bases those are pertinent to the development of plant, derived drugs in plant: The potentials for extracting Protein, Medicines, and Other Useful Chemicals. Workshop Proceedings. OTA, BP, F, 23.U.S. Congress, Office of Technology Assessment, Washington, DC., pp 178-195. Publisher

3. Sergio, H.M., Gloria, A., 2003. Antimicrobial peptide: A natural alternative to chemical antibiotics and potential for applied biotechnology. *Electronic Journal of Biotechnology* 2(3) (cited 2003) ISSN: 0717-3458

4. Paul, M.D., and John, W., 2002. Medicinal natural products, a biosynthetic approach. 2nd edition. p. 515. School of pharmaceutical Sciences, University of Nottingham, UK.

5. Fyhrquist, P., Mwasumbi, L., Haeggstrom C.A., Vuorela H., Hiltunen, R., Vuorela, P., 2002, Ethnobotanical and antimicrobial investigation of some species in Terminalia and Combretum (Combretaceae) growing in Tanzania. *J. Ethnopharmacol.* 79: 169-177. Publisher

6. WHO, 2005. Shigellosis: disease burden, epidemiology and case management. *Weekly Epidemiological Record*, **80** (11): 94-9.

7. Siegel, J.D., Rhinehart, E., Jackson, M., Chiarell L., 2006. Management of multidrug resistant organisms in healthcare settings. pp. 1-74. Centers for Disease control and Prevention (CDC).

8. Domin, M.A. (1998). Highly virulent pathogens-a post antibiotic era. *Br J theater Nurs.*; 8: 14-18.

9. Bruneton, J. (1999), Pharmacognosie, Phytochimie et Plantes medicinales. *In Lavoisier, Plantes Medicinales :* 309-801. Paris

10. Newman, J., 2007. Natural Products as Sources of New Drugs over the last 25 Years. *Journal of Natural Products*, 461- 477

11. Sharma A.S., Mann V., Gajbhiye and Kharya, M.D., 2007. Phytochemical profile of *Boswelia serrata*: an overview. *Pharmacognosy Reviews*, 1(1): 137-142. Publisher

12. Orwa, C., Mutua, A., Kindt, R., Jamnadass, R., Simons, A., 2009. Agroforestry database, accessed on February 22nd, 2013 from http:www.worldagroforesrty.org/af/treedb/

13. Randy, C., Ploetz, Kay Kepler, A., Jeff Daniells, and Scot, C., Nelson, 2007. Banana and Plantain – An overview with emphasis on Pacific island cultivars. Musaceae banana family. Species profile for Pacific Islands Agroforestry: p 2, 3-7

14. Njogu, P.M., Thoithi, G.N., Mwangi, J.W., Kamau, F.N.,. Kibwage, I.O., Kariuki, S.T., Yenesew, A., Mugo, H.N., and Mwalukumbi, J.M., 2011. Phytochemical and Antimicrobial Investigation of *Girardinia diversifolia* (Urticaceae). *East and Central African Journal of Pharmaceutical Sciences* Vol. 14: 89

15. Epiphanie, N., 2012. Chemical Screening of plants used in traditional medicine to treat diarrheic cases of *Acacia hockii* (Umugenge). Memoire, NUR.

16. Arias M., *et al.*, 2004. Antimicrobial activity of ethanolic and aqueous extracts for *Acacia aroma* Gill.ex Hook ate Am. *Journal of Life Sciences*, 191-202.

17. Mukazayire, M.J., Minani, V., Ruffo, C.K., Bizuru, E., Bigendako, M.J., Stevigny, C. and Duez P. 2010. Plants and folk medicine used in liver pathologies in Rwanda and biochemical evaluation of patients treated by *Ocimum lamiifolium* in traditional medicine. Université Libre de Bruxelles (ULB), Laboratory of Pharmacognosy, Bromatology and Human Nutrition, CL 205,9,B,1050 Brussels, Belgium: 59.

18. Huma, S., Ghazala, H.R., Shaukat, M., Raheela, K., Hina Z., 2012. *In vitro* antimicrobial and phytochemical analysis of *Cardiospermum halicacabum L. Pak. J. Bot.*, 44(5): 1677-1680.

19. Matthew A. and Dempsy B., 2011. *Anatomical and morphological responses of Cardiospermum halicacabum L.* Pp.1-56. Master thesis, University of North Texas.

20. Shekhawat M. 2012. *In vitro* Clonal propagation of *Cardispermum Halicacabum* Through nodal segment cultures. *The Pharma innovation*, 1-7.

21. Adeniyi, A. J., Olatunde, R.O., 2012. Systematic studies in some *Ipomoea linn* Species using pollen and flower morphology. *Annals of West University of Timi°oara*, ser. Biology, XV (2): 177-187.

22. Dennis I. M., 2009. 95 POLYGONACEAE, Flora of Tasmania, version 2009: 1.

23. Department of Agriculture and Food, Government of Western Australia, 2009. Harmful garden plants in Western Australia. Bulletin No. 4641 ISSN 1448-0352: 8.

24. Orwa, C., Mutua, A., Kindt, R., Jamnadass, R., Simons, A., 2009. *Psidium guajava*, Agroforestry Database 4.0: 1.

25. Kokwaro, J.O., 2009. Medicinal plants of East Africa: Third Edition. Nairobi: University of Nairobi Press. pp. 478.

26. Teklehaymanot, T.; Giday, M., 2010. Ethnobotanical study of wild edible plants of Kara and Kwego semi-pastoralist people in Lower Omo River Valley, Debub Omo Zone, SNNPR, Ethiopia. *J. Ethnobiol. Ethnomedi.*, 6 (1): 23.

27. Yang, F., 2005. Genome dynamics and diversity of Shigella species, the etiologic agents of bacillary dysentery". *Nucleic Acid Research*, 5. (33)9: 6445-6458.

28. Harris, L.G., Foster, S.J., and Richards, R.G., 2002. An Introduction to *Staphylococcus aureus* and Techniques for Identifying and Quantifying *S. aureus* Adhesins In Relation to Adhesion to Biomaterials: Review. L.G. Harris *et al.*, *European Cells and Materials* (4): 39- 60.

29. O'Sullivan, J., Bolton, D. J., Duffy, G., Baylis,C., Tozzoli, R., Wasteson, Y., and Lofdahl, S., 2006. Methods for Detection and Molecular Characterisation of Pathogenic *Escherichia coli*. Co-ordination action food-ct-2006-036256 *Pathogenic Escherichia coli*, Network.

30. Chhetri, H.P., Yogel, N.S., Sherchan, J., Anupa, K., and Mansoor, S., 2008. Phytochemical and antimicrobial evaluation of some medicinal plant of Nepal. *Science Engineering and Technology*, (1): 49-54.

31. Gnan, S.O., Demello, M.T., 1999. Inhibition of *Staphylococcus aureus* by aqueous Goiaba extracts. *J. Ethnopharmacol.*, 68: 103-108.

32. Fleisher, T.C., Ameade, E.P.K., Sawer, I.K., 2003. Antimicrobial activity of the leaves and flowering tops *of Acanthospermum hispidum. Filotherapia*, (**74**): 130-132.

33. Calixto, J.B., 2000. Efficacy, safety, quality control, marketing and regulatory guidelines for herbal medecines (Phytotherapeutic agents), *Braz J. med. Biol. Res.*, 33(2): 179-189.

34. Okigbo, R.N., Igwe, D.I. (2007). The antimicrobial effects of *Piper guineense* 'uziza' and *Phyllantus amrus* 'ebe-benizo' on *Candida albicans* and *Streptococcus faecalis*. *Acta Microbiologica et Immunologica Hungarica*, 54(4): 353-366 (Free).

35. Ogkibo, R.N., Omodamiro, O.D., 2006. Antimicrobial Effects of leaf extracts of Pigeon pea (*Cajanus cajan* (L.) Millsp.) on some human pathogens. *J. Herbs, Spices Medicinal Plants*, 12(1/2): 117-127.

Chapter 12

Antibacterial and Antifungal Potential of Plant Seeds

Tijen Talas Ogras

TUBITAK, Marmara Research Center,
Genetic Engineering and Biotechnology Institute,
P.O. Box 21, 41470 Gebze, Kocaeli, Turkey,
E-mail: Tijen.ogras@tubitak.gov.tr

ABSTRACT

Plants have a wide array of molecules including proteins with antibacterial and/ or antifungal activities to cope with the attacks of microbial pathogens. Antimicrobial peptides (AMPs) constitute a heterogenous class of low molecular weight proteins which are recognized as important components of innate defence system of both animals and plants. In this study, the antifungal and antibacterial activities of low molecular weight cationic peptide fractions of dry and germinated wheat seeds were analyzed using disk diffusion and turbidity measurement assays. The peptides exhibited effective *in vitro* antifungal activity against four plant pathogenic fungi including *B. cinerea, F. oxysporum R. solani, V. dahlia*. The peptides also demonstrated inhibitory activity against Gram-positive bacterium *Clavibacter michiganensis* subsp. *sepedonicus*, however they were less effective against Gram-negative bacterium *Erwinia carotovora* subsp. *carotovora*. The antimicrobial activity of germinated wheat seed peptides was more effective than dry wheat seed peptides. The activities were also tested in the presence of $CaCl_2$.

Keywords: Antibacterial, Antifungal, Peptides, Seed, Peptide, Fungi, Wheat (Triticum durum).

Abbreviations

WAP: Wheat antimicrobial protein;

GWAP: Germinated wheat antimicrobial protein;

AMPs: Antimicrobial peptides.

1. INTRODUCTION

Antimicrobial peptides are diverse group of molecules that are produced by animals, plants and microorganisms. These peptides are capable of inhibiting microorganisms and they are interesting issue for medicine and food industry. AMPs constitute a heterogenous class of low molecular weight proteins, which are recognized as important components of innate defense system of both animals and plants. They directly interfere with the growth, multiplication and spread of microbial organisms (Lehrer and Ganz, 1999). AMPs are found in a wide range of secondary structures such as α-helices, β-strands with one or more disulphide bridges, loop and extended structures. Most of AMPs are cationic, and only a few of them are anionic (Carvalho *et al.*, 2009). More than 2000 AMPs have been reported in antimicrobial peptide database (http://aps.unmc.edu/AP/main.php/). Plants produce a wide array of defense proteins in their different tissues to anticipate and cope with attacks of microbial pathogens. They are different from regular antibiotics that they are coded by genes. Several classes of proteins with antibacterial and/ or antifungal properties with different structures have been isolated from various plants (Gao *et al.*, 2000). Some of these proteins are classified as: thionins, lipid transfer proteins, plant defensins, chitinases, 2S albumins and ribosome inactivating proteins (Castro and Fontes, 2005). Plant extracts having antimicrobial activity have been of great interest to researchers for the discovery of new drugs. Consequently, a number of studies on the antimicrobial screening of plant extracts for antimicrobial activities have appeared in medicinal research. In this report, basic, water soluble, low molecular weight cationic peptides were isolated from dry and germinated wheat seeds and antimicrobial activities were investigated.

2. MATERIALS AND METHODS

Purification of Small Basic Proteins of Wheat Seeds

Wheat (*Triticum durum* L. Altintoprak-98) seeds were obtained from the Faculty of Agricultural, the University of Trakya. The wheat seeds were surface sterilized with 70 per cent alcohol for 5 min, rinsed in sterile distilled water three times and germinated between sterile wet filter papers for 48 h in the dark. The dry and germinated wheat seeds were ground finely and the seed meal (60 g) was extracted with four volumes of cold extraction buffer (10 mM Na_2HPO_4, 15 mM NaH_2PO_4, 100 mM KCl, 1.5 mM EDTA) with constant stirring for 2 h and followed by centrifugation at 15 000 g for 30 min. During extraction phenylmethylsulfonyl fluoride (PMSF) was added at a level of 1 mM to inhibit proteolysis. The clear supernatant (total crude extract) was subjected to ammonium sulphate $((NH_4)_2SO_4)$ precipitation to obtain a final 30 per cent saturation and was gently stirred for 2 h. The mixture was centrifuged at 15 000 g for 30 min and, the supernatant was adjusted to 70 per cent relative ammonium sulphate saturation. The slurry was stirred for 2 h and centrifuged as previously done. The pellet was resuspended in minimal amount of distilled water and the protein suspension was kept at 80 °C for 10 min. The soluble fraction obtained after a final centrifugation at 13 000 g for 15 min and dialyzed against distilled water for 20 h using a dialysis tubing with a molecular cut-off 1 kDa. Elimination of the protein samples greater than 10 kDa was performed using

Centricon YM-10 filter (Millipore Corporation) as described by manufacturer. The sample was applied directly to a diethylaminoethylcellulose (DEAE-52) column (1.0 x 10 cm) which had been previously equilibrated with 20 mM Tris-HCl buffer (pH 8.0). Unadsorbed protein fractions were collected with the equilibration buffer while adsorbed proteins were eluted by addition of 0.5 M NaCl in the buffer. The protein eluates were collected at a flow rate of 22.5 ml h^{-1} and the elution profile was monitored at 280 nm (A$_{280}$). All steps were performed at 4°C.

The eluates were sterilized with 0.22 μm syringe-filter and submitted to bioassay for detection of antimicrobial activity. The active fractions of peak D1 were combined and dried in a vacuum concentrator. The dried material were dissolved in water and further submitted to HPLC- gel filtration chromatography on a Bio-Sil Sec 250 (Bio-Rad) 300 x 7.8 mm column. The column was equilibrated in 150 mM NaCl, 50 mM potassium phosphate buffer (pH 8.0). The column was calibrated with several proteins of known molecular masses (24.000, 14.200, 6.500, 1.35 kDa) for the estimation of molecular weights of the peptides. Absorbance at 280 nm was also used to monitor elution profiles during chromatography and to determine the protein content of the column eluates. The absolute protein concentrations of the eluates were determined as described by Bradford (Bradford, 1976) using bovine serum albumin (BSA) as a standard. Sodium dodecyl sulphate-polyacrylamide gel electrophoresis (SDS-PAGE) was used to check the purity of the protein fractions. SDS-PAGE was carried on 15 per cent acrylamide separating gel (Laemmli, 1970) and after electrophoresis the protein bands were stained with silver reagents (Nesterenko *et al.*, 1994).

Microorganisms and Antimicrobial Activity Assays

Four strains of fungi, *Botrytis cinerea, Fusarium oxysporum, Rhizoctonia solani,Verticillium dahliae* and two bacterial strains *Corynebacterium michiganense* CM 10, *Erwinia carotovora* subsp. *Carotovora* NCPPB 929 were used in the bioassays. The fungi were maintained on potato dextrose agar (PDA). For the preparation of spores, sterile distilled water (5 ml) was added to the dishes of fungal culture and these were gently agitated for 1 min for spore liberation. The spore densities were determined microscopically by using a Thoma counting chamber. Paper disk diffusion and microtiterplate based turbidity assays were performed to determine the antifungal/antibacterial activities of the small, basic peptides of the dry and germinated wheat seeds.

The assay of antifungal activity towards fungal species was assessed on basis of hypal extension assay of filamentous fungi by paper disk diffusion assay. An agar plug containing mycelia of the test fungus was harvested from actively growing fungal plates and placed in the center of 9.1 mm petri plate containing 15 ml PDA. The plates were incubated at 26 °C for 7 daysin the darkuntil the mycelial colony reached a diameter of 3 cm. Sterile filter paper disks (Whatman No.3; 6 mm in diameter) were placed on the agar surface at a distance of 0.5 cm away from the growing front of the mycelial colony and then various amounts of the protein solution was applied to each disk in 10 μl. Control disks were prepared by replacing the protein sample with the same volume of 10 mM Tris-HCl (pH 7.5). The test plates were incubated at room temperature and the diameter of the inhibition zone around each disk was measured after 72 h. If the material being tested had antifungal activity, a transparent zone was observed around each disk. The fungal concentration of 2×10^4 spores mL^{-1} was used for the antifungal assays.

In vitro inhibitory effect of the basic peptides of wheat seeds were quantified using flat bottom 96 well microtiterplate based turbidity measurement assay and the percentage of growth inhibition was determined. The growth medium for the antifungal assay was either half-strength potato dextrose broth (PDB) (medium I), or supplemented with 5 mM CaCl$_2$ (medium II).The turbidity assay for bacteria was carried out in two media; half strength LB broth (medium I) and medium I supplemented with 5 mM CaCl$_2$ (medium II). Each well contained twenty μl of various concentrations of the filter sterilized protein solution and 110 μl of appropriate growth medium containing 2×10^4 fungal spores mL^{-1} or 2×10^5 colony forming units ml^{-1} of bacterial suspension, in a total of volume 130 μl. A control well without the addition of protein solution was also used. The microplate was incubated at 26 °C and the degree of growth inhibition was determined by measuring culture turbidity over time intervals (24 h for bacteria and 48 h for fungi in the dark) at 595 nm using an automatic microplate reader. All assays were performed in triplicate.

3. RESULTS AND DISCUSSION

Purification of the Antimicrobial Proteins

The heat-resistant low molecular weight protein fractions of the wheat seeds were submitted to ion-exchange chromatography on DEAE-52 column. The flow-through from the column represented the basic protein fractions (peak D1) and adsorbed protein fractions (peak D2) were eluted with 0.5 M NaCl wash of the column (Figure 12.1). The protein eluates were monitored spectrophotometrically at 280 nm. The arrow indicates addition of 0.5 M NaCl to the column.

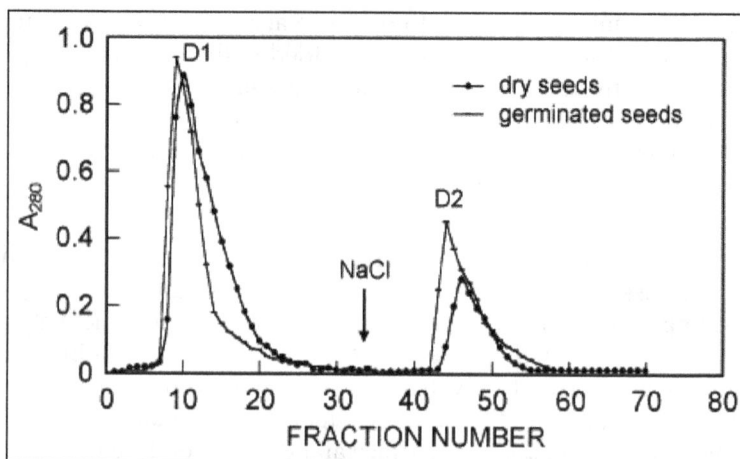

Figure 12.1: Elution Profile of an Ion-Exchange Column to Fractionate Basic Proteins of Dry and Germinated Wheat Seeds. Ammonium sulphate precipitated proteins of the seeds were heat denatured, fractionated with YM-10 filter and applied onto column of DEAE-52.

Protein fractionations from the dry and germinated wheat seeds showed similar chromatographic profiles on the column. The overall purification stages of the proteins were summarized in Table 12.1.

Table 12.1: Purification Steps of Small Protein Fractions from the Germinated and Non-germinated Wheat Seed Meal and Inhibition Concentrations (IC_{50}) of the Fractions were Determined against *V. dahlia*

Purification Step	Total Protein (mg)[D]	Yield[D] (per cent)	IC_{50} Values[D]	Total Protein[G] (mg)	Yield[G] (per cent)	IC_{50} Values[G]
Crude seed extract	320	100	400	300	100	400
$(NH_4)_2SO_4$ precipitation (70 per cent)	68.5	21.4	360	62.0	20.8	350
80 ºC denaturation	54	16.8	250	49.5	16.5	250
Dialysis	49	15.3	220	42	14.0	210
YM-10 filtration	21	6.5	180	18	6.0	160

[D] Dry wheat seed, [G] Germinated wheat seed.

The purity and apparent molecular weight of the peptides were analyzed by 15 per cent SDS-PAGE acrylamide gel under reducing conditions (Laemmli 1970) and they gave bands with a molecular mass of approximately 5 kDa (Figure 12.2). Molecular weight estimation of the WAP and GWAP from the active fractions of two peak D1 yielded two closest peaks by gel filtration chromatography with a value of 5 kDa.

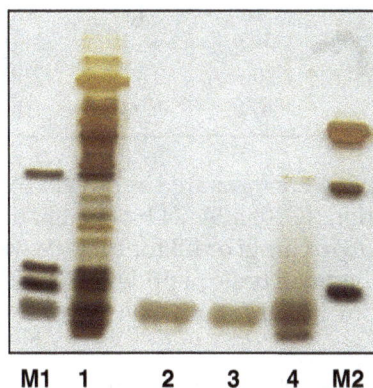

Figure 12.2: SDS-PAGE of different Protein Fraction(s) from the Dry and Germinated Wheat Seeds.

Lane 1: dialyzed protein fraction of the dry seed. Lanes 2 and 3: 10 µg WAP and GWAP. Lane 4: YM-10 filtration of the dry seeds. M1 shows Sigma molecular weight standards (from top downward, 26.600 kDa, 17.000 kDa, 14.200 kDa, 6.500 kDa, 3.496 kDa). M2 shows the molecular weight marker proteins (from top downward, bovine serum albumin, 67 kDa; trypsinogen, 24.000; α-lactalbumin, 14.2 kDa)

Various concentrations of the protein fractions were sterilized and assayed for the presence of antimicrobial activity against the selected of pathogenic fungi and bacteria. The active peptides of the dry and germinated seeds were named WAP (wheat antimicrobial peptide) and GWAP (germinated wheat antimicrobial peptide), respectively.

Antimicrobial Properties of the Wheat Seed Proteins

In vitro antimicrobial susceptibility tests and inhibition concentrations(IC_{50}) of the basic wheat proteins were performed against four agronomically important phytopathogenic fungi and two bacterial strains. The basic protein fractions of DEAE-52 (peak D1) showed effective antifungal activity against the tested fungi while no such activity was detected with second peak (D2) (data not shown). *In vitro* antimicrobial susceptibility tests of the proteins were performed by turbidity measurement and paper disk diffusion assay (Cole 1994). The data of the assay was used to obtain the dose response curves (percent growth inhibition versus protein concentration) where effective protein concentrations required for 50 per cent inhibition of bacterial growth (IC_{50}: inhibition concentration) were determined. The results were quantified and shown in Table 12.2.

Table 12.2: Antifungal Activity of WAP and GWAP

Test organism	Medium I		Medium II	
	WAP	GWAP	WAP	GWAP
B. cinerea	20	20	200	200
F. oxysporum	30	20	200	200
R.. solani	20	15	200	200
V. dahliae	40	30	200	200
C. michiganensis	>200	180	>200	>200
E. carotovora	>200	>200	>200	>200

The antifungal activities were measured as protein concentrations required for 50 per cent growth inhibition (IC50) after 24 h of incubation. Half strength growth medium (Medium I) of PDB for fungi or LB for bacteria were supplemented with 5 mM $CaCl_2$ (Medium II). The small basic peptides of the dry and germinated wheat seeds inhibit growth of four tested fungi when assayed in the low ionic strength growth medium.

The IC_{50} values of WAP and GWAP against the four fungi ranged from 20 to 40µg/ml and 15 to 30 µg/ml, respectively. The basic peptide GWAP from the germinated seed of the wheat seems to be more effective antifungal protein than the peptide of the dry wheat seed WAP. The minimum IC_{50} value of the peptides towards fungi was 15 µg protein ml^{-1}, so GWAP affected growth of *R. solani* at this concentration in the medium I. The basic peptides of the wheat did not influence the growth of the Gram-negative bacteria *E. carotovora* when added in the growth media at up to 200 µg/ml. Likewise WAP did not affect the growth of *C. michiganense* at this concentration but GWAP inhibited the growth of this bacterium with 180 µg protein ml^{-1}. The antimicrobial activity of the wheat peptides is more potent against the tested fungi than the tested bacteria. At concentrations of up to IC_{50} value of 180 µg/ml protein WAP and GWAP did not affect the viability of the bacteria.

In the presence of 5 mM $CaCl_2$ in the assay medium, growth inhibition of microorganisms was barely detectable at concentration of 200 µg protein ml^{-1}. Addition of calcium ions to the growth media significantly reduced the activity of the WAP and GWAP. Thus, IC_{50} values rose more than 10 fold. This phenomenon has been identified for a number of small antimicrobial peptides and reported previously (Cammue *et al.*, 1992, Terras *et al.*, 1992).

Two basic, small-sized (around 3000 kDa) cysteine-rich antimicrobial polypeptides of *Amaranthus caudatus* inhibited the growth of different plant pathogenic fungi at low doses and inhibited growth of two Gram-positive bacteria but, these peptides were did not show activity against two Gram-negative bacteria (Broeakaert *et al.*, 1992). The antimicrobial activities of the peptides were also determined by paper disk diffusion assay. The mycelial growth formed transparent zone of inhibition around disks which contain plant extract with antifungal activity.

4. CONCLUSIONS

The isolated peptides were shown to be active *in vitro* against four fungi and one bacterium of economically important plant hosts. The WAP and GWAP were also heat stable as the antifungal activity of these peptides was not affected by heat treatment at 80 °C.

The molecular weights of the basic peptides of the dry and germinated seed of wheat were determined by SDS-PAGE and gave approximately 5 kDa of protein bands which are similar to the molecular weight of known AMPs. Gel filtration analysis of the peptides are also correlated with the SDS-PAGE analysis. The majority of the antimicrobial proteins isolated from other plants (Cammue *et al.*, 1995) is relatively small, cysteine-rich basic peptides (>10 kDa) (Koo *et al.*, 2002, Garcia-Olmedo *et al.*, 1998). Their low molecular masses coupled with their high antifungal potency, should make them strong candidates for exploitation of its biological activities.

In vitro antimicrobial screening allows the selection of plant extracts with potentially useful properties to be used for further studies for plant protection. The bioassays showed that WAP and GWAP as well as other identified antimicrobial peptides (Koo *et al.*, 2002) exhibited effective inhibitory activity as antimicrobial agents especially against the tested pathogenic fungi. The WAP and GWAP can inhibit the growth of all examined pathogenic fungi including *B. cinerea, F. oxysporum, R. solani* and *V. dahliae*.

The antagonistic effect of calcium ion was examined on the activity of the peptides by adding 5 mM $CaCl_2$ to the growth media. Addition of the $CaCl_2$ to the growth media completely reduced the activity of the peptides at least 13 fold. It has been reported that the activity of some small cationic antimicrobial proteins, including plant defensins (Terras *et al.*, 1992, Osborn *et al.*, 1995), and thionin is reduced or blocked in the presence of monovalent and divalent cations, especially $CaCl_2$, $MgCl_2$ and KCl. However, divalent cations like $CaCl_2$ affect the antimicrobial activity more strongly than do monovalent cations (Terras *et al.*, 1992, Harrison *et al.*, 1997).

The antimicrobial spectrum of the WAP and GWAP is more effective against fungi than bacteria when assayed in the growth medium I. The antibacterial activity of the basic peptides were not so effective on the growth of Gram-positive bacterium *C. michiganense* and Gram-negative bacterium *E. carotovora*. The GWAP exhibited inhibitory effect against *C. michiganense* with the IC_{50}180 µg protein ml^{-1}. Growth of Gram-negative bacteria *E. carotovora* was suppressed by radish 2S albumins (with IC_{50} 250 µg/ml) (Terras *et al.*1992). Generally Gram-positive bacteria are more sensitive to antimicrobial agents because they lack an outer membrane that serves as an effective barrier for penetration of large molecules (Zang and Lewis, 1997).

The mechanisms of action of the antimicrobial peptides are not known in detail however, it is assumed that the antimicrobial proteins are interacting directly with microbial membranes or they have a protein/receptor target. It is also believed that the antifungal property of small peptides might be related to their small size which would allow them to penetrate fungal wall (Money 1990). They are believed to generate pores in the microbial membranes resulting in leakage of the cytoplasmic material which then leads to the death of microorganisms (Giudici *et al.*, 2000). Inhibitory effect of antimicrobial peptides on bacteria requires an electrostatic interaction with negatively charged membrane phospholipids, followed by either pore formation and a special interaction with a domain in the membrane (Floerack and Stiekema, 1994).

In this study, effect of germination on the antimicrobial activity of the wheat proteins was also examined. During germination period of the plant seed, there is a drastic onset of antimicrobial activities that are not present in dormant phase of the seeds. It is believed that there is a mechanism in the seed to generate an array of antimicrobial proteins for the protection of embryo and young seedlings during earlier stage of development in highly microbial environment of soil (De Bolle *et al.*, 1996). It is also reported that there is an increase of antifungal activity during plant seed germination and in seed protein extract *in vitro* of some plants including cheeseweed, cigar tree and wheat (Wang *et al.*, 2002). Assays with radish seeds have demonstrated that some antimicrobial proteins (defensin) represent over 30 per cent of the proteins released during germination (Terras *et al.*, 1995). The studies have shown that the secondary metabolite content varies during germination of seeds. Purification and antifungal properties of barley seed chitinases released during early stages of imbibition have also been reported by Swegle and colleagues (1992).

The study revealed that the small, heat stable, basic protein fractions of the dry and germinated seeds of wheat contained potential antimicrobial activity profiles on the growth of some agronomically important phytopathogenic bacteria and fungi. The detailed analysis of the peptides should be clarified by amino acid analysis and protein database searches to find out identity with the known antimicrobial proteins and further studies are being conducted to elucidate the properties of the peptides for antimicrobial activity. Isolation of active antimicrobial proteins from wheat seed provides a tool for the use of gene transformation studies aimed to obtain transgenic cotton plants displaying enhanced tolerance to fungi especially for

Verticillium dahliae. In conclusion, it is shown that wheat seed contains antimicrobial proteins which inhibit growth of a range of fungi and bacteria *in vitro*.

5. ACKNOWLEDGEMENTS

This work was in part by the project No. 5003301 TUBITAK, Genetic Engineering and Biotechnology Institute and financially supported by Turkish Textile Association.

6. REFERENCES

1. Bradford, M.M., 1976. A rapid and sensitive method for the quantitation of microgram quantities of protein utilizing the principle of protein-dye binding. *Anal. Biochem*. 72: 248-254.

2. Broekaert, B.P.A., Marien, W., Terras, F.R.G., De Bolle, M.F.C., Proost, P., Van Damme, J.,Dillen, L.,Claeys, M., Rees, S. B., Vanderleyden, J., Cammue, B.P.A., 1992. Antimicrobial peptides from *Amaranthus caudatus* seeds with sequence homology to the cysteine/glycine rich domain of chitin-binding proteins. *Biochemisrty* 31: 4308-4314.

3. Broekaert, W.F., Cammue, B.P.A., De Bolle, M.F.C., Thevissen, K., De Samblanx, G.W., Osborn, R.W., 1997. Antimicrobial peptides from plants. *Crit. Rev. Plant Sci*. 16: 297-323.

4. Cammue, B.P.A., De Bolle, M.F., Terras, F.R.G., Proost, P., Van Damme, J., Rees, S.B., Vanderleyden, J., Broekaert, W. F., 1994. Isolation and characterization of a novel class of plant antimicrobial peptides from *Mirabilis jalapa* L. seeds. *J. Biol. Chem*. 267: 2228-2233, 1992.

5. Cammue, B.P.A.,Thevissen, K., Hendriks, M.; Eggermont, K., Goderis, I.J., Proost, P., Van Damme, J., Osborn, R.W., Guerbette, F., Kader, J.C., Broekaert, W.F., 1995. A potent antimicrobial protein from anion seeds showing sequence homology to plant lipid transfer proteins. *Plant Physiol*. 109: 445-455.

6. Carvalho, A. D. O., Gomes, V. M., 2009. Plant defensins-prospects for the biological functions and biotechnological properties, *Peptides*, vol. 30(5): 1007–1020.

7. Castro, M. S. and Fontes, W., 2005 Plant defense and antimicrobial peptides, *Protein and Peptide Letters*, 12(1): 13–18.

8. Cole, M.D., 1994. Key antifungal, antibacterial and anti-insect assays. A critical review. *Biochemical Systematics and Ecology* 22: 837-856.

9. De Bolle, M.F., Osborn, R.W., Goderis, I.J., Noe, L., Acland, D., Hart, C.A., Torrekens, S., Van Leuven, F., Broekaert, W. F., 1996. Antimicrobial peptides from *Mirabilis jalapa* and *Amaranthus caudatus*: expression, processing, localization and biological activity in transgenic tobacco. *Plant Mol. Biol*. 31: 993-1008.

10. *FEBS Letters* 368: 257-262.

11. Floerack, D. E. A., Stiekema, W, J.: Thionins: properties, possible biological roles and mechanisms of action. *Plant Mol. Biol*. 26: 25- 37, 1994.

12. Gao, A.G., Hakim, S.M., Mittanck, C.A., Wu, Y., Waerner, B.M., Stank, D.M., Shah, D.M., Liang, J., Rommrns, C.M.T., 2000. Fungal pathogen protection in potato by expression of a plant defensin peptide. *Nature Biotechnol*. 18: 1307- 1310.

13. Garcia-Olmedo, F., Molina, A., Alamillo, J.M., 1998. Plant defense peptides. *Biopolymers* 47: 479-491.

14. Giudici, A.M., Regente, M.C., De La Canal, L., 2000. A potent antifungal protein from helianthus annuus flowers is a trypsin inhibitor. *Plant Physiol. Biochem*. 38: 881-888.

15. Harrison, S.J., Marcus, J.P., Kenneth C. Goulter., Green, J.L., Maclean, D.J., Manners, J.M., 1997. An antimicrobial peptide from the Australian native *Hardenbergia violacea* provides the first functionally characterised member of a subfamily of plant defensins. *Aust. J. Plant Physiol*. 24: 571-578.

16. Koo, J.C., Chun. H.J., Park, H.C., Kim, M.C., Koo, Y.D., Koo, S.C., Ok, H.M., Park, S.J., Lee, S-H., Yun, D-J., Lim, C.O., Bahk, J.D., Lee, S.Y., Cho, M.J., 2002. Over-expression of a seed specific hevein-like antimicrobial peptide from *Pharbitis nil* enhances resistance to a fungal pathogen in transgenic tobacco plants. *Plant Molecular Biology* 50: 441-452.

17. Laemmli, U.K. 1970. Cleavage of structural proteins during the assembly of the head bacteriophage T4. *Nature*, 227: 680-685.

18. Lehrer, R. I., Ganz, T., 1999.Antimicrobial peptides in mammalian and insect host defence. *Curr. Opin. Immunol*. 11: 23-27.

19. Money, N.P., 1990. Measurement of pore size in hypal cell wall of *Achyla bisexualis*. *Exp. Mycol*. 14: 234-242.

20. Nesterenko, M. V., Tilley, M., Upton, S.J., 1994. A simple modification of Blum's silver stain method allows 30 minute detection of proteins in polyacrylamide gels. *J. Biochem. Biophys. Methods* 28: 239-242.

21. Osborn, R.W., De Samblanx, G. W., Thevissen, K., Goderis, I., Torrekens, S., Van Leuven, F., Attenborough, S., Rees, S. B., Broekaert, W. F., 1995. Isolation and characterization of plant defensins from seeds of *Asteraceae, Fabaceae, Hippocastanaceae* and *Saxifragaceae*.

22. Swegle, M., Kramer, K.J., Muthukrishnan, S., 1992. Properties of barley seed chitinases and release of embryo-associated isoforms during early stages of imbibition. *Plant Physiol*. 99: 1009-1014.

23. Terras, F.R.G., Schoofs, H.M.E., De Bolle, M.F.C., Van Leuven, F., Rees, S.B., Vanderleyden, J., Cammue, B.P.A., Broekaert, W.F., 1992. Analysis of two novel classes of plant antifungal proteins from radish (*Raphanus sativus* L.) seeds. *Journal of Biological Chemistry* 267: 15301-15309.

24. Terras, F.R.G., Eggermont, K., Kovaleva, V., Raikhel, N.V., Osborn, R.W., Kester, A., Rees, S.B., Torrekens, S., Van Leuven, F., Vanderleyden, J., Cammue, B.P.A, Broekaert, W.F., 1995. Small cysteine-rich antifungal proteins from radish: their role in host defense. *Plant Cell* 7: 573-588.

25. Wang, X., Thoma, R.S., Carroll, J.A., Duffin, K.L., 2002. Temporal generation of multiple antifungal proteins in primed seeds. *Biochem. Biophys. Res. Commun.* 292 (1): 236- 242.

26. Zhang,Y., Lewis, K., 1997. Fabatins: new antibicrobial plant peptides. *FEMS Microbiology* 149: 59-64.

III. Beneficial Effects on Health and Toxicity Studies of Herbal Drugs

Chapter 13

Sedative Activity of Essential Oil from *Heracleum afghanicum* KITAMURA Seeds

Abdul Ghani Karimi and Michiho ITO

Department of Pharmacognosy,
Faculty of Pharmacy, Kabul University, Afghanistan
E-mail: karimiabg@gmail.com

ABSTRACT

Heracleum afghanicum (Apiaceae) is a perennial plant indigenous to Afghanistan. Phytochemical and pharmacological analyses of *H. afghanicum* seeds essential oil were carried out to investigate its possible sedative effects on mice spontaneous locomotor activity. The essential oil was analyzed by GC and GC/MS, and thirty three constituents were identified. Hexyl butyrate (34.3 per cent) and octyl acetate (21.1 per cent) were found as its principal constituents. The sedative effect of *H. afghanicum* essential oil was confirmed using an open field test with ddY mice. The essential oil significantly decreased the locomotor activity of mice suggesting its sedative effect. Hexyl butyrate and octyl acetate were found to be responsible for the sedative activity of *H. afghanicum* seeds essential oil.

Keywords: *Heracleum afghanicum, Apiaceae, Essential oil composition, Sedative effect, Aliphatic esters, Hexyl butyrate, Octyl acetate.*

1. INTRODUCTION

Essential oils (EOs) are used for their therapeutic action (*e.g.* eucalyptous oil), for flavoring (*e.g.* lemon oil), in perfumery (*e.g.* rose oil) or as starting materials for the synthesis of other compounds (*e.g.* turpentine oil). The medicinal use of EOs began in ancient Egypt and has been continued until present. EOs are believed to be important for their effectiveness in treating various illnesses but the lack of scientific basis for their use remains an obstacle (Umezu T., 2006). Aromatherapy which uses essential oils for various treatments has attracted much attention as an alternative medicine, and the inhalation of EOs is one of these treatments.

The genus *Heracleum* includes ca.70 species distributed all around the world (Evans W.C., 2002). Many kinds of secondary metabolites such as coumarins, anthraquinones, and flavonoids have been isolated from several species of this genus and their structures were identified (Sayyah M., 2005). The EO compositions of these species have been studied and many monoterpenes, oxygenated monoterpenes, sesquiterpenes and different aliphatic esters were reported (Mojab F., 2003). *H. afghanicum* (known as Balderghan in Persian) is a perennial plant endemic to Afghanistan. Its seeds are used as spice, leaves are for pain killer and anti-fever, and the young stems are edible. The anticonvulsant, analgesic, anti-inflammatory and antimicrobial effects of other members of *Heracleum* such as *H. persicum* have been reported in recent studies (Tosun F., 2008). However, scientific studies on the phytochemical and pharmacological aspects of *H. afghanicum* are not established. The objective of this study was to identify the chemical composition of the EO of this plant, to investigate its possible effects on locomotor activity in mice, as well as to analyze the active components of the EO.

2. MATERIALS AND METHODS

Materials

Seeds of *H. afghanicum* were collected at Syakhark in Darrah-i-Fringel (Ghorband district, Parwan province, Afghanistan) in July, 2008 and air dried. The plant was identified by Mr. Kh. A. Yarmal, a professor of botany, Faculty of Science Kabul University and three voucher specimens (No.s 4968, 4969 and 4970) were deposited in the herbarium of Experimental Station for Medicinal Plants, Graduate School of Pharmaceutical Sciences, Kyoto University. Lavender oil (Nacalai Tesque Co., Ltd.) and benzylacetone (Tokyo Kasei) were used as positive controls. Triethyl citrate (Merck), an odorless solvent, was used for dissolving the fragrant components. Hexyl butyrate, octyl acetate, octyl butyrate, and butyl butyrate were purchased from Wako. All chemicals used in this study were of the highest grade available.

Isolation of EO and Fractionation

The EO of *H. afghanicum* was prepared by distillation of seeds for 2 hours using an apparatus designated in Japanese Pharmacopeia (JP XV) and captured in hexane. Obtained pale yellow EO was dried over anhydrous sodium sulfate and stored in sealed vials at 4°C before analysis. The EO was subjected to silica gel column chromatography (Wakogel C-200) to be fractionated. The column was eluted with hexane : acetone (3:1) to give fractions 1–2, and then washed with absolute acetone to give Fr.3.

Qualitative and Quantitative Analyses of Essential Oil

The qualitative analysis of essential oil was carried out on a Hewlett Packard 5890 series gas chromatograph connected to AUTOMASS 50 (JEOL) with operation conditions as follows; column: fused silica capillary column, TC–wax (HP), 60 m x 0.25 mm x 0.25 μm; column temperature program: 40 –130°C increasing at a rate of 2°C/min, holding at 130°C for 25 min, 130° – 140°C at 2°C/min, holding at 140°C for 15 min, 140° – 200°C at 15°C/min, ending at 200°C for 30 min; injector temperature: 180°C; carrier gas: helium, 25 cm/sec; ionization energy: 70 eV; injection volume: 1.0 μL.

Quantitative analysis was carried out on Hitachi G-5000 equipped with flame ionization detector with the following condition; column: fused silica capillary column, TC- wax (HP), 60 m x 0.25 mm x 0.25 μm; column temperature: same as GC/MS; injector: 180°C, detector: 200°C; carrier gas: helium, 0.8 ml/min; split: 33:1; injection volume: 1 μl. Compounds were identified by comparison of their mass spectra with those in the library (NIST 2) and of authentic samples.

Animals

Male 4-week old ddY mice were purchased from Japan SLC (Shizuoka, Japan). The animals were housed in colony cages with a 12 h light/dark cycle at 25±2°C and relative humidity of 50±10 per cent with free access to food and water before being used for experiments. All experiments were conducted between 10:00 AM and 05:00 PM under the same conditions. Animal experiments were designed following recommendations by the Animal Research Committee of Kyoto University, Kyoto, Japan (approval number 2010-23). Experimental procedures involving animals and their care were conducted in conformity with the institutional guidelines in compliance with the Fundamental Guidelines for Proper Conduct of Animal Experiment and Related Activities in Academic Research Institutions under the jurisdiction of the Ministry of Education, Culture, Sports, Science and Technology, Japan (2006).

Evaluation of Spontaneous Motor Activity

The sedative effects of *H. afghanicum* EO was evaluated on mice spontaneous motor activity in an open-field test. Doses administered were expressed as milligrams of the EO in 400 μl of triethyl citrate (TEC) per cage. Four pieces of filter-paper soaked with EO dissolved in triethyl citrate were placed in the four corners of the inner walls of the glass cage (W 60 cm × L 30 cm × H 34 cm) using adhesive tape so that the vapor pervaded the cage by natural diffusion. Sixty minutes after charging the sample, a mouse was placed in the center of the cage and was monitored by a video camera for 60 min. The frequency of each mouse crossing the lines drawn on the floor of the cage (at10 cm intervals) was counted every five min for 60 min. Area under the curve (AUC) which represented total locomotor activity was calculated by trapezoidal rule.

Statistics of Data

Data were expressed as mean value±standard error of mean (SEM) of six mice. Statistical analyses were done by one way ANOVA followed by Dunnett's multiple comparison test using GraphPad Instat (GraphPad Software, San Diego, CA, USA). A probability level of $P < 0.05$ was interpreted to be statistically significant.

3. RESULTS AND DISCUSSION

Analyses of Essential Oil

The seeds of *Heracleum afghanicum* afforded 1.5 per cent (v/w) EO with a pale yellow color and sharp characteristic odor. Table 13.1 shows 33 constituents identified in the EO which were listed in order of their elution from the TC-wax column. It was found that *H. afghanicum* EO consists mainly of aliphatic esters. Hexyl butyrate (34.3 per cent) and octyl acetate (21.1 per cent) were found to be principal constituents of the EO. These 2 compounds were also reported as principal components of *H. persicum* EO (Sheffer, 1984).

Table 13.1: Chemical Composition of the Essential Oil from *Heracleum afghanicum* Seeds

Compounds	[1]RI	[2]PA (per cent)	Compounds	RI	PA (per cent)
Isopropyl isobutyrate	980	2.1	(*E*)-5-Decen-1-ol acetate	1512	3.4
Isopropyl 2-methylbutanoate	1053	2.5	Cis-4-Decenal	1536	T
Isopropyl 3-methylbutanoate	1077	1.9	Octyl isobutyrate	1545	1.5
Butyl butyrate	1221	2.5	Linalool	1558	1.0
Isopropyl 3-methyl-2-butanoate	1235	0.7	1-Octanol	1567	3.6
Terpinene	1238	0.7	1,4-Octadaiene	1598	T
Isopentyl butyrate	1264	t	Cis-5-octen-1-ol	1611	0.6
Hexyl acetate	1276	2.5	Hexyl caproate	1613	1.5
Octanal	1288	t	Octyl butyrate	1620	6.9
Unknown	1299	0.7	Octyl 2-methylbutyrate	1624	T
Hexyl isobutyrate	1344	0.7	Octyl isovalerate	1630	1.7
n-Hexanol	1362	2.3	1-octen-3-ol, butyrate	1653	1.0
Hexyl butyrate	1419	34.3	Tridecen-1-ol, acetate	1699	T
Hexyl 2-methylbutyrate	1428	1.1	Octyl hexanoate	1813	0.6
Hexyl isovalerate	1444	t	Anethol	1839	2.3
Octyl acetate	1479	21.1	Unknown	1876	1.5
Botanoic acid, 4 hexenyl ester	1489	t	Octanoic acid	1962	0.7
Decanal	1505	0.8			

Chemical composition of *H. afghanicum* EO.

[1]Retention indices. [2]Peak area. Compounds listed in order of their elution from TC-wax column.

Effect of Vapor Inhalation of *H. afghanicum* EO on Mice Locomotor Activity

The result of administration of whole essential oil is shown in Figure 13.1. The essential oil from *H. afghanicum* seeds was administered to mice by inhalation at doses ranging from 0.004 – 4 mg dissolved in 400 µl of triethyl citrate. Sedative activity was observed in a dose dependent manner and the strongest effect was observed at the dose of 0.4 mg which calmed mice after 10 min. The AUC values of treated groups were significantly smaller than control and decrease in locomotor activity produced by doses of 0.4 and 4 mg were statistically significant (Figure 13.1, $P < 0.01, P < 0.05$). This suggested the potential sedative activity of *H. afghanicum* EO.

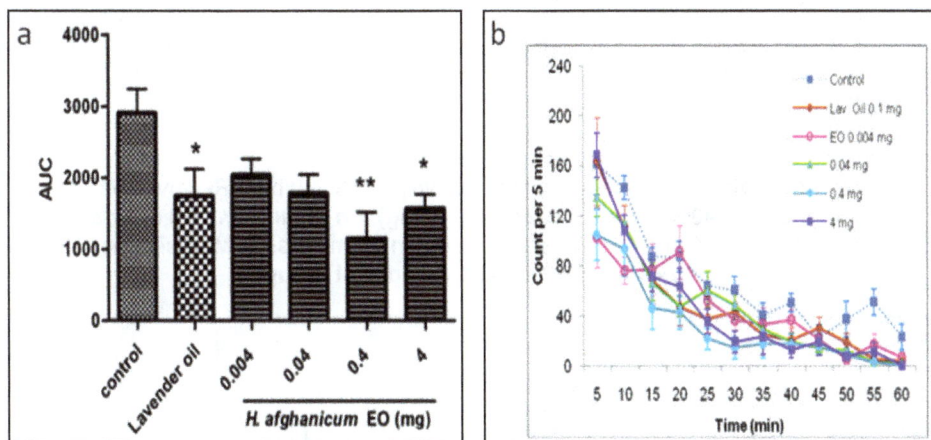

Figure 13.1: Total Spontaneous Motor Activity (a) and Locomotor Activity Transition (b) of Mice which Received Vehicle (Triethyl citrate 400μl), Lavender Oil (0.1 mg), and *H. afghanicum* EO (0.004, 0.04, 0.4 and 4 mg). Data are shown as means±SEM of six mice (*p<0.05 and **p<0.01 vs. control group).

Effect of Vapour Inhalation of Fractions on Mice Locomotor Activity

The EO fractions were administered to mice by inhalation at doses of 0.004, 0.04 and 0.4 mg. Among these fractions, Fr.1 was effective which significantly decreased the locomotor activity of mice (Figure 13.2a) but fractions 2 (Figure 13.2b) and 3 (Figure 13.2c) were not effective. This suggested that Fr.1 contained active ingredients and it was then analyzed for its chemical composition using GC/MS. The results revealed that hexyl butyrate and octyl acetate were the main compounds of fraction 1 along with other minor constituents such as octyl butyrate and (*E*)-5-decen-1-ol acetate.

Effect of Hexyl Butyrate, Octyl Acetate and Octyl Butyrate on Locomotor Activity of Mice

The main compounds of Fr.1, hexyl butyrate, octyl acetate and octyl butyrate were assayed for their sedative activity. Hexyl butyrate administered to mice at doses of 0.04, 0.4 and 4 mg. The mice sedated 10 min after administration of 0.4 and 4 mg of the compound. The AUC values of the 0.04, 0.4 and 4 mg treated groups were 66, 42.7 and 53 per cent of the control, respectively (Figure 13.3a). Octyl acetate also significantly decreased the locomotor activity of mice and the strongest effect was observed at dose of 4 mg. The AUC values of treated groups with this compound were 75.4, 63.2, 41.5 and 80 per cent of the control (Figure 13.3b). However, octyl

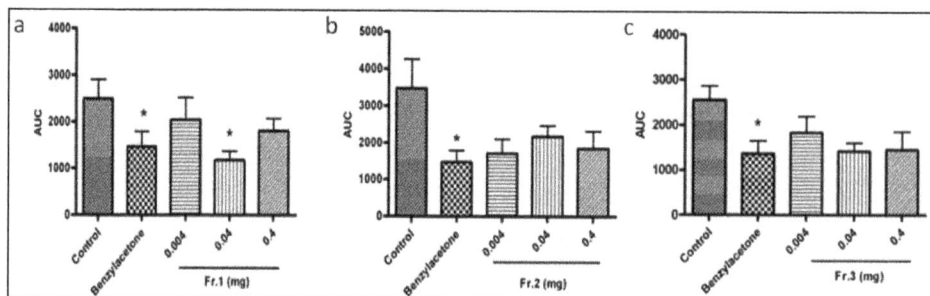

Figure 13.2: Total Spontaneous Motor Activity of Mice which Received Vehicle (Triethyl citrate 400µl), Benzylacetone (0.4 mg), Frl (a) and Fr.2 (b) or Fr.3 (c) (0.004, 0.04, 0.4 and 4 mg). Data are shown as means±SEM of six mice (*$p<0.05$ and **$p<0.01$ vs. control group).

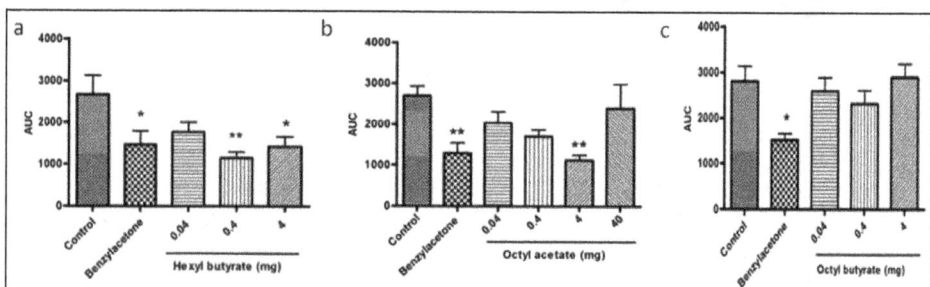

Figure 13.3: Total Spontaneous Motor Activity of Mice which Received Vehicle (Triethyl citrate 400µl), Benzylacetone (0.4 mg), Hexyl Butyrate (a), Octyl acetate (b) or Octyl butyrate (c) (0.04, 0.4 and 4 mg). Data are shown as means±SEM of six mice (*$p<0.05$ and **$p<0.01$ vs. control group).

butyrate did not alter the locomotor activity of mice (Figure 13.3c). This confirmed that hexyl butyrate and octyl acetate were responsible for the sedative activity of Fr.1.

Comparison of the Effects of Butyl Butyrate and Octyl Butyrate with Hexyl Butyrate

It was observed that hexyl butyrate decreased the locomotor activity of mice whereas octyl butyrate which differed from hexyl butyrate only in carbon chain length did not alter the locomotor activity of mice. The structure-activity relationship of the aliphatic esters present in *H. afghanicum* EO was thus investigated. Figure 13.4 shows a comparison of the effects of butyl butyrate, hexyl butyrate, and octyl butyrate on locomotor activities of mice. Hexyl butyrate (C=10) and butyl butyrate (C=8) showed sedative activity while octyl butyrate (C=12) was not effective. The sedative effect of hexyl butyrate was stronger than that of butyl butyrate. These three compounds structurally differ only in carbon chain length. It is thought that the length of carbon chain may affect sedative activity of these aliphatic esters. However, the discrepancy between the sedative activity of hexyl butyrate and butyl butyrate and their carbon chain length suggests that carbon chain length alone may not account for the sedative activities of these compounds. It may be due to

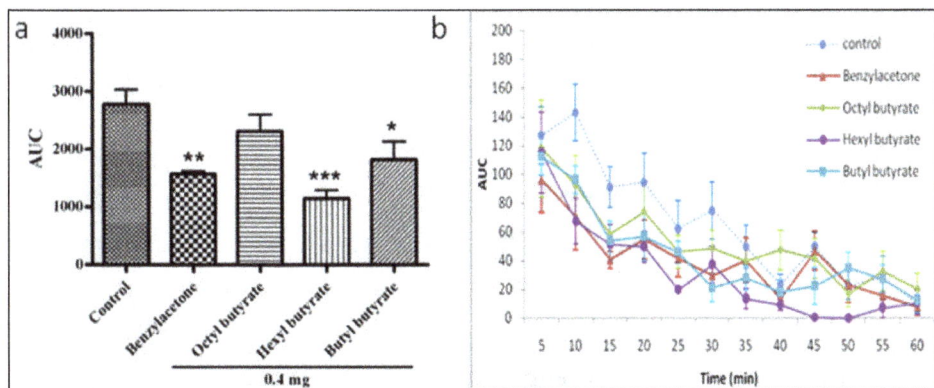

Figure 13.4: Total Spontaneous Motor Activity (a) and Locmotor Activity Transition (b) of Mice which Received Vehicle (Triethyl citrate 400µl), Benzylacetone, Octyl Butyrate, Hexyl Butyrate and Butyl Butyrate (0.4 mg). Data are shown as means±SEM of five mice (*$p<0.05$, **$p<0.01$ and ***$p<0.001$ vs. control group).

differences in their physical properties such as vapor pressure and lipophilicity, in addition to carbon chain length. It has been reported that each increase in chain length of two carbons resulted in a decrease in vapour pressure at 20°C by a factor of ~ 4 (Jocelyn, 1997). Also, Hexyl butyrate (log P = 3.28) is more lipophilic than butyl butyrate (log P = 2.39) and may easily penetrate the cell membrane. It is possible that the sedative effects of the aliphatic esters of *H. afghanicum* may be influenced by carbon chain length, vapor pressure and lipophilicity.

The effective doses of hexyl butyrate and octyl acetate in this study were in the range of 0.04 – 4 mg. According to literature, the oral toxicity (LD50) of these compounds are 5000 and 3000 mg/kg, respectively (Golberg L., 1974; Weatherston I., 2002). This means that our applied doses are much smaller than the toxic doses and the observed effects may not be from toxic doses.

The number of patients with psychiatric disease such as insomnia, anxiety, depression and Alzheimer's diseases are increasing. Current therapeutic goal in the treatment of psychiatric diseases are to improve quality of life, normalizing mood, increasing awareness of personal pleasures and interests, and to reverse the functional and social disabilities. Currently available synthetic drugs used for this purposes sometimes have sever adverse effects and even adduction or dependency. Therefore there is a need for more effective and safe drugs. Thus EOs exhibiting anti anxiety and sedative effects are very important. In the present study *H. afghanicum* EO exhibited significant sedative activity by vapour inhalation system which is rather safe and acceptable for patients especially children and elderly people.

4. CONCLUSIONS

This study revealed that *H. afghanicum* EO consists mainly of aliphatic esters with hexyl butyrate and octyl acetate as principal compounds. The EO significantly decreased the locomotor activity of mice at 0.4 mg suggesting the potential sedative activity of this oil. Hexyl butyrate is shown to be the main active compound of *H. afghanicum* EO. However further studies are required in order to elucidate the exact mechanism of action as well as to investigate the structure-activity relationship. In this paper, the phytochemical and pharmacological activity of the EO from *H. afghanicum* seeds is being reported for the first time.

5. ACKNOWLEDGEMENTS

The authors would like to express their deepest thanks to the Takeda Foundation for financial support.

6. REFERENCES

1. Anil J. (2007). Chemical Composition of Leaf and Fruit Oils of *Heracleum candolleanum*. *J. Essent. Oil Res.* 19, pp. 358-359.

2. Evans W. C. (2002). *Trease and Evans Pharmacognosy* (15th ed.), W. B. Saunders Company, London, Chepter 5 and 23, pp. 31, 253.

3. Golberg, L. and Seeley A. M. (1974). Fragrance raw materials monographs: Acetate C-8, *Food and Cosmetics Toxicology*, 12, pp. 815-816.

4. Hajhashemi V. (2009). Anti-inflammatory and analgesic properties of *Heracleum persicum* Essential oil and hydroalcoholic extract in animal models. *Journal of Ethnopharmacology*, 124, pp. 475–480.

5. Jocelyn G. (1997). Sex pheromone of the Mirid bug. *Journal of Chemical Ecology*, 23, pp. 1743-1754.

6. Kuljanabhagavad T. (2010). Chemical Composition and Antimicrobial Activity of The Essential Oil from *Heracleum siamicum*. *J Health Res* 24, pp. 55-60.

7. Mojab, F. (2003). Composition of the essential oil of the root of Heracleum persicum from Iran. *I.J. Pharm. Res.*, pp. 245-247.

8. Papageorgiou V.P. (2000). Composition of the Essential Oil from *Heracleum dissectum*, NATEC Institute, Hamburg 50, West Germany, pp. 851-853.

9. Sayyah, M. (2005). Anti convulsant activity of *Heracleum persicum* seeds. *Journal of Ethnopharmacology*, pp. 209-211.

10. Scheffer J. C. (1984). Composition of the essential oil of *Heracleum persicum* fruits. *Planta Medica*, **50**, pp. 56–60.

11. Sefidkon F. (2004). Analysis of the oil of *Heracleum persicum* L. (leaves and flowers). *Journal of Essential Oil Research* 16, pp. 295–297.

12. Tosun, F. (2008). Anticonvulsant activity of furanocoumarins and the essential oil obtained from the fruit of *Heracleum crenatifolium*. *Food and Chemistry*, pp. 990-993.

13. Umezu T. (2006). Anticonflict effects of lavender oil and identification of its active constituents. *Pharmacology, Biochemistry and Behavior*, pp. 713-721.

14. Weatherston I. (2002). Regulatory issues in the commercial development of phermones and other semiochemicals. *IOBC wprs Bulletin*, pp. 1-10.

Chapter 14

Mechanism of Action of Active Constituents from *Zingiber officinale* Roscoe var. *rubrum* (Halia Bara) on Psoriasis

Nurul Izza Nordin[1], Simon Gibbons[2], David Perrett[3] and Rizgar A. Mageed[4]

[1]Industrial Biotechnology Research Centre, SIRIM Berhad, 40700 Shah Alam, Malaysia
E-mail: izza@sirim.my
[2]Department of Pharmaceutical and Biological Chemistry, University College London, United Kingdom
[3]Translational Medicine and Therapeutics,
[4]Bone and Joint Research Unit, Queen Mary University of London, United Kingdom

ABSTRACT

Psoriasis is an autoimmune skin disease whose manifestation involves multifaceted interaction between genetic and environmental factors (1–5). The disease is associated with aberrant activation of phagocytes (such as macrophages), lymphocytes and the production of pro-inflammatory cytokines and chemokines. To-date, multidimensional treatment for psoriatic patient is required due to the paradoxical immunopathology of psoriasis (6-8). Our studies reveal the therapeutic efficacy and mechanism of action of the ginger species *Zingiber officinale* Roscoe var. *rubrum* (ZOR), on key immunopathogenic mechanism in psoriasis.

It is evidenced in our studies that two identified gingerol-related constituents isolated from an active fraction of ZOR (F6) have been proven to possess potent effects against relevant inflammatory responses in psoriasis. Those are phenylpropanoids and characterized as 6-shogaol and 1-dehydro-6-gingerdione. The identified active constituents strongly modulated activated immune responses by macrophages, polymorphonuclear leukocytes (PMN), keratinocytes and T-lymphocytes that are key mediators in psoriasis pathogenesis (9). In-depth experiments showed that ZOR chloroform extract (HBO_2), its active fraction (F6) and the two identified constituents effectively inhibited nitric oxide (NO) and prostaglandin E_2 (PGE_2) production by activated macrophages, comparable to dexamethasone and indomethacin. Interestingly, the samples strongly down-regulated mRNA level of iNOS, IL-12p40 and IL-23p19 in pre-treatment experiments of activated macrophages. Further, studies of immune cell migration showed that F6 and the constituents inhibited the migration of polymorphonuclear neutrophils (PMNs) through human vascular endothelial cells (HUVEC). An *in vitro* 2D model of epidermal inflammation indicated that ZOR samples directly inhibited keratinocyte proliferation and the production of key psoriasis-promoting cytokines, IL-20 and IL-8. Moreover, the structural orientation of both constituents satisfies the criteria of a 'druggable' compound according to Lipinski Rule of Five (10). Conceivably, ZOR constituents showed convincing regulatory effects on inflammatory responses relevant to psoriasis, thus, demonstrate potential in the treatment of psoriasis.

Keywords: *Psoriasis, Zingiber officinale Roscoe var. rubrum (Halia Bara), 6-shogaol, 1-dehydro 6-gingerdione.*

1. INTRODUCTION

Psoriasis is an autoimmune disease mediated by aberrant T-lymphocyte and B-lymphocyte responses. The aetiology of psoriasis is unknown but there is good evidence that susceptibility to the disease involves the interaction between genetic and environmental factors (1-4). Clinically, the disease is characterised by the appearance of red scaling plaques that range from a few lesions to total coverage of the skin. The disease is non-contagious and while symptoms can be managed by medical intervention there is, at present, no cure.

Interactive cellular responses in psoriatic lesions appear to be due to an imbalance between the activation of both innate and acquired immune systems and factors produced by keratinocytes that directly influence T-cells and DCs (5). PMNs, mDCs and CD11c+ mDCs are effector innate immune cells that are stimulated by chemokines produced by keratinocytes in the epidermis. Consequently, immune cell-derived cytokines produced through this interaction activate the transcription of genes encoding pro-inflammatory proteins and those that promote keratinocyte proliferation.

Evidence implicating Th1- and Th17-induced inflammatory pathways in the pathogenesis of psoriasis is demonstrated by a linear relationship between proximal inducer cytokines, such as IL-23, IL-12, IFN-γ, TNFα and activation of many IFN-responsive genes through signal transducer and activator of transcription (STAT) -1 and/or -3 (5). Induction of the cascades leads ultimately to the production of keratinocyte-derived cytokines such as ECGF, VEGF and PDGF as well as IL-1α, IL-7, IL-19, IL-20 and IL-22 which play parts in keratinocytes proliferation.

Th1 cells are activated by APCs that produce IL-12. In contrast, differentiation of naive T-cells to Th17 cells is promoted by the presence of IL-23. The infiltration of Th17 cell leads to the production of IL-17 and IL-22. IL-22 is a critical cytokine in the establishment of an inflammatory environment that promotes autoreactive T-cell activation. Production of IL-17 is also facilitated by IL-22. Most recently, the pathogenesis of psoriasis was associated with another T-cell subset, Th22 cells. The notion that expansion of Th1 and Th17 cells are indirectly influenced by IL-22 suggested that IL-22 could play a specific role in psoriasis. The link between Th17 and Th22 is that both are activated by IL-23, denoting IL-23 as the possible key cytokine in psoriasis pathogenesis.

In epidermal keratinocytes, IL-22 induces IL-20, an upstream inflammatory product in the epidermis of psoriatic skins, which specifically cause phenotypic skin changes including hyperproliferation and aberrant differentiation. Although IL-22 has been shown to induce hyperproliferation and altered differentiation in epidermal keratinocytes *in vitro* that resemble acanthosis, its role in the manifestation of psoriatic lesions might be magnified by presence of IL-20 (6). Nevertheless, detailed studies on the role of the Th22 pathway in psoriasis might be an interesting area to be explored in the future.

Plant-derived active compositions and extracts have been shown to possess multiple functions in disease therapy. Although, natural products and plant-derived constituents have not been the subject of recent drug discoveries, their potential should not be ignored since important drugs such as Ciclosporin and dithranol were derived from natural product/resources. Numerous studies have shown that ginger has a broad spectrum of anti-inflammatory activity involving multiple mechanisms of action. Several studies have demonstrated that ginger extracts possess similar effects to NSAID in inhibiting PG synthesis by inhibiting COX activity (7). Interestingly, some ginger constituents were found to have dual inhibitory effects against COX and LOX (lipooxygenase), thus, reducing the occurrence of gastrointestinal and renal side effects while expressing their anti-inflammatory effects (7). Various studies have been carried out to understand the pathogenic and therapeutic roles of NO and PGE_2. iNOS and COX-2 are induced and regulated through a number of mechanisms and by a number of mediators. A number of studies indicate that NO could activate and regulate COX enzyme activity and, thus, regulate PGE_2 production. Crosstalk between these two pro-inflammatory enzymes has attracted much interest but data generated so far are conflicting. In addition, ginger extracts and its components have been shown to suppress expression of genes related to inflammation including genes encoding TNFα, IL-1β and the transcription factor, NF-κB (7, 8). Intriguingly, this highlights the potency of ginger as an anti-inflammatory agent with minimum side effects. This study aims to identify the key mechanism of action of a ginger species, Halia Bara or *Z.officinale* Roscoe var. *rubrum* (ZOR) against inflammatory responses relevant to psoriasis.

2. MATERIALS AND METHODS

Samples

Rhizomes of *Z.officinale* Roscoe var. *rubrum* were obtained from Pahang, Malaysia. Two active compounds, 6-shogaol (6S) and 1-dehydro-6-gingerdione (GD) were isolated from the active fraction (F6) of a crude chloroform extract from *Z.officinale* Roscoe var. *rubrum* (HB02), and tested for anti-inflammatory effects relevant to psoriasis pathogenesis. The samples were dissolved in dimethyl sulfoxide (DMSO) prior to experiments. Positive controls used were (L-NAME) (SIGMA Aldrich), dexamethasone (SIGMA Aldrich), indomethacin (Fluka) and cyclosporine (Neoral®).

Cell Culture

Murine macrophages *Abelson;* (RAW 264.7) was obtained from cell ATCC. Polymorphoneutrophil (PMN) isolated from blood from human volunteer, human vascular endothelial cell (HUVEC) monolayer and primary normal human keratinocytes (NHEK) from patient (courtesy of Biochemical Pharmacology Department and Cutaneous Centre, Queen Mary University of London, respectively) were used for the experiments.

Measurement of Nitrite/Nitrate and PGE_2 Production

RAW 264.7 cell was treated with ZOR's extract (HB02), fraction 6 (F6) and active compounds, prior to and after stimulation of the cells with LPS (SIGMA Aldrich). The NO content of the supernatants was measured using the Griess reagent, which measures nitrite. 100 µl of culture supernatants were dispensed into 96 well plates followed by the addition of 100 µl of Griess reagent (1 per cent sulphanilamide in 2.5 per cent phosphoric acid and 0.1 per cent naphthylenediamine in dH_2O). Absorbance at 560 nm was measured within 10 minutes using a microplate reader (TECAN GENios). Nitrite contents were calculated by extrapolation from a standard curve constructed using known concentrations of sodium nitrite, 0-100 µM. PGE2 was measured using a PGE2 ELISA kit (R and D Systems) and an HTRF PGE2 kit (CISBIO).

RNA Isolation and Quantitative Polymerase Chain Reaction (qPCR)

RAW 264.7 cells were harvested and lysed using lysis/binding solution of RNAqueous®-4PCR kit (AMBION). 1 µg of mRNA was obtained from $1x10^5$ cells. The yield of mRNA (ng/µl) was determined using a NanoDrop 1000 spectrophotometer, at 260 nm. 20 ng of total RNA extracted from cell lysates were reverse transcribed to single-stranded cDNA in a 20µl reaction mixture using High-Capacity cDNA Reverse Transcription Kit (Promega). The generated cDNA was then used for realtime-PCR for assessing the level of pro-inflammatory cytokines transcripts using glyceraldehydes 3-phosphate dehydrogenase (GAPDH) gene as a control. The relative expression or level of mRNA for iNOS (Mm01309898_m1), IL-12p40 (Mm00434174_m1*) and IL-23p19 (Mm00518984_m1*) in the stimulated RAW 264.7 cells was determined using TAQMAN gene expression assay. GAPDH (Mm99999915_g1) was used as the housekeeping gene.

Western Blot

Cells were washed with PBS and lysed using RIPA buffer (SIGMA Aldrich) and 10 μl/mL protease inhibitor cocktail (SIGMA Aldrich). Cells were left on ice for 10-15 minutes and then centrifuged (1700 g) at 4°C for ~10 minutes to eliminate any protein residues. Equal amount of proteins (30 μg) were mixed with reducing loading buffer (LB) in a volume ratio of 1:3 (LB: sample). Sample LB mixtures were loaded in each well of the NuPAGE gel (Invitrogen) alongside with 10-250kDal Precision Plus protein standard. The gel was run for 2 hours at 150V in 1x NuPAGE running buffer (Invitrogen). The blotting was carried out for 2 hours at 30V. Once proteins were transferred onto the nitrocellulose membrane. The membrane was blocked with 5 per cent fat-free milk in PBS for 1 hour at room temperature (RT) and then incubated with the primary antibody, mouse iNOS specific rabbit antibody (Cell Signaling) in 5 per cent milk (ratio of 1:1000) to detect the level of total iNOS protein. The membrane was incubated overnight at 4pC. The blot was washed with PBS/T (0.05 per cent Tween-20 in 1x PBS) and incubated with the secondary antibody, sheep anti-rabbit IgG polyclonal antibody conjugated with horseradish peroxidase (AbD Serotec). The membrane was washed PBS/T, soaked in enhanced chemiluminescence (ECL) solution for 1 minute and exposed to autoradiography film using SRX-101A (Konica Minolta). These steps were repeated for the detection of Actin using rabbit anti-Actin antibody (SIGMA Aldrich).

PMN Migration

Peripheral blood from three healthy individuals was collected into 50 mL falcon tubes containing sodium citrate at a 1:10 ratio (citrate: blood) and then diluted 1:1 in RPMI1640 medium. Prior to the PMN isolation process, a mixture of 3 mL 11191 histopaque (Sigma) and 3 mL of 10771 histopaque (SIGMA) was dispensed and prepared in 15 mL falcon tube. The pellet was re-suspended in PBS without calcium (containing 0.1 per cent BSA) to a final volume of 2 mL. The number of cells was counted using Turk's solution.

The migration assay was carried out using a flow chamber assay. Prior to the experiment, a confluent human vascular endothelial cell (HUVEC) monolayer (maximum of passage 4) was stimulated with 10 ng/mL TNFα for 4 hours. PMNs at 1×10^6 cell/mL were incubated with $HB0_2$, F6, 6S and GD at 1:6 dilution in PBS (supplemented with calcium, magnesium and 0.1 per cent BSA) at 37 °C for 10 minutes in a water bath. PMNs were incubated with HBO_2, F6, 6S and GD at a final concentration of 3.3 μg/mL. The treated PMNs were then perfused over the TNFα-stimulated HUVEC monolayers in the flow chamber at a constant rate at 1dyne/cm^2 using a syringe pump.

After 8 minute of perfusion, six random fields were captured for 10s each at 40x magnification using JVC TK-C1360B digital colour video camera. The number of PMNs captured, rolling and migrating was counted and analysed using ImagePro Plus software (Media Cybemetic, Wokingham, Berkshire). The number of PMNs was determined as an initial cell capture and then categorised as rolling, adhering and migrating. Among the captured PMNs, those, which have firmly adhered remained stationary for the 10s of observation while others rolled or migrated through the endothelial junction.

In vitro Model of Inflammatory Epidermis

NHEK cells were obtained from 3 healthy individuals (A-page, SFK1 and SFK2) and each at their third passages. The cells were cultured onto irradiated 3T3 cells in cell culture medium (DMEM F12) containing RM++ supplement, 2 mM L-glutamine, 10 per cent FBS and 100 U/mL penicillin and 100 µg/mL streptomycin.

NHEK cells were cultured to 80 per cent of confluence in DMEM F12 medium supplemented with RM++ growth factor cocktail. Complete Epilife medium with HKGS that contains specific and necessary growth factors for undefined NHEK cell was used. 1×10^5 cells/well NHEK cells were seeded into 24 well plates in Epilife medium supplemented with Human Keratinocyte Growth Supplement (HKGS) and cultured for 24 hours. The cells were then starved for another 24 hours in Epilife medium without HKGS. The purpose of cell starvation is to retard the progression of cell proliferation and production of growth factors by the cells. After the 24-hour-starvation, the cells were stimulated with, or without, cytokine cocktail (CT) containing recombinant IL-17A, OSM, TNFα, IL-22, and IL-1α at 10 ng/mL each and incubated for 6 hours. It is anticipated that after 6 hours incubation, gene transcription necessary for secretion of the related cytokines has taken place (9). After the 6-hour stimulation period, the cells were treated either with F6, 6S, GD at 20 µg/ml or positive control, Ciclosporin at 20 ng/mL. The treated cells were incubated for another 48 hours. The collected supernatants were tested for the level of IL-8 and IL-20 using human IL-8 Tissue Culture kit (Mesoscale) and IL-20 ELISA kit (R and D Systems) respectively. The protocol was carried out according to the manufacturer's (MSD®) instructions. IL-8and IL-20 concentrations were determined with Softmax Pro Version 4.6 software, using a curve fit model (log-log or four-parameter log-logit) as suggested by the manufacturer.

Statistical Analysis

Data was analysed using GraphPad Prism software (GraphPad Prism, San Diego, California). Means and standard deviation were used to describe normally distributed data. The significance of differences between groups was assessed using One-way ANOVA (Tukey's or Dunn's multiple comparison test) or Two-way ANOVA (Bonferroni posttest). Comparison between two selected groups was performed using Student t-test. Difference of $p<0.05$ is considered significant.

3. RESULTS and DISCUSSION

ZOR's Active Compounds Display Dual Inhibitory Effects on NO and PGE$_2$ Production

To address the possible link between the iNOS and COX pathways, the effect of ZOR's extract (HB0$_2$), fraction 6 (F6) and the compound 6S and GD on NO and PGE2 production were studied in comparison to known NSAID; L-NAME, indomethacin and dexamathasone.

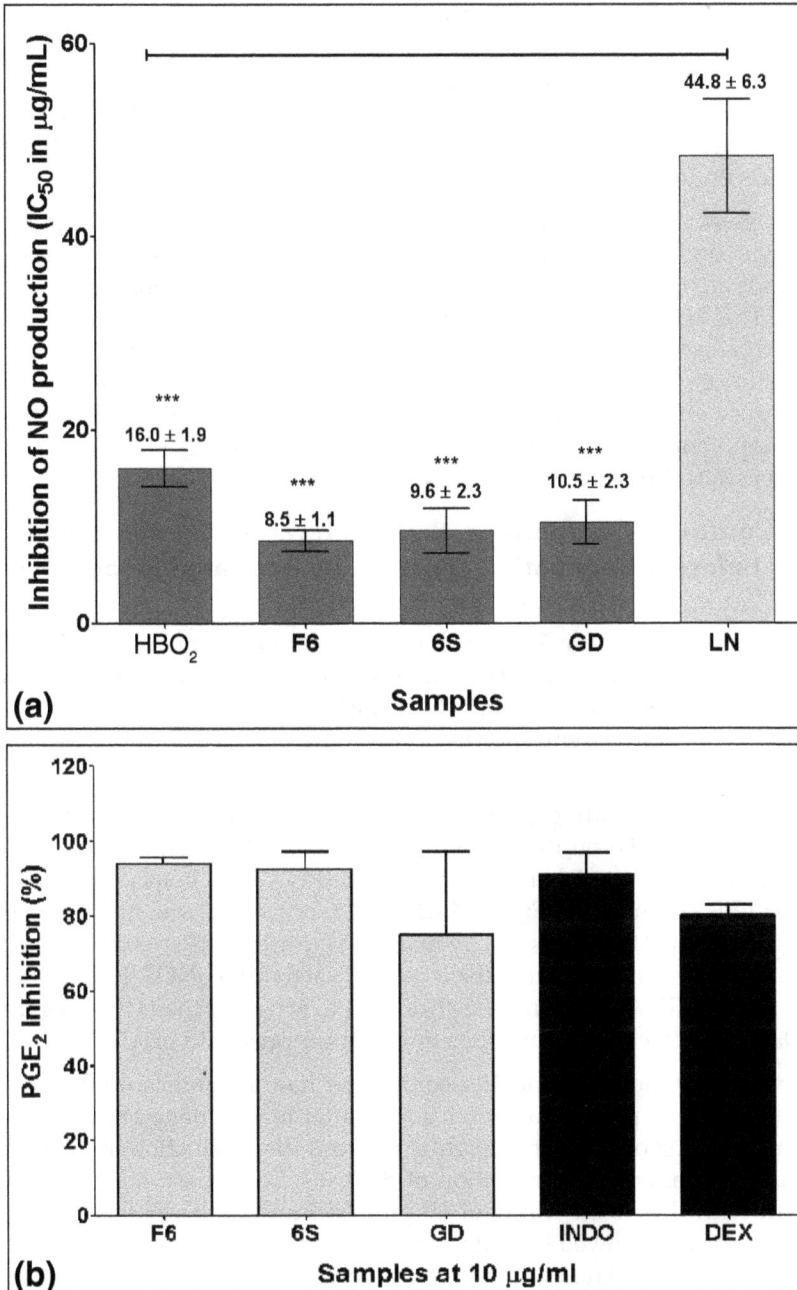

Figure 14.1: The Inhibitory Effects of the Crude Extract (HBO$_2$), Active Fraction (F6) and Two Compound; 6-shogaol(6S) and 1-dehydro-6-gingerdione (GD) from *Z.officinales* Roscoe var *rubrum* against Nitric Oxide (NO) and Prostaglandin E$_2$ (PGE$_2$) Production. The effects of samples were compared to NO inhibitor L-NAME (LN) and PGE$_2$ inhibitors, Indomethacin (INDO) and dexamethasone (DEX). The data is the mean±SEM of at least five independent experiments.

Figure 14.1(a) shows the HB02, F6 and the compound 6S and GD have significantly lower IC_{50} than the IC_{50} of O inhibitor, L-NAME. F6 was initially identified among several fraction from HB02 based on it potent effect in inhibiting NO production, whereby it was also two-fold stronger than HB02. Since F6 and its compound have variable but significant effects on NO production at 8-12 ug/mL, their effects on PGE_2 at 10 ug/mL were tested (Figure 14.1b).

The assessment shows that 6S has similar potent inhibitory effects to F6 on PGE_2 production while GD has lower inhibitory effects. Interestingly, F6 and both compounds are comparable to PGE_2 inhibitors; indomethacin and dexamethasone ($p>0.05$). This observation complements theirs effects on NO production (Figure 14.1a) and reflects the dual inhibitory effects of the samples against NO and PGE_2. These results suggest that F6/6S may affect iNOS and COX pathways. The anti-inflammatory effects of 6S has been studied before and there is evidence to show that 6S has potent inhibitory effects on iNOS and COX (*30, 33*). However, it remains unclear whether either F6 or 6S are specific in their action on iNOS and/or COX1/2.

HB02 Contains Compounds that Preferably Act in Prophylactic Manner before Inflammation Initiated by Activated Macrophages

To determine the basis for the inhibition of NO production and extent of effects, the expression of iNOS at mRNA level before and after stimulation of RAW 254.7 cells with LPS were determined. The rationale is to determine whether the inhibition was at the transcriptional level of iNOS or during the release of NO by stimulated RAW 264,2 cells. The concentration used were based on the determined IC_{50} of each samples (Figure 14.1a).

Based on these findings, the contribution of these two compounds to the biological effects of F6 on iNOS mRNA and protein levels was investigated. 6S showed similar effects to F6 in reducing iNOS mRNA level in macrophages before and after LPS stimulation (Figure 14.2b). GD, however, was more effective in reducing iNOS mRNA expression by macrophage after LPS stimulation. GD may have tendency towards reducing the enzymatic activity of iNOS. Hence, the effect of F6 on iNOS mRNA levels may be attributed to 6S rather than GD. Therefore, 6S and GD have distinct effects on iNOS but both suppress NO production.

Interestingly, however, neither compound had a significant effect of iNOS protein level or NO production when added after macrophage activation (Figure 14.3). The results showed that 6S slightly reduced the production of iNOS protein before and also after LPS stimulation of the cells. This observation, however, is consistent with the noted effects of 6S on iNOS mRNA level (Figure 14.2b). In contrast, the results showed that GD marginally decrease the level of iNOS protein when the cells were treated with the compound before and after LPS stimulation (Figure 14.3b), which is similar to the effect of L-NAME (200 µM).

Hence, GD does not have a significant effect on iNOS mRNA and protein levels but is likely to have an inhibitory effect on its enzymatic activity. In contrast, 6S may have modulating effects on iNOS mRNA and protein levels. These results suggest that HB02 contains compounds that preferably act in a prophylactic manner before inflammation initiated by activated macrophage.

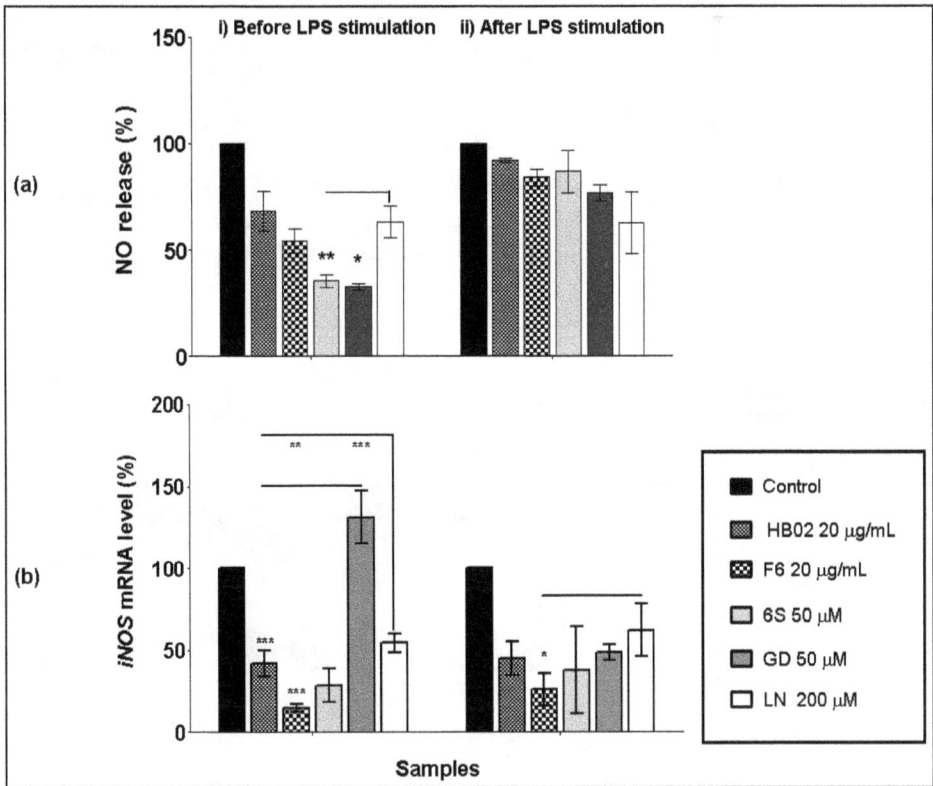

Figure 14.2: Cultured RAW264.7 Cells were Either Treated with HB02, F6 6S, GD and LN Before or After Stimulation with LPS. The crude extract and fraction were tested at 20 μg/mL whilst the compounds were tested at 50 μM. The data represent data from at least 3 sets of independent experiments.

ZOR's Extract and Active Constituents Reduce the Expression of IL-12 and IL-23 with p40 Subunit

IL-23 and IL-12 share a common protein subunit, p40 that forms a heterodimeric complex with p19 and p35, respectively. The heterodimeric complexes are biologically active in the differentiation of Th17 and Th1 cells, and important in the pathogenesis of psoriasis. Experiments to assess the effects of ZOR on the IL-12/IL-23 axis showed that the ZOR samples strongly suppressed IL-12p40 as well as IL-23p19 mRNA levels when added to the macrophages before LPS stimulation (Figure 14.4). HB02, F6 and 6S showed very significant inhibition of IL-12/IL-23p40 when added after LPS stimulation ($p<0.001$) whilst GD showed contradictive effect. This observation could be relevant to psoriasis since IL-12 and IL-23 in key cytokines in the differentiation Th1 and Th17 cells, respectively (*5, 10, 11*). These findings, therefore, suggest that HB02, F6 and 6S could modulate the generation of Th1 and Th17 T-cell functions. These findings are important since IL-23 is established as one of the prime inflammatory mediators in psoriasis (*12, 13*) and targeting p40

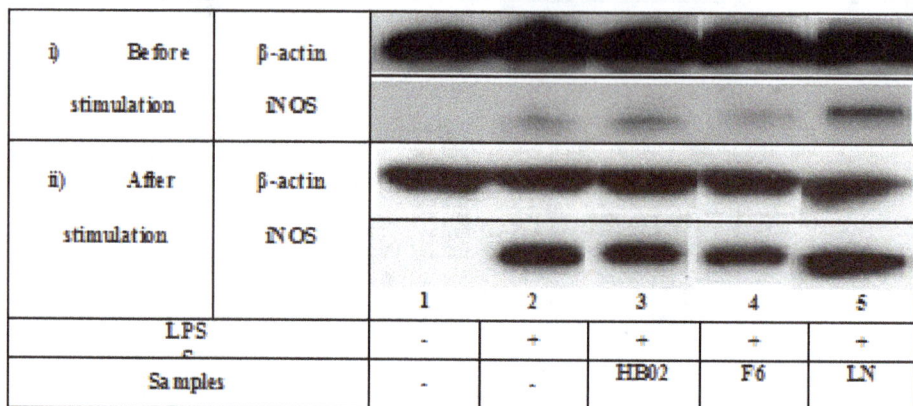

(a)

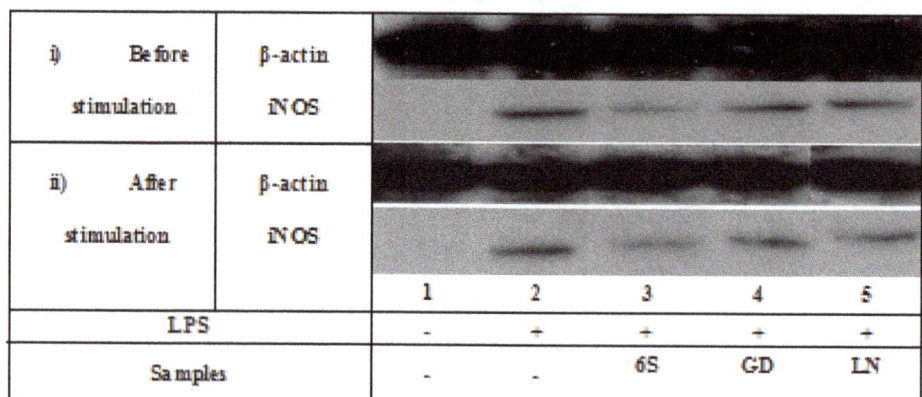

(b)

Figure 14.3: Western Blotting Analysis of iNOS Protein Levels in RAW 264.7 Cells Treated with HB02, Fraction 6 (F6) (a) and the Compounds 6-shogaol and 1-dehydro-6-gingerdione (b). RAW 264.7 cells were treated with 20 µg/mL of HB02, F6, 50 µM 6S or GD either (i) before or (ii) after stimulation with LPS. L-NAME (LN) was used at 54 µg/mL or 200 µM.

subunit has been used for the treatment of psoriasis (*14*). These results could, thus, suggest that ZOR and, particularly F6 and its compounds, could strongly modulate macrophage responses during inflammation.

HB02, F6 and 6S at Low Concentration Modulate PMN Migration

Study on PMN migration revealed that even at these lower concentrations (3.3 µg/mL), HB02, F6 and 6S reduced PMN capture, rolling, adhesion and migration by up to ~46 per cent relative to control (containing 0.03 per cent DMSO) (Figures 14.5 and 14.6). GD, however, augmented PMN adhesion by 34.2 per cent (Figure 14.6c). It is indicated in Figure 14.5(e) wherein higher margination of PMNs in medium containing GD. Nevertheless, GD reduced PMN rolling by 27.6 per cent

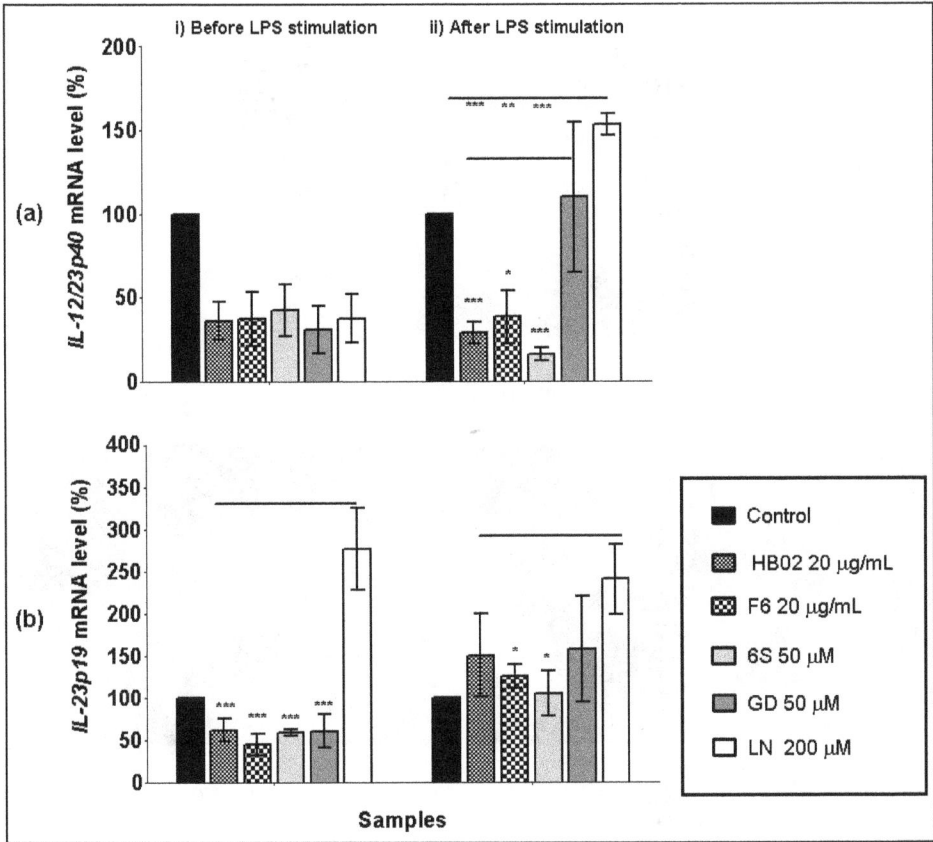

Figure 14.4: The Effect of 6S and GD on the Level of IL-12/IL23p40 and IL-23p19 mRNA.

The level of IL-12/IL-23p40 and IL-23p19 mRNA were determined using real-time PCR. The effect of treatment with 6S and GD on the level of IL-12/IL-23p40 and IL-23p19 mRNA were determined using LPS-stimulated RAW 264.7 cells. The cells were either treated with HB02, F6, 6S and GD 30 minutes before stimulation with LPS or treated 12 hours after stimulation. Statistical comparisons were carried out using Two-Way ANOVA (Bonferroni posttest). Data are presented as the mean±SD of per cent of IL-12/IL-23p40 mRNA level relative to control. The results for the effect of 6S vs. GD on IL-12/IL-23p40 mRNA are based on at least three independent experiments.

and migration by 29.2 per cent. Thus, GD has some inhibitory effect against PMN migration suggesting that this compound may act to inhibit the adhesion of PMN to HUVEC.

Figure 14.6 shows that F6 and 6S at the lower concentration of 3.3 μg/mL had comparable effects on PMN capture, rolling, adhesion and migration, that were more effective than HB02 (Figure 14.6). 6S had significant inhibitory effects (p<0.05) on PMN capture (~34.2 per cent inhibition), whilst F6 significantly affect PMN rolling (44 per cent inhibition) compared to control (Figures 14.6a and 14.6b). These data

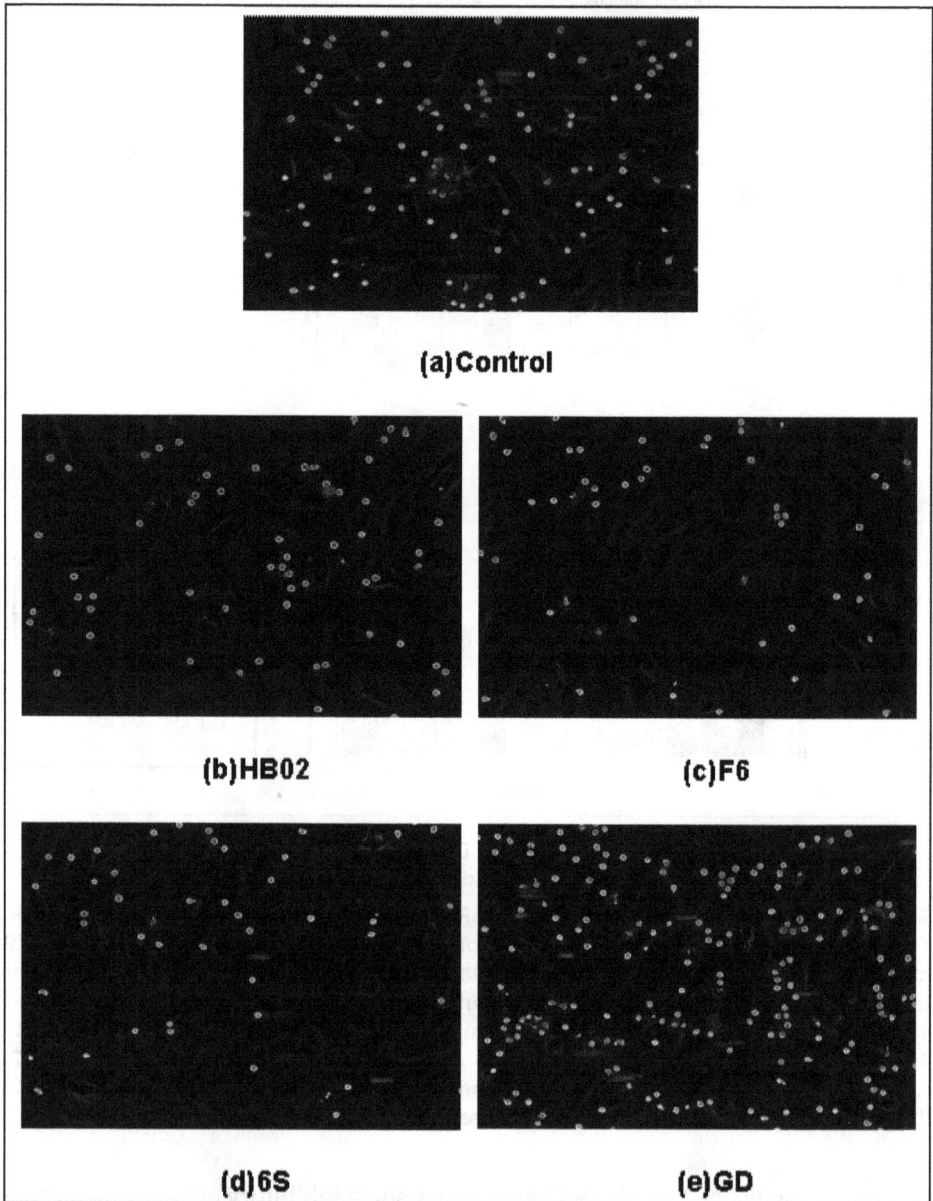

Figure 14.5: The Interaction of Treated PMNs with HUVEC during Perfusion.

The interaction between PMNs and TNFα-activated HUVEC during the 8 minutes perfusion into the flow chamber. Prior to perfusion, the PMNs were treated with (a) DMSO control (at 0.03 per cent), (b) HB02, (c) F6, (d) 6S and (e) GD at 3.3 µg/mL, for 10 min. HB02, F6 and 6S reduced PMNs margination and recruitment on the endothelial cell, except GD. This observation represents three independent experiments.

may suggest that both F6 and 6S may have inhibitory effects during early stages of inflammation through preventing PMN tethering. 6S and GD have different, but possibly, complimentary modes of action on PMN migration.

In addition to the effects in modulating iNOS and IL-12/IL-23 by activated macrophages, these results highlight their modulatory effects on the response of PMNs to inflammation. This property could be important since PMN activation is linked to the balance of interaction between iNOS, IL-12/IL-23 and PGE2 (15, 16). Therefore, although these observations are based on limited assay systems (macrophages and PMNs), it is sufficient to indicate that the tested samples possess important anti-inflammatory properties.

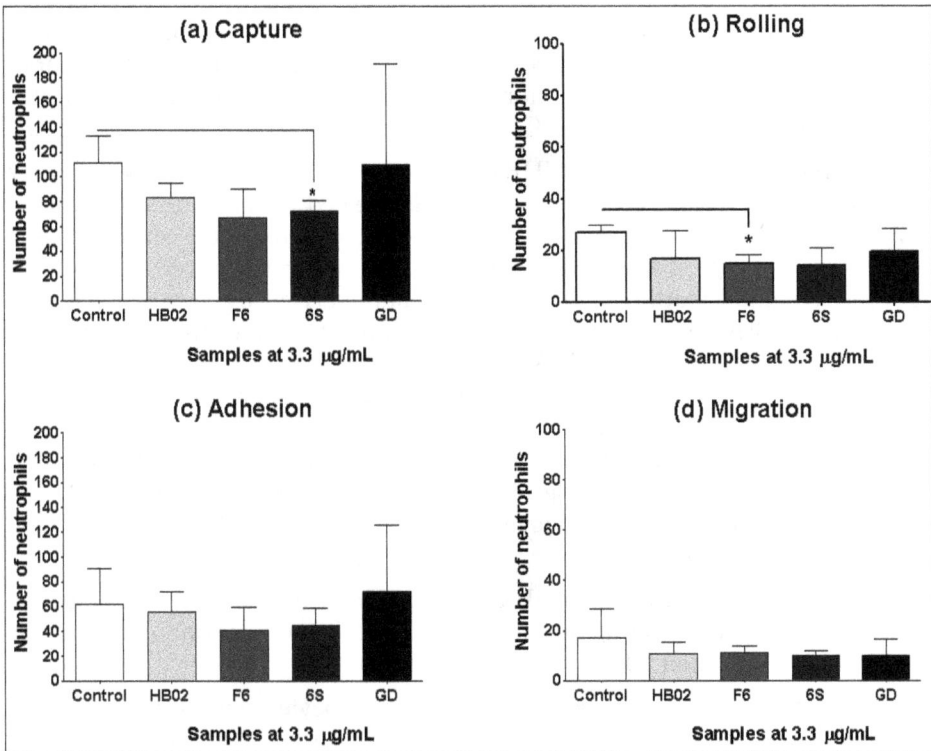

Figure 14.6: The Modulatory Effects of Low Concentrations of HB02, F6, 6S and GD Relative to Control (DMSO 0.03 per cent) on PMN Biology were Tested at 3.3 µg/mL.

The effects shown are for: a) capture, b) rolling, c) adhesion, and d) migration of PMN using TNFα-activated HUVEC in flow chambers. The interactions were assessed in six random fields from three sets of independent experiments. Statistical comparison between the samples and control were carried out using Student t-test. * =p<0.05

HB02 Contains Compounds that Influence Keratinocyte Biology and Onhibited Production of IL-20 and Il-8

In this model, human primary keratinocytes are stimulated with IL-17, IL-22, TNFα, IL-1β and OSM (a member of IL-6 family), all established key cytokines in psoriasis pathogenesis (*17-21*). The effect of the samples on keratinocyte proliferation and cytokine/chemokine production in this system were then investigated. Cytokines studied included IL-20, which is known to promote cell hyperproliferation in psoriasis (*22*) and IL-8 that plays a important role in inflammatory responses, particularly as a chemoattractant (*23*). IL-20 is produced exclusively by keratinocytes in response to Th22 and Th17-related cytokines (*24*) and is expressed on the entire layer of the epidermal keratinocytes in patients with psoriasis (*22*). IL-8 production by keratinocytes is triggered by the synergistic effect of TNFα and IL-17 (*20*). Therefore, the potential of Z.officinale Roscoe var rubrum (ZOR) to influence keratinocyte biology and contribution to inflammation, by evaluating the effect of HB02, F6, 6S and GD on the production of IL-20 and IL-8 by the activated NHEK, CsA was chosen as a positive control based on its potent inhibitory effect on keratinocytes proliferation and chemotactic activity as reported (*31, 32*) and for its use in treating psoriasis. The inhibitory effects of HB02, F6, 6S and GD were tested in comparison with Ciclosporin A (CsA) at a concentration range of 2-20000 ng/mL

The results show that the NHEK cells from individual SFK1 which were stimulated with cytokine cocktail (CT) and then treated with HB02, F6, 6S and GD at 20 µg/mL shrank and became less confluent compared with control (Figure 14.7). The phenotype changes were similar to the appearance of cells undergoing apoptosis. Visually, apoptotic bodies are characterised by a condensed and 'invaginated' cell membrane. F6 and 6S had more potent effects compared with HB02 and GD. Treatment with F6 and 6S at 20 µg/mL resulted in a dramatic reduction of NHEK proliferation. GD, however, did not appear to affect cell morphology at 20 µg/mL. The morphology and cell density were comparable to cells which were only stimulated with CT (Figure 14.7b). Stimulated cells are presented as polygonal-shaped cells with distinctive nuclei. These results, therefore, confirm that GD was less affective in regulating keratinocytes proliferation compared with HB02, F6 and 6S at this concentration.

To further explore the potential of ZOR to influence keratinocyte biology and contribution to inflammation, the effect of HB02, F6, 6S and GD on the production of IL-20 by the activated NHEK was studied (Figure 14.8).

IL-20 has been shown to be involved in keratinocyte hyperproliferation and the alteration of terminal differentiation process in psoriatic skin (*34*). By assessing the effect of selected ZOR composition (HB02, F6, 6S and GD) on IL-20 production, it would be possible to confirm their potential therapeutic effects on keratinocytes proliferation and differentiation. HB02, F6, 6S and GD significantly down-regulated IL-20 production at 20µg/mL (90.8±6.0 per cent, 99.9±0.3 per cent, 99.4±0.5 per cent and 88.9±7.0 per cent, respectively) (Figure 14.8). The inhibitory effects were significantly higher than seen with Ciclosporin A (CsA) at 20 ng/mL ($p < 0.01$). However, despite this concordance in the effect of the test compounds on IL-20 production and on proliferation and morphology, it remains to be established whether these effects are connected at the molecular level.

Further, the effect of HB02, F6, 6S and GD on IL-8 by the activated NHEK cells was studied. IL-8 is a key chemoattractant cytokines in inflammatory responses and is known to be involved in psoriasis. At the concentration of 20 µg/mL, HB02, F6 and

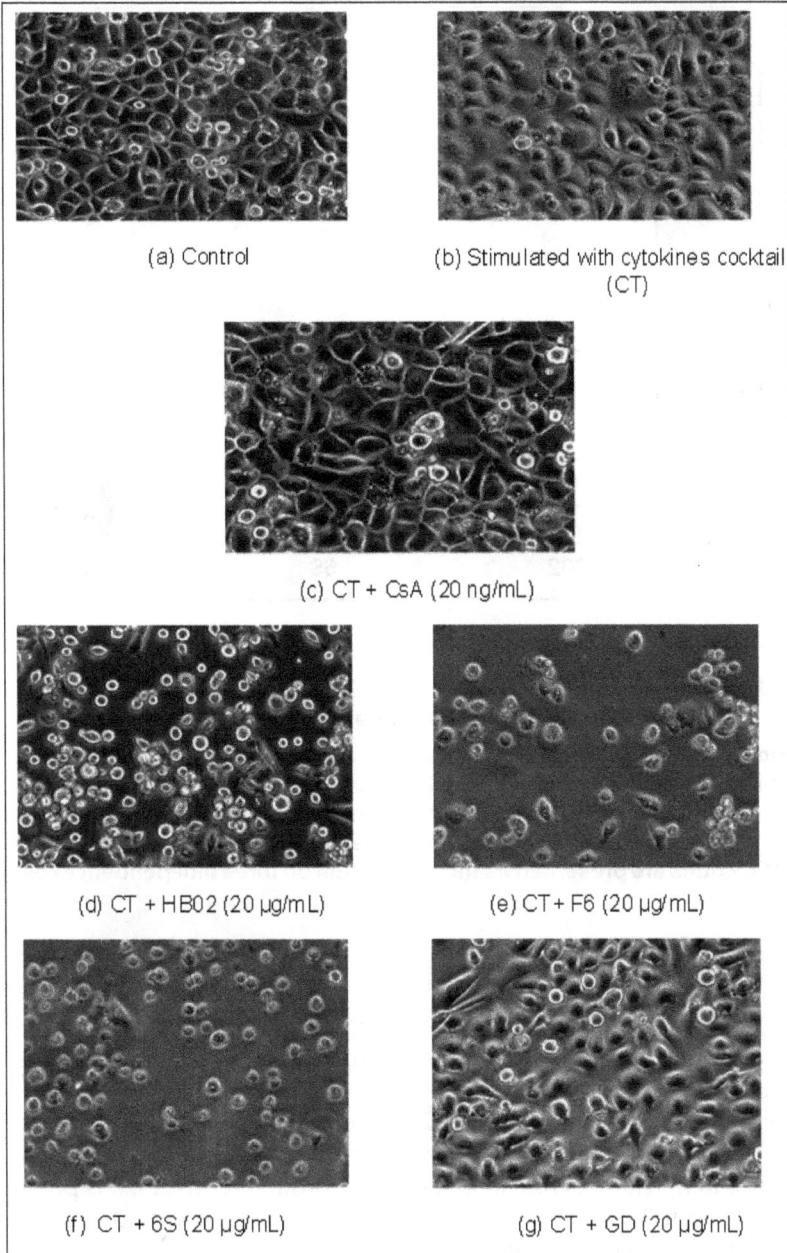

(a) Control

(b) Stimulated with cytokines cocktail (CT)

(c) CT + CsA (20 ng/mL)

(d) CT + HB02 (20 µg/mL)

(e) CT + F6 (20 µg/mL)

(f) CT + 6S (20 µg/mL)

(g) CT + GD (20 µg/mL)

Figure 14.7: Phenotype Modulation of NHEK Cells Treated with Selected ZOR Samples (HB02, F6, 6S and GD) at 20 µg/mL.

Changes in the NHEK cells were assessed after stimulation with the CT and treatment with (a) HB02 (b) F6 (c) 6S and (d) GD at 20 µg/mL for 48 hours. Cell morphology was assessed at 40x magnification and captured using JVC TK-C1360B digital colour video camera.

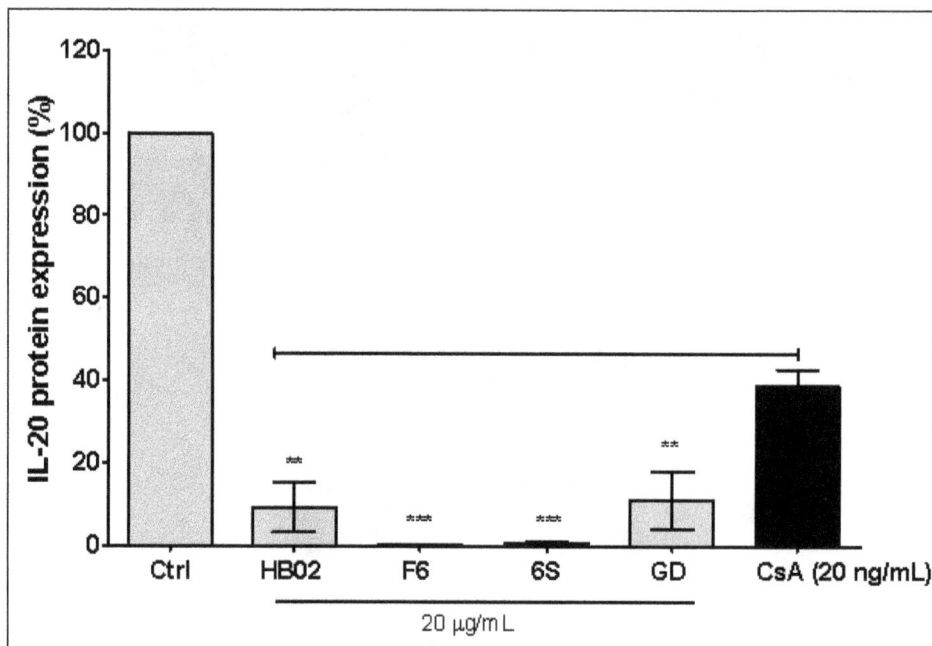

Figure 14.8: The Effect of HB02, F6, 6S and GD on IL-20 Production by Activated Keratinocytes.

The samples were assessed for their inhibitory effects on IL-20 production by NHEK cells stimulated with the cytokine cocktail (CT) containing IL-17A, OSM, TNFα, IL-22, and IL-1α, each at 10 ng/mL. HB02, F6, 6S and GD were tested at 20 μg/mL and the results analysed using One-way ANOVA (for comparison of the effect of the samples to CsA (20 ng/mL). Values are presented as the mean±SEM on three independent experiments.

GD increased IL-8 expression (Figure 14.9). However, 6S reduced IL-8 production by 68.0±17.0 per cent relative to the control. This inhibition was statistically significant when compared with CsA at 20 ng/mL ($p < 0.05$).

The results revealed that 6S had a profound effect on inhibiting IL-20 and IL-8 production by activated keratinocytes. This effect was significantly more pronounced compared with the effect seen for HB02, F6 and GD. The effect observed for 6S was also higher than seen for the positive control CsA. In view of the reported increase in IL-20 production and its role in keratinocyte proliferation (22), it is possible that the ability of 6S to inhibit keratinocyte proliferation may partly due to its potent inhibitory effect on IL-20 production.

This beneficial therapeutic effect would further be bolstered by the ability of 6S to strongly inhibit IL-8 production. A previous study proposed that IL-8 production is primarily localised to the suprabasal layer of the psoriatic epidermis, suggestive of the role IL-8 plays in keratinocyte differentiation rather than proliferation (25).

6S also promoted changes in the phenotype of activated keratinocytes suggestive of inducing apoptosis in these cells. This effect was also noted for HB02

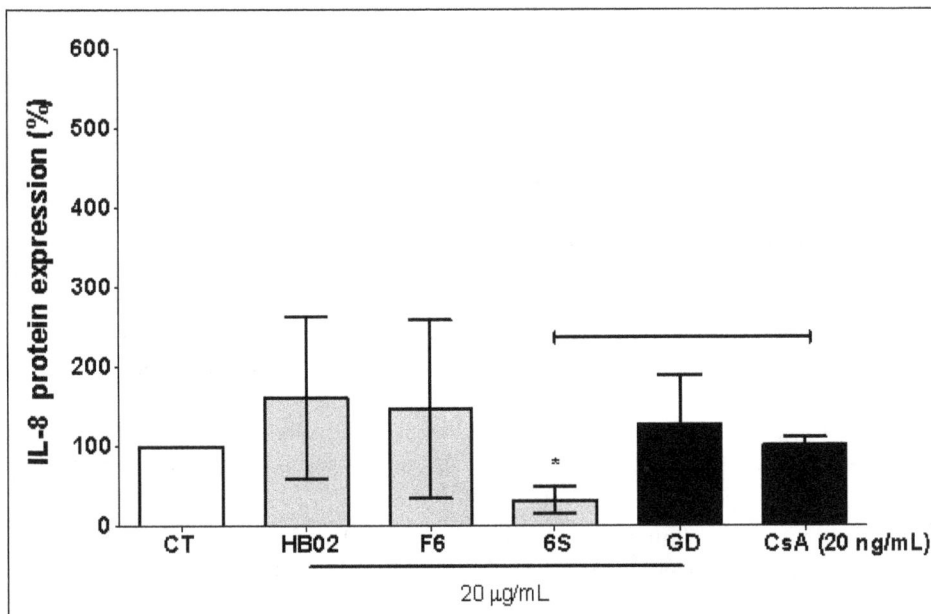

Figure 14.9: The Effect of HB02, F6, 6S and GD on IL-8 Production by Activated Keratinocytes.

The effect of HB02, F6, 6S and GD on IL-8 production by NHEK was studied by ELISA. The NHEK cells were stimulated with the cytokine cocktail (CT) then cultured either with or without HB02, F6, 6S, GD or CsA at the indicated concentrations. One-way ANOVA (Tukey's test) analysis was used to compare the effect of the test samples relative to that obtained with the control (CT). No significant differences were observed for the samples at 20 µg/mL relative to control (CT) or CsA (20 ng/mL) except for 6S (p<0.01). Values are presented as the mean±SEM on three independent experiments.

and F6 and, thus, highlights the potential of ZOR in psoriasis. GD, in contrast, did not affect the phenotype of the keratinocytes nor had an effect on IL-8 production despite its effect on IL-20 production.

4. CONCLUSIONS

It has been shown that ZOR extract, fractions and two identified compounds have modulatory effects on a range of pro-inflammatory pathways including the production of cytokines, cell migration and proliferation. Interestingly, however, the study also revealed that the two identified and fully characterised from one of the most relevant ZOR fractions, 6S and GD displayed differential effects on activated macrophage, PMNs, and keratinocytes. Overall, the data presented provide a detailed insight into the multi-target effects of 6S and its modulatory effects on inflammatory responses mediated by immune cells and by keratinocytes. Furthermore, the similarities between the biological effects revealed for 6S and, the rather unrefined HB02 and F6, highlights the influence of 6S on the activity of the HB02 and, by inference, ZOR.

The study has also revealed a number of interesting mechanisms through which 6S and GD could influence inflammation. Thus, the results have suggested that the suppressive effects of the two compounds on NO and PGE_2 production may indirectly reduce IL-12/IL-23 mRNA expression by macrophages. The reduction in IL-12 and IL-23 levels, in turn, will reduce the activation and differentiation of Th1 and Th17 which have been shown to be important for the development of psoriasis. Further, the ability of the two compounds to inhibit PMN migration indicates that the compounds would have strong anti-inflammatory effects during the initiation of inflammation. In addition to the inhibitory effects of 6S and GD on immune cells, the two compounds also modulated IL-20 production by keratinocytes. IL-20 is a member of the IL-10 family of cytokines and was recently shown to be involved in inflammation and bone resorption in patients with inflammatory conditions (29). In addition to the relevance of this observation to the pathogenesis of psoriasis, the data suggest that the compounds could have therapeutic benefits in psoriatic arthritis which is associated with joint damage and bone loss. However, further studies will be required to test this possibility.

The structural orientation of 6-shogaol and 1-dehydro-6-gingerdione are differentiated by the orientation at B (Figure 14.10), which is the substitution pattern of the hydroxyl and carbonyl moiety on the side chain. Compared with 6-shogaol,

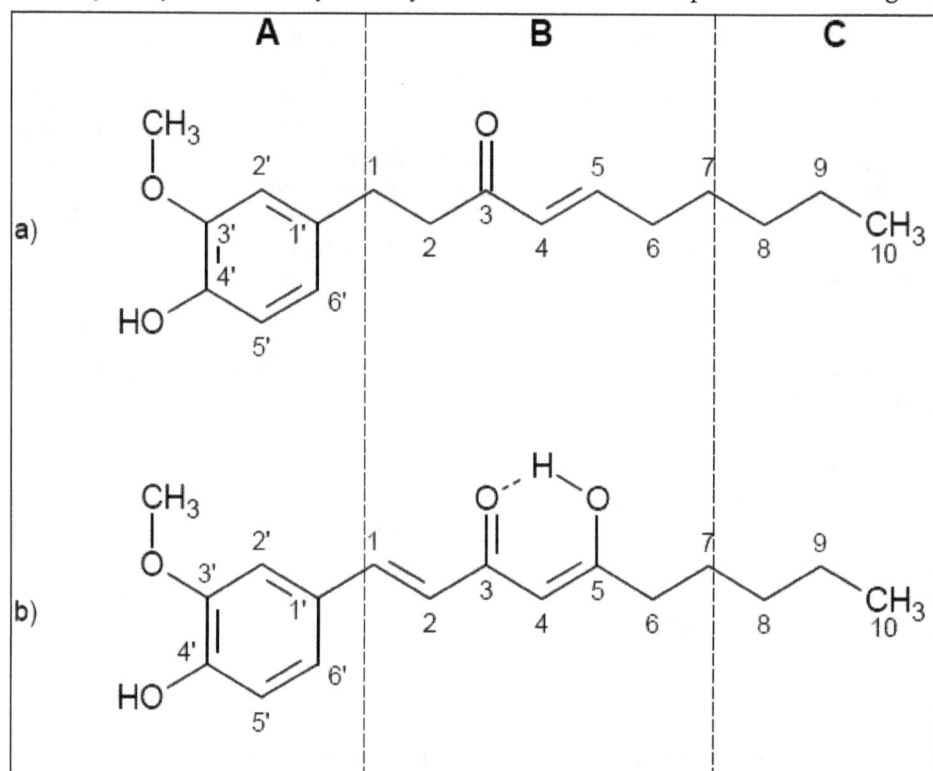

Figure 14.10: The Structural Orientation of a) 6-shogaol and b) 1-dehydro-6-gingerdione. Structural orientation of 6-shogaol and -dehydro-6-gingerdione denotes by A, B and C, may contribute to the activity of the compounds.

GD has a hydroxyl group as part of an enolic moiety which could be easily chelated. 6-shogaol only features one free hydroxyl whilst GD has a free hydroxyl and enol-hydroxyl groups in the alkyl chain. Therefore, the activity of both compounds could be differentiated by the presence of an enolic moiety which is more reactive as an enolate ion in basic solution. Both compounds contain one oxygen atom and at least one hydroxyl group which are hydrogen acceptors and donors, respectively.

This satisfies the criteria of a 'druggable' compound (hydrogen-donor≤5 and hydrogen-bond acceptor≤10). However, Lipinski's Rules of five is not an absolute indicator of the efficacy of a compound as a drug, instead, it is used as a preliminary guideline to identify drug candidates in natural product research and drug discovery. Thus, in terms of function-structural relationships, it is tempting to speculate that these differences might be due to structural differences between 6S and GD. The two compounds are differentiated by the presence of an enolic-hydroxyl moeity at C-5 on the side chain of GD which is absent in 6S. Instead, at the same position in 6S, there is a double bond. These observations raise issues as to whether it would be preferable to use the compounds individually or in combination. However, further detailed studies are required.

Overall, the study provides a new overview of potential mechanisms of action of ZOR in modulating inflammatory cascades, potentially relevant to psoriasis. Both compounds, albeit at different doses, possess complimentary effects in regulating NO and PGE_2 production, reduce level of IL-12p40 and IL-23p19 mRNA, inhibit PMN migration and inhibit IL-20 production by keratinocytes. The studies suggest that the compounds could work synergistically to modulate a range of inflammatory pathways in psoriatic plaque. Thus, further work that arises from this study is to identify the mechanistic interaction of both compounds in modulating the inflammatory events mentioned above.

5. ACKNOWLEDGEMENTS

The authors gratefully thank Prof. R.A.Mageed, Prof. D. Perrett and Prof. S. Gibbons for their expert assistance in chemical elucidation and biological studies. We thank Prof. M. Philpott, Dr. Fulvio and Dr, Diane for the willingness to provide the cells and kind guidance in the study. The work was supported by study grant from SIRIM Berhad and Malaysian Ministry of Science, Technology and Innovation (MOSTI).

6. REFERENCES

1. Chandran, V., and Raychaudhuri, S. P. (2010). Geoepidemiology and environmental factors of psoriasis and psoriatic arthritis, *Journal of Autoimmunity* 34, J314-J321.

2. Grjibovski, A. M., Olsen, A. O., Magnus, P., and Harris, J. R. (2007). Psoriasis in Norwegian twins: contribution of genetic and environmental effects, *Journal of the European Academy of Dermatology and Venereology* 21, 1337-1343.

3. Naldi, L., Peli, L., and Parazzini, F. (1999). Association of early-stage psoriasis with smoking and male alcohol consumption: evidence from an Italian case-control study, *Archieves of Dermatology 135*, 1479-1484.

4. Naldi, L., and Mercuri, S. R. (2009). Smoking and Psoriasis: From Epidemiology to Pathomechanisms, *Journal of Investigative Dermatology 129*, 2741-2743.

5. Lowes, M. A., Bowcock, A. M., and Krueger, J. G. (2007). Pathogenesis and therapy of psoriasis, *Nature 445*, 866-873.

6. Wenjun, O. (2010). Distinct roles of IL-22 in human psoriasis and inflammatory bowel disease, *Cytokine and Growth Factor Reviews 21*, 435-441.

7. Grzanna, R., Lindmark, L., and Frondoza, C. G. (2005). Ginger-an herbal medicinal product with broad anti-inflammatory actions, *J Med Food 8*, 125-132.

8. Nievergelt, A., Marazzi, J., Schoop, R., Altmann, K. H., and Gertsch, J. (2011). Ginger Phenylpropanoids Inhibit IL-1β and Prostanoid Secretion and Disrupt Arachidonate-Phospholipid Remodeling by Targeting Phospholipases A_2, *Journal of Immunology 187*, 4140-4150.

9. Otkjaer, K., Kragballe, K., Johansen, C., Funding, A. T., Just, H., Jensen, U. B., Sorensen, L. G., Norby, P. L., Clausen, J. T., and Iversen, L. (2007). IL-20 Gene Expression Is Induced by IL-1β through Mitogen-Activated Protein Kinase and NF-κB-Dependent Mechanisms, *Journal of Investigative Dermatology 127*, 1326-1336.

10. Bastos, K. R. B., Marinho, C. R. F., Barboza, R., Russo, M., Ãlvarez, J. M., and D'Imperio Lima, M. R. (2004). What kind of message does IL-12/IL-23 bring to macrophages and dendritic cells?, *Microbes and Infection 6*, 630-636.

11. Di Cesare, A., Di Meglio, P., and Nestle, F. O. (2009). The IL-23/Th17 axis in the immunopathogenesis of psoriasis, *Journal of Investigative Dermatology 129*, 1339-1350.

12. Chen, X., Tan, Z., Yue, Q., Liu, H., Liu, Z., and Li, J. (2006). The expression of interleukin-23 (p19/p40). and inteleukin-12 (p35/p40). in psoriasis skin, *Journal of Huazhong University of Science and Technology Medical Sciences 26*, 750-752.

13. Lee, E., Trepicchio, W. L., Oestreicher, J. L., Pittman, D., Wang, F., Chamian, F., Dhodapkar, M., and Krueger, J. G. (2004). Increased Expression of Interleukin 23 p19 and p40 in Lesional Skin of Patients with Psoriasis Vulgaris, *The Journal of Experimental Medicine 199*, 125-130.

14. Yeilding, N., Szapary, P., Brodmerkel, C., Benson, J., Plotnick, M., Zhou, H., Goyal, K., Schenkel, B., Giles-Komar, J., Mascelli, M. A., and Guzzo, C. (2011). Development of the IL-12/23 antagonist ustekinumab in psoriasis: past, present, and future perspectives, *Annals of the New York Academy of Sciences 1222*, 30-39.

15. Kalinski, P., Vieira, P. L., Schuitemaker, J. H. N., de Jong, E. C., and Kapsenberg, M. L. (2001). Prostaglandin E_2 is a selective inducer of interleukin-12 p40 (IL-12p40). production and an inhibitor of bioactive IL-12p70 heterodimer, *Blood 97*, 3466-3469.

16. Lemos, H. P., Grespan, R., Vieira, S. M., Cunha, T. M., Verri, W. A., Fernandes, K. S. S., Souto, F. O., McInnes, I. B., Ferreira, S. H., Liew, F. Y., and Cunha, F. Q. (2009). Prostaglandin mediates IL-23/IL-17-induced neutrophil migration

in inflammation by inhibiting IL-12 and IFN-γ production, *Proceedings of the National Academy of Sciences 106*, 5954-5959.

17. Boniface, K., Bernard, F.-X., Garcia, M., Gurney, A. L., Lecron, J.-C., and Morel, F. (2005). IL-22 Inhibits Epidermal Differentiation and Induces Proinflammatory Gene Expression and Migration of Human Keratinocytes, *The Journal of Immunology 174*, 3695-3702.

18. Boniface, K., Diveu, C., Morel, F., Pedretti, N., Froger, J., Ravon, E., Garcia, M., Venereau, E., Preisser, L., Guignouard, E., Guillet, G. r., Dagregorio, G., Pène, J., Moles, J.-P., Yssel, H., Chevalier, S., Bernard, F. o.-X., Gascan, H., and Lecron, J.-C. (2007). Oncostatin M Secreted by Skin Infiltrating T Lymphocytes Is a Potent Keratinocyte Activator Involved in Skin Inflammation, *The Journal of Immunology 178*, 4615-4622.

19. Harper, E. G., Guo, C., Rizzo, H., Lillis, J. V., Kurtz, S. E., Skorcheva, I., Purdy, D., Fitch, E., Iordanov, M., and Blauvelt, A. (2009). Th17 Cytokines Stimulate CCL20 Expression in Keratinocytes *In Vitro* and *In Vivo*: Implications for Psoriasis Pathogenesis, *Journal of Investigative Dermatology 129*, 2175-2183.

20. Chiricozzi, A., Guttman-Yassky, E., Suarez-Farinas, M., Nograles, K. E., Tian, S., Cardinale, I., Chimenti, S., and Krueger, J. G. (2011). Integrative Responses to IL-17 and TNF-α in Human Keratinocytes Account for Key Inflammatory Pathogenic Circuits in Psoriasis, *Journal of Investigative Dermatology 131*, 677-687.

21. Guilloteau, K., Paris, I., Pedretti, N., Boniface, K., Juchaux, F., Huguier, V., Guillet, G., Bernard, F. X., Lecron, J. C., and Morel, F. (2010). Skin Inflammation Induced by the Synergistic Action of IL-17A, IL-22, Oncostatin M, IL-1α, and TNFα Recapitulates Some Features of Psoriasis, *Journal of Immunology 184* 5263.

22. Wei, C.-C., Chen, W.-Y., Wang, Y.-C., Chen, P.-J., Lee, J. Y.-y., Wong, T.-W., Chen, W. C., Wu, J.-c., Chen, G.-y., Chang, M.-S., and Lin, Y.-c. (2005). Detection of IL-20 and its receptors on psoriatic skin, *Clinical Immunology 117*, 65-72.

23. Jiang, W. Y., Chattedee, A. D., Raychaudhuri, S. P., Raychaudhuri, S. K., and Farber, E. M. (2001). Mast cell density and IL-8 expression in nonlesional and lesional psoriatic skin, *International Journal of Dermatology 40*, 699-703.

24. Wolk, K., Witte, E., Warszawska, K., Schulze-Tanzil, G., Witte, K., Philipp, S., Kunz, S., Döcke, W.-D., Asadullah, K., Volk, H.-D., Sterry, W., and Sabat, R. (2009). The Th17 cytokine IL-22 induces IL-20 production in keratinocytes: A novel immunological cascade with potential relevance in psoriasis, *European Journal of Immunology 39*, 3570-3581.

25. Beljaards, R. C., Beek, P. V., Nieboer, C., Stoof, T. J., and Boorsma, D. M. (1997). The expression of interleukin-8 receptor in untreated and treated psoriasis, *Archives of Dermatological Research 289*, 440-443.

26. Ovigne, J. M., Baker, B. S., Brown, D. W., Powles, A. V., and Fry, L. (2001). Epidermal CD8+ T cells in chronic plaque psoriasis are Tc1 cells producing heterogeneous levels of interferon-gamma, *Experimental Dermatology 10*, 168-174.

27. Hamada, H., Garcia-Hernandez, M. d. l. L., Reome, J. B., Misra, S. K., Strutt, T. M., McKinstry, K. K., Cooper, A. M., Swain, S. L., and Dutton, R. W. (2009). Tc17, a Unique Subset of CD8 T-Cells That Can Protect against Lethal Influenza Challenge, *The Journal of Immunology 182*, 3469-3481.

28. Res, P. C. M., Piskin, G., de Boer, O. J., van der Loos, C. M., Teeling, P., Bos, J. D., and Teunissen, M. B. M. (2010). Overrepresentation of IL-17A and IL-22 Producing CD8-T Cells in Lesional Skin Suggests Their Involvement in the Pathogenesis of Psoriasis, *PLoS ONE 5*, e14108.

29. Hsu, Y.-H., Chen, W.-Y., Chan, C.-H., Wu, C.-H., Sun, Z.-J., and Chang, M.-S. (2011). Anti-IL-20 monoclonal antibody inhibits the differentiation of osteoclasts and protects against osteoporotic bone loss, *The Journal of Experimental Medicine 208*, 1849-1861.

30. Jolad, S. D., Lantz, R. C., Solyom, A. M., Chen, G. J., Bates, R. B., and Timmermann, B. N. (2004). Fresh organically grown ginger (*Zingiber officinale*): composition and effects on LPS-induced PGE2 production. *Phytochemistry*, 65(13), 1937–54. doi:10.1016/j.phytochem.2004.06.008

31. Kaplan, A., Matsue, H., Shibaki, A., Kawashima, T., Kobayashi, H., and Ohkawara, A. (1995). The effects of cyclosporin A and FK506 on proliferation and IL-8 production of cultured human keratinocytes. *Journal of Dermatological Science*, 10(2), 130–138. doi:10.1016/0923-1811(95)00395-9

32. Karashima, T., Hachisuka, H., and Sasai, Y. (1996). FK506 and cyclosporin A inhibit growth factor-stimulated human keratinocyte proliferation by blocking cells in the G0G1 phases of the cell cycle. *Journal of Dermatological Science*, 12(3), 246–254. doi:10.1016/0923-1811(95)00480-7

33. Sang, S., Hong, J., Wu, H., Liu, J., Yang, C. S., Pan, M.-H., Ho, C.-T. (2009). Increased growth inhibitory effects on human cancer cells and anti-inflammatory potency of shogaols from *Zingiber officinale* relative to gingerols. *Journal of Agricultural and Food Chemistry*, 57(22), 10645–50. doi:10.1021/jf9027443

34. Stenderup, K., Rosada, C., Worsaae, A., Dagnaes-Hansen, F., Steiniche, T., Hasselager, E., Dam, T. N. (2009). Interleukin-20 plays a critical role in maintenance and development of psoriasis in the human xenograft transplantation model. *The British Journal of Dermatology*, 160(2), 284–96. doi:10.1111/j.1365-2133.2008.08890.x

Chapter 15

Assessment of Enzyme Activities and Heamatological Parameters in Male Wister Rats Administered with Aqueous Extract of *Massularia acuminata* Root.

O.S Awotunde[1] and M.T. Yakubu[2]

[1]Department of Biochemistry,
Habib Medical School, IUIU, Kampala, Uganda
E-mail: derockng@gmail.com
[2]Phytomedicine, Toxicology, Reproductive and
Developmental Biochemistry Research Laboratory,
Department of Biochemistry, University of Ilorin, Ilorin, Nigeria

ABSTRACT

The aqueous extract of *Massularia acuminata* root at the doses of 50, 100 and 200 mg/kg body weight was investigated for its toxicological implication on selected enzyme activities and heamatological parameters *in vivo and in vitro* in male wister rats. Phytochemical screening reveals the presence of alkaloids, anthraquinones, saponins, phenolics, flavonoids and tannins in the extract. It caused labilization in some organs whereas it induced the synthesis of enzymes by the increase and decrease ($p<0.05$) in the activities of alkaline phosphatase, acid phosphatase and gamma-glutamyl transferase. Alterations in the normal levels ($p<0.05$) of the haematological parameters such as PCV, RBC, WBC, platelets, neutrophils and lymphocytes showed that the aqueous extract exhibited erythropoeitin potential enhancing the rate of oxygen transport but suppressed the immune system by reduction of the white blood cells. Therefore the aqueous extract of *M. acuminata* root at 50, 100 and 200 mg/kg body weight caused alterations in the normal levels of the haematological parameters and tissue enzymes which might adversely affect the normal functioning of biomolecules and by extension, the organs.

Keywords: Massularia acuminata, Enzyme activities, Heamatological parameter.

1. INTRODUCTION

Many of the effects of drugs and side effects caused by drug interaction with tissues are of a pharmacologic nature that modifies normal responses. Drugs and chemicals can produce, in the human body, the same spectrum of tissue changes that occur in diseases not caused by drug and chemicals, therefore there is hardly a drug or chemical that will not cause an adverse reaction (Irey, 1982). In the same vein, there is no organ or tissue that may not be the site of an "adverse drug reaction" (ADR). Particularly likely locations of ADR are organs and tissues related to absorption, metabolism, storage and excretion of particular drugs and chemicals such as the gastrointestinal system, the liver and the kidney. However, no organ, system or tissue is exempted from the possibility of being an ADR site (Irey, 1982).

In our previous study of the aqueous extract of *Massularia acuminata* root, a plant also known as pako ijebu (Yoruba), a small tropical plant found undergrowth of closed moist forest growing up to 5m high and is distributed from Sierria Leone through Nigeria to Zaire, It was concluded that the aqueous root extract has androgenic potentials which may stimulate male sexual maturation and enhance normal testicular functions.

However, apart from the androgenic potential studies of *Massularia accuminata* root, information on its toxicological implications appears scanty to the best of our knowledge. Therefore this study was undertaken to provide information on the possible inhibitory effect of the aqueous root extract of the plant or their metabolites on enzyme activities and heamatological parameters of some selected tissues.

2. MATERIALS AND METHODS

Plant Material and Authentication

The sample of the plant was obtained from herb sellers at Ijebu-ode, Ogun state, Nigeria. It was authenticated at the Forestry Research Institute of Nigeria (FRIN), Ibadan, Nigeria where a voucher specimen (FHI107644) was deposited.

Experimental Animals

Male female white albino rats (*Rattus novergicus*) weighing between 250-280 g and 170-200 g respectively, were obtained from the Animal Holding Unit of the Department of Biochemistry, University of Ilorin, Ilorin, Nigeria.

Drugs and Kits

The assay kits for gamma-glutamyl transferase, alkaline phosphatase, acid phosphatase, bilirubin and urea were products of human Gessellscchaft fiir

Biochemica und Diagnostica mbh Max-planck-Ring21-D65205 Wiesbaden-Germany. All other reagents used were of analytical grade and were prepared in all glass distilled water.

Preparation of Aqueous Extract of *Massularia accuminata* Root

The method as described by Yakubu (2006) was used.

Animal Grouping and Extract Administration

For the toxicity studies, male rats were randomly grouped into four groups (A, B, C and D) of 15 animals each. Rats in groups B, C and D were administered with the plant extract once daily at 24hr interval at the dose of 50, 100, 200 mg/kg body weight respectively for 21 days. Group A, which serves as the control, received 1.0 cm^3 of the vehicle (distilled water) and was treated exactly like the test groups. All administrations were done daily at the same point time of between 0900hrs-1000hr. The experimental rats were allowed free access to rat pellets and tap water after the daily dose of the extract/distilled water.

Collection of Biological Fluids

The rats were anaesthetized in a jar containing cotton wool saturated with ether. Blood samples were quickly collected by jugular incision into separate bottles with anticoagulant and clean, dry centrifuge tubes (Akanji and Ngaha, 1989).

Determination of Specific Enzyme Activity

Hillman method and Kinetic colorimetric method were used to determine alkaline phosphatase, acid phosphatase, gamma-glutamyl transferase (GGT).

W.B.C. (Total Count)

Procedure for counting of white blood cell (WBC), 0.02ml of blood sample was added into 0.38ml of WBC diluting fluid to make 1/20 blood dilution charged and placed on microscopy, using x10 objective lens to count the WBC in all the 64 small squares Thin blood film was made and stained with Lieshman stain, before the counting of different type of white blood cells was achieved (Purves, 2004), (Ganong, 1977).

Differential Count

A drop of immersion oil was added to stained thin blood film and was examined with x100 objective lens. The different white blood cells like lymphocytes, neutrophils eosinophils, monocytes and basophils were counted with different counter. The count was recorded (Ganong, 1977).

Statistical Analysis

All significant differences were determined by ANOVA and were complimented by Duncan's multiple range tests. Significant differences were reported at $p < 0.05$.

3. RESULTS

Enzyme Activities Studied

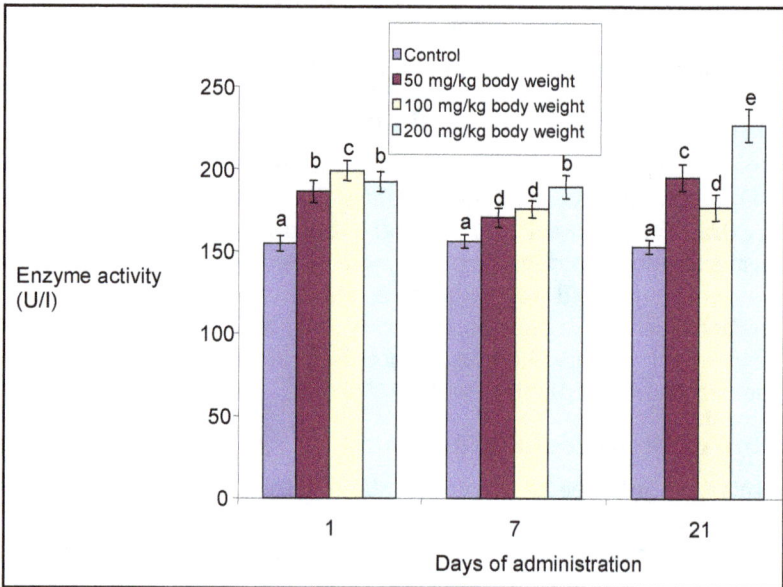

Figure 15.1: Effect of Administration of Aqueous Extract of *Massularia acuminata* Root on the Male Rat Liver Alkaline Phosphatase Activity.

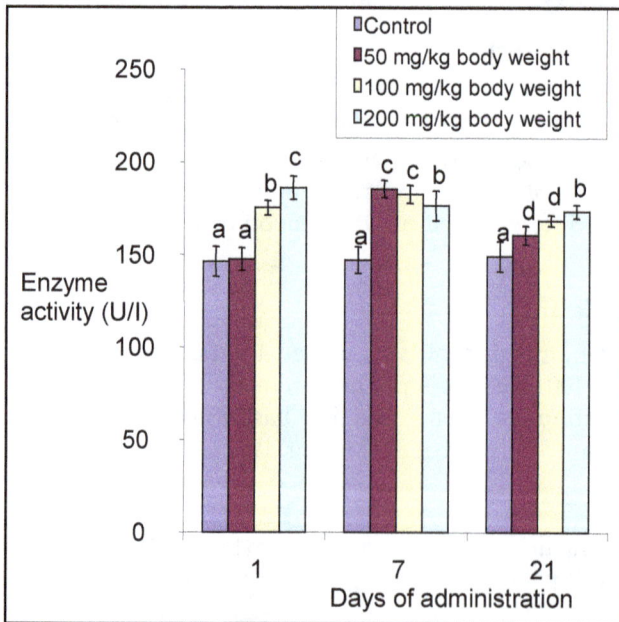

Figure 15.1: Effect of Administration of Aqueous Extract of *Massularia acuminata* Root on the Male Rat Testicular Alkaline Phosphatase Activity.

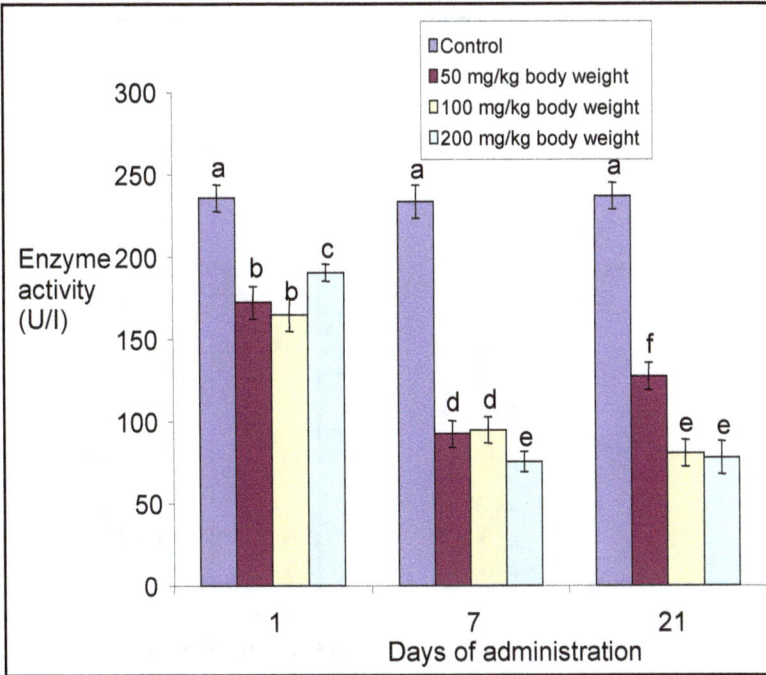

Figure 15.3: Effect of Administration of Aqueous Extract of *Massularia acuminata* Root on the Male Rat Kidney Alkaline Phosphatase Activity.

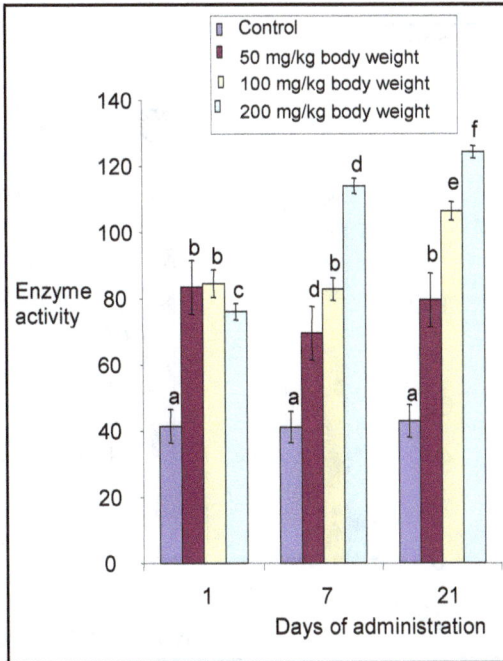

Figure 15.4: Effect of Administration of Aqueous Extract of *Massularia acuminata* Root on the Male Rat Serum Alkaline Phosphatase Activity.

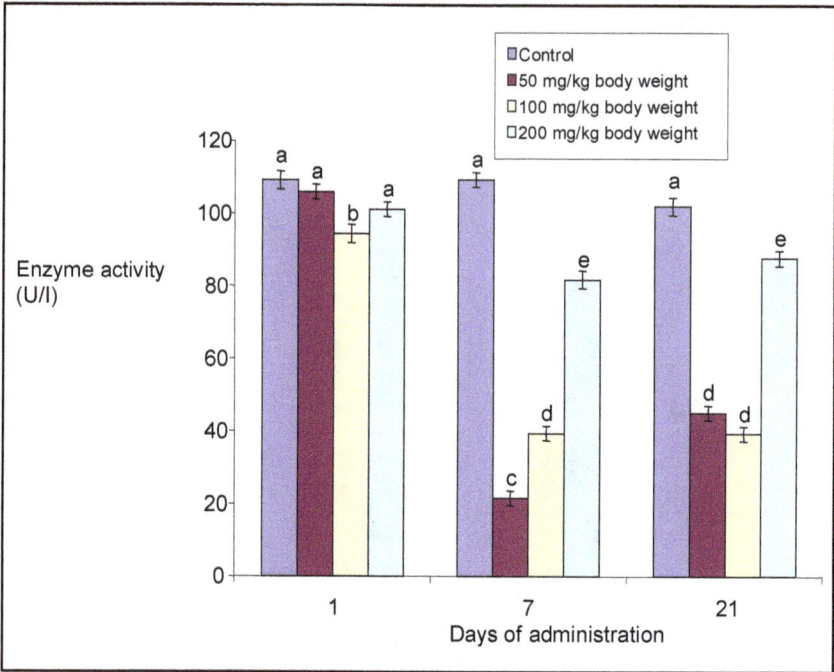

Figure 15.5: Effect of Administration of Aqueous Extract of *Massularia acuminata* Root on Gamma Glutamyl Transferase Activity of Male Rat Liver.

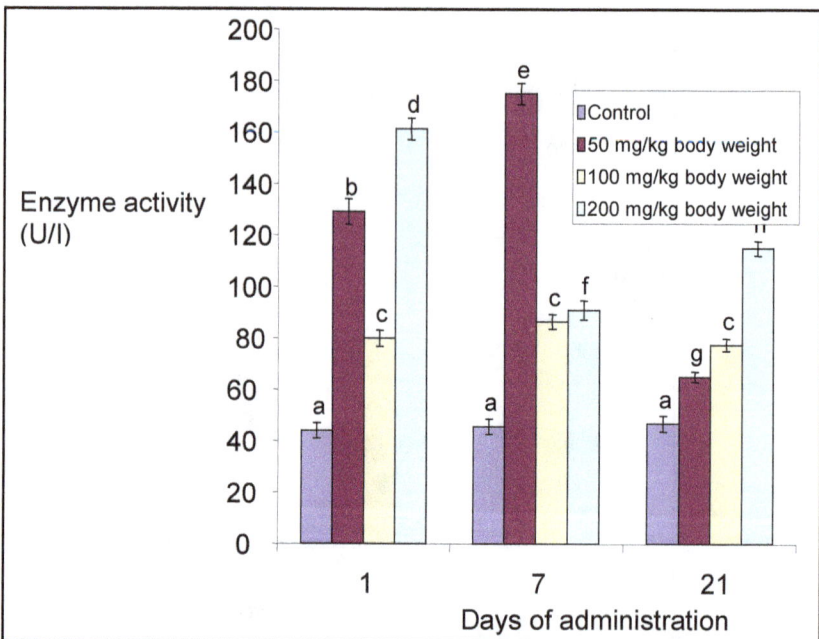

Figure 15.6: Effect of Administration of Aqueous Extract of *Massularia acuminata* Root on Gamma Glutamyl Transferase Activity of Male Rat Testes.

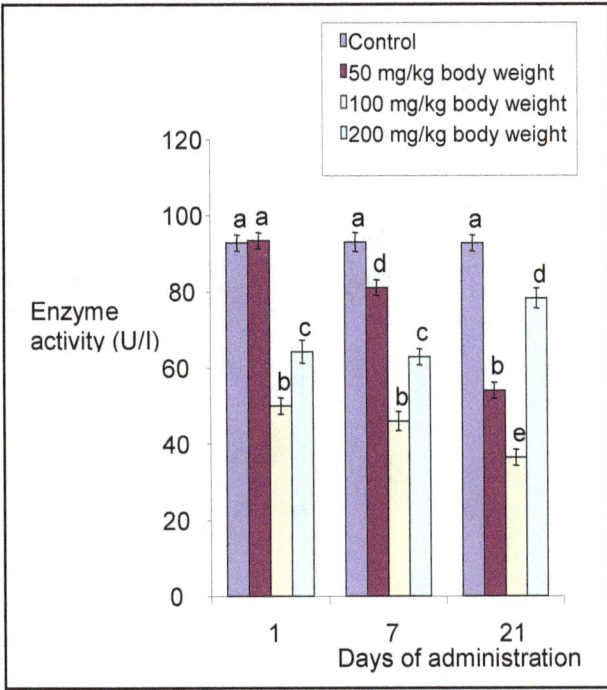

Figure 15.7: Effect of Administration of Aqueous Extract of *Massularia acuminata* Root on Gamma Glutamyl Transferase Activity of Male Rat Kidney.

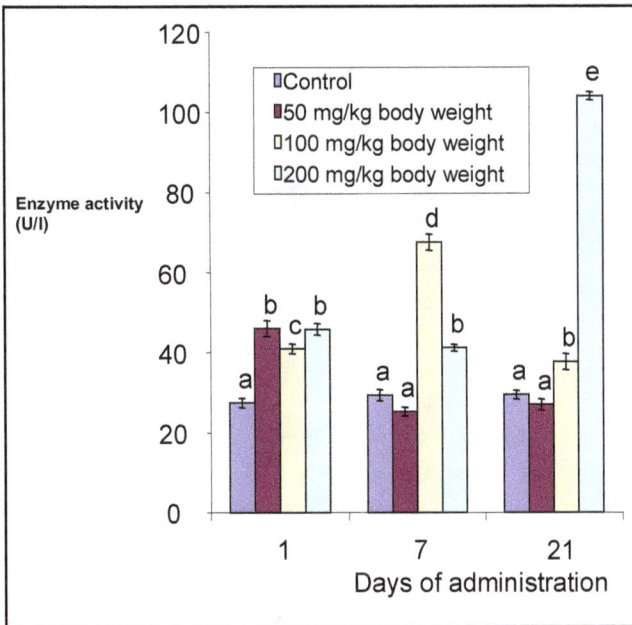

Figure 15.8: Effect of Administration of Aqueous Extract of *Massularia acuminata* Root on Gamma Glutamyl Transferase Activity of Male Rat Serum.

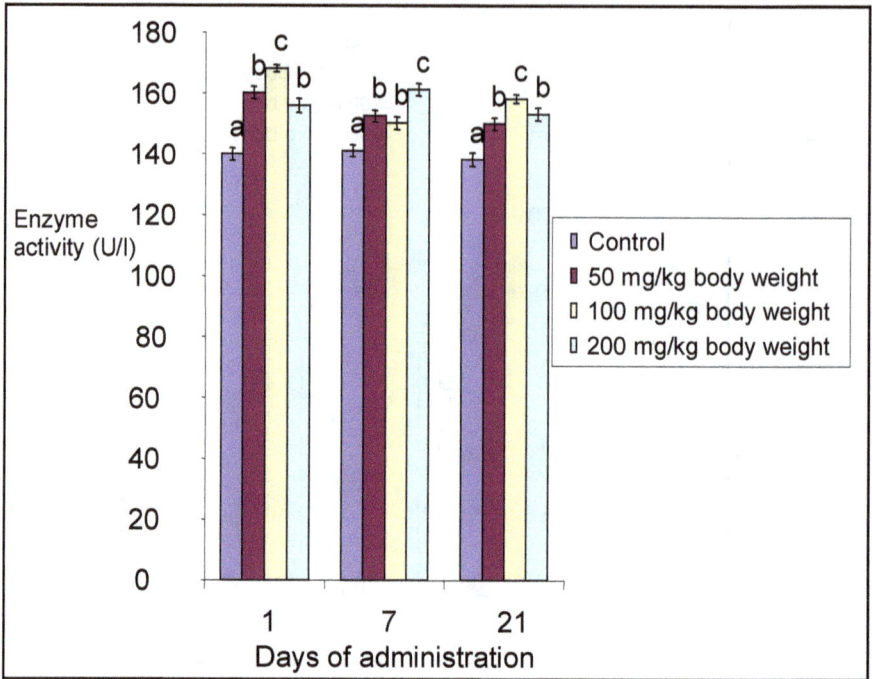

Figure 15.9: Effect of Administration of Aqueous Extract of *Massularia acuminata* Root on the Male Rat Testicular Acid Phosphatase Activity.

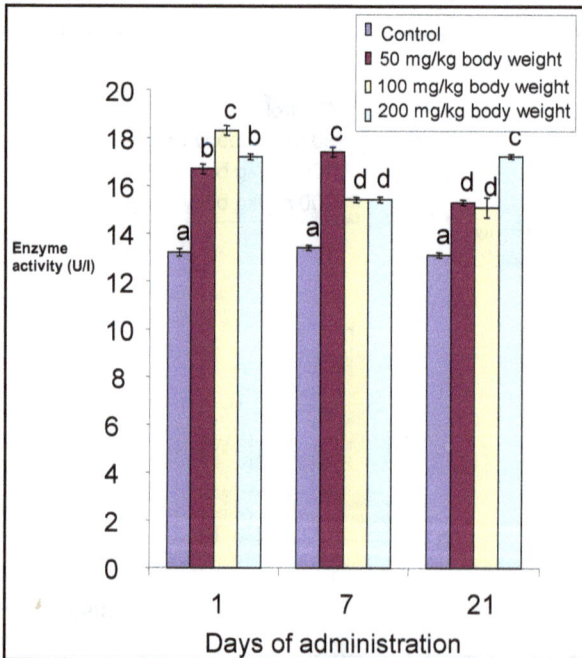

Figure 15.10: Effect of Administration of Aqueous Extract of *Massularia acuminata* Root on the Male Serum Acid Phosphatase Activity.

4. DISCUSSION

Cellular Enzymes

The measurement of the activities of 'marker' enzymes in tissues and body fluids can be used in assessing the degree of assault and the toxicity of a chemical compound on organ/tissues (Malomo, 2000), (Yakubu *et al.*, 2003). Measurement of enzyme activities can also be used to indicate tissue cellular damage caused by a chemical compound long before histological changes (Akanji, 1986).

Alkaline Phosphatase

Alkaline phosphatase is often used to assess the integrity of the plasma membrane (Akanji *et al.*, 1993). It may also be used to indicate bone or liver disease. It is a 'marker' enzyme for the plasma membrane and endoplasmic reticulum (Wright, 1974), (Shahjahan *et al.*, 2004). It is frequently used to assess the integrity of the plasma membrane (Akanji *et al.*, 1993), such that any alteration in the activity of the enzyme in the tissue and serum would indicate likely damages to the external boundary of cells (plasma membrane) (Akanji *et al.*, 1993).

The significant increase in the alkaline phosphatase activity of the male rat liver and testes (Figures 15.1 and 15.2) may be attributed to induction in the enzyme synthesis probably by *de novo* (Umezawa, 1982). Such increase in alkaline phosphatase activities can constitute threat to the life of cells that are dependent on a variety of phosphate esters for their vital process since there may be indiscriminate hydrolysis of phosphate ester of the tissues. The reduction in the ALP activity in the kidney (Figure 15.3) might be attributed to either loss of membrane components (including ALP) into the extracellular fluid, the serum (Malbica, 1971), inhibition of the enzyme activity at the cellular level or inactivation of the enzyme molecule *in situ* (Umezawa, 1982). The reduction in kidney alkaline phosphatase activity might be the resultant effect of compromise of the plasma membrane as this was further confirmed by the concomitant increase in the serum ALP activity (Figure 15.4). Such reduction in the kidney alkaline phosphatase activities would hinder adequate transportation of required ions or molecules across their cell membrane and this may lead to starvation of the cells (Akanji *et al.*, 1993). and might also adversely affect other metabolic processes where the enzyme is involved such as the synthesis of nuclear proteins, nucleic acids and phospholipids as well as in the cleavage of phosphate esters (Akanji *et al.*, 1993).

Gamma Glutamyl Transferase

Gamma glutamyl transferase is a marker enzyme that can be used for detecting liver or bone disease. It is the most sensitive enzymatic indicator of hepatobiliary disease (Ramalingman, 2002). It is also another membrane-localized enzyme that plays a major role in glutathione metabolism and resorption of amino acids from the glomerular filtrate and intestinal lumen. The decrease in GGT activity of the liver and kidney (Figures 15.5 and 15.7) might be attributed to leakage from the organs to the extracellular fluid since there was corresponding increase in the serum enzyme (Figure 15.8). The decrease in GGT activity level may also be due to liver and kidney damage. The alterations brought on the liver and kidney GGT activity

Table 15.1: Effect of Administration of Aqueous Extract of *Massularia acuminata* Stem on some Haematological Indices of Male Rats

Indices	Control 1ml	50mg/kg	100mg/kg	200mg/kg	Control 1ml	50mg/kg	100mg/kg	200mg/kg	Control 1ml	50mg/kg	100mg/kg	200mg/kg
PCV	38.0±0.89[a]	38.8±2.6[a]	37.2±1.5[a]	38.6±2.3[a]	32.6±2.4[a]	52.0±3.5[b]	53.6±3.0	53.8±3.9[b]	38.2±1.1[b]	38.0±0.9[a]	38.0±2.0[a]	36.0±0.6[a]
RBC	8.2±0.4[a]	8.8±0.4[a]	7.4±0.8[a]	8.4±0.8[a]	7.6±0.5[a]	11.0±0.6[b]	11.0±0.4[b]	10.0±0.6[b]	7.0±0.63[a]	7.0±0.52[a]	8.0±0.6[a]	7.9±0.21[a]
WBC	35.4±5.8[a]	29.4±2.1[b]	17.60±1.0[c]	11.6±3.3[d]	34.0±1.4[a]	17.0±1.0[b]	19.0±1.4[b]	7.0±1.8[c]	35.2±0.7[a]	18.0±2.0[b]	19.2±1.6[b]	14.4±0.6[a]
NEUTROPHIl	16.6±1.2[a]	14.8±4.5[a]	16.30±2.2[a]	14.2±3.0[a]	15.0±1.2[a]	15.0±1.6[a]	16.0±1.6[a]	17.0±1.0a	17.0±0.63[a]	10.4±0.2[b]	8.4±0.4[b]	5.2±0.1[c]
LYMPHOCYTE	66.0±6.8[a]	87.0±2.7[b]	83.6±2.7[b]	80.4±2.8[b]	65.0±7.7[a]	82.0±2.5[b]	86.0±0.5[b]	86.0±1.0[b]	65.2±4.1[a]	84.6±4.3[b]	102.0±0.6[c]	85.6±3.2[b]

PCV: Packed Cell Volume; RBC: Red Blood Cell; WBC: White Blood Cell.

by the extract may adversely affect the metabolism of glutathione and resorption of amino acids from the glomerular filterate and intestinal lumen. However, the increase of the GGT activity of the testes (Figure 15.7) might indicate impaired function of sertoli cells (Ramalingman, 2002).

Acid Phosphatase

Acid phosphatase in its own case is a 'marker' enzyme for the lysosomal membrane Elevated acid phosphatase activity of male rat testes (Figure 15.9) could be due to increase in functional activity of the testes. It can also be attributed to the effect of the component of the extract leading to *de novo* synthesis of the enzyme molecule *in situ* (Umezawa, 1982), (Yakubu *et al.*, 2003) or loss of other protein from the tissue (Yakubu *et al.*, 2003) (Akanji, 2000). However, the increase in the serum acid phosphatase activity (Figure 10) might be due to contribution from other organs of the rat (Latner, 1961).

Haematological Parameters

Assessment of haematological parameters can be used to determine the extent of deleterious effect of foreign compound including plant extract on the blood. It can also be used to explain blood relating functions of chemical compound/plant extract (Yakubu *et al.*, 2007). Haematological parameters can also be used to investigate and determine blood diseases that affect the production of blood and its components, such as PCV, RBC, WBC, platelets, neutrophils, lymphocytes *etc*. The increase in the RBC and PCV might be due to increase in the rate of production of erythrocytes. It may also shows that that the extract exhibited erythropoeitin potential (Ganong, 1977). This might be that the oxygen-carrying capacity of the blood and amount of oxygen delivered to the tissues was enhanced by the extract administration. The decrease in WBC at all dose levels (Table 15.1) is an indication that there is no challenge on the immune system, while further decrease might indicate depression of the immune system by the plant extract. The presence of tannins might have also contributed to the decrease in the production of the WBC because of its potential antiviral, anti-inflammatory, antibacterial and antiparasitic effect and their ability to stop infection while they continue to heal wound internally with little cytotoxicity.

5. CONCLUSIONS

Results from enzyme and haematological studies showed that the aqueous extract of *Massularia acuminata* root might affect the organs/tissues. It caused labilization in some organs whereas it induced the synthesis of enzymes in some other. The effect of the extract was also pronounced at the various dose levels. Such alterations in the normal levels of the tissue and serum enzymes and haematological parameters might adversely affect the normal functioning of the biomolecules and by extension, the organs.

6. ACKNOWLEDGMENTS

Part of this research was supported by the internation Foundation for Science (IFS), Stockholm, Sweden, through a grant to Dr. M.T. Yakubu (grant number F/3977-1 and F/3977-2).

7. REFERENCES

1. Akanji, M.A and Ngaha, E.O (1989). Effect of repeated administration of berenil or urinary excretion with corresponding tissue pattern in rats. *Pharmacol. Toxicology.* 64, 272—275.

2. Akanji, M.A and Yakubu, M.T (2000). Tocopherol protects against metabisulphite-induced tissue damage in rats. *Nig. J. Biochem. Mol. Biol.*, 15(2); 179- 183.

3. Akanji, M.A. (1986). A comparative biochemical study of the interaction of some tyrpanocides with rat tissue cellular system. *Ph.D Thesis*, University of ile-ife, Nigeria.

4. Akanji, M.A.,Olagoke, O.A and Oloyede, O.B (1993). Effect of chronic consumption of metabisulphate on the integrity of rat cellular system. *Toxicol.* 81; 173- 179.

5. Collins A.J and Lewis D.A (1971). Lysosomal enzyme level in blood of arithritic rats. *Biochem. Pharmacol.* 28, 251-253.

6. Ganong, W.F. (1977). *Review of Medical Physiology* 9th Edition. Lange Medical Books/McGraw-Hill Medical Publishing Division, London. pp. 214-517.

7. Irey,N.S (1982). When drug is induced? In; Pathology of drug induced and toxic disease. Robert H. Ridell (ed) Churchill Livingstone, pp. 1-18.

8. Latner,A.L and Skillen, A, (1961). Clinical application of the dehydrogenase isoenzyme, A simple method for their detection. *Lancet* 1286-1288.

9. Malomo,S.O (2000). Toxicological implication of ceftriaxone administration in rats. *Nig. J. Bichem. Mol. Biol.* 15(1); 33-38.

10. Malbica,J.O and Hart, L.G (1971). Effect of adenosine triphosphate (ATP) and some anti-inflammatory agents on purified lysosomal fraction having high acid phosphate and labile gluconosidase activity. *Biochem. Pharmacol.*, 20; 2017-2026.

11. Purves, William, K., David Sadava, Gordon,.H. Orians, H and Craig Heller (2004). *Life; the Science of Biology*, 7th edition, Sunderland, Mass; Sinauer Associates, 954.

12. Ramalingman,V and Vimaladevi,V (2002). Effect of mercuric chloride on membrane-bound enzyme in rats testis. *Asian J. Androl.*, 4;309-311.

13. Rotimi, V.O and Mosadomi, H.A (1987). The effect of crude extract of nineAfrican chewing sticks on oral anaerobes. *Pharmacological J.*, 150-153.

14. Shahjahan, M, Sabitha, K.E, Mallika,J. and Shyamala-Devi, C.S (2004).Effect of Solanum trilobatum against tetrachloride carbon induced hepatic damage in albino rats. *Indian J. Med. Res.*, 120: 194-198.

15. Umezawa, H and Hooper, R (1982). Amino-glycoside antibiotic. Springer-Verlag- Berlin, Hadelberg, New York.

16. Wright, P.J and Plummer, D.T (1974). The use of urinary enzyme measurement to detect renal damages caused by nephrotoxic compounds. *Biochem. Pharmacol.*, 12: 65.

17. Yakubu, M.T (2006). Aphrodisiac Potentials and toxicological evaluation of aqueous extract of *Fadogia agrestis* (Schweinf, ExHiern) stem in male albino rats. *Ph.D. Thesis*, University of Ilorin, Ilorin.

18. Yakubu, M.T., Bilbis, L.S., Lawal, M.A (2003). Effect of repeated administration of sildenafil citrate on selected enzyme activities of liver and kidney of male albino rats. *Nig. J. Pure and Appl. Sci.*, 18: 395-400.

19. Yakubu M.T.Akanji and A.T., 2007. Heamatological evaluation in male albino rats following chronic administration of aqueous extract of *Fadogia agrestis* stem. *Pharmacog. Mag.*, 3(1): 34-38.

Chapter 16

Sub-chronic Toxicity of Detogen-B: A Herbal Detoxifying Product

Akinbosola Jibayo Philips, Orgah Emmanuel Adikwu,
Salihu Timothy, Oguntoyinbo Abel and Idowu Ifeoluwa

Product Development and Quality Assurance Department,
Nigeria Natural Medicine Development Agency (FMST),
9, Kofo Abayomi Street, Victoria Island Lagos, Nigeria
E-mail: jakinbosola@gmail.com

ABSTRACT

Subchronic toxicity of Det-B, a Polyherbal formulation used for the management of various debilitating conditions was evaluated.

Keywords: Detogen-B, Acute toxicity and Sub chronic toxicity.

1. INTRODUCTION

Plants derived medicine or herbal medicine are generally regarded as safe based on their long standing use in various cultures (Mosihuzzaman and Iqbal Choudhary, 2008). Herbal medicines remain the mainstay of the health care system in developing countries and are also gaining increasing popularity in developed countries where orthodox medicines are predominant. Herbal medicine has the advantage of being effective as well as being a cheap source of medical care (Ogbonnia *et al.*, 2010a). There is a growing global disillusion with orthodox medicines, coupled with the misconception that herbal products being natural were devoid of adverse effects associated with conventional and allopathic medicines. This is fuelling the rising demand for herbal medicine.

With increasing herbal medicine consumption, concerns have been raised over the lack of quality control and scientific evidence of the efficacy and safety of herbal medicine (Firenzuoli and Gori 2007, Seef 2007 and Tang *et al.*, 2008). Herbal medicines are most often poly-herbal, being prepared from mixtures of different plant parts obtained from various plant species and families and may contain multiple bioactive constituents that could be difficult to characterize (Ogbonnia *et al.*, 2010b). The bioactive principles in most herbal preparations are not always known and there are possibilities of interaction within the mixture which could be deleterious, synergistic or antagonistic. The quality as well as the safety criteria for herbal drugs may be based on a clear scientific definition of the raw materials used for such preparations. Also herbal medicine may have multiple Physiological activities and could be used in the treatment of various disease conditions (Pieme *et al.*, 2006).

Traditionally herbal medicine might be administered in some disease states over a long period of time without proper dosage monitoring and consideration of toxic effects that might arise from such prolonged usage (Ogbonnia *et al.*, 2010c). This implies a long term deposition of would be toxic constituents of the drug. The danger associated with the potential toxicity of herbal therapies employed over a long period of time demands that healthcare institutions be kept abreast of the possible incidences of renal and hepatic toxicity resulting from the ingestion of such herbal drugs (Tedong *et al.*, 2007; Ogbonnia *et al.*, 2008).

Detogen-B is an herbal formulation constituted of multiple herbal materials and used by the formulator as systemic detoxifier and diuretic agents. Being a formulator product with no data on safety and quality, it is essential to conduct an evaluation of sub-chronic toxicity to establish its safety and obtain data on any potential adverse effects of this formulation considering its use in a primary health care environment. The study examined the safety of the oral use of Detogen-B and was conducted in accordance with the guidelines established by the Organization for Economic Cooperation and Development (OECD, TG 408) for the testing of chemicals and recent Good Laboratory Practice (GLP) Regulations.

2. MATERIALS AND METHODS

Test Material

Detogen-B herbal drug was supplied by a natural healthcare centre in Nigeria in a powdered form. The therapeutic label claimed that the formulation was prepared with unspecified quantities of some listed plants and the prescribed daily recommended dose for human adult was 1 teaspoon 3-4 times daily and children between 5 to 12 years is 1/2 teaspoon 3-4 times daily. The drug is used as systemic detoxifier and as diuretic agent. 35g of Detogen-B powder was dissolved in 500 ml water, which gave a stock concentration of 70mg/ml. The drug extract was kept in a refrigerator until the time of drug administration. It was withdrawn from the fridge and upon attainment of room temperature (28°C±2), appropriate volumes of the preparation were administered directly to the experimental animals via a stainless steel canula.

Animal

Swiss albino mice (40±2) g of either sex was used for the acute toxicity studies, while Adult albino rats (98 – 174) g were used for sub chronic profiling. The animals were supplied by the Animal Facility Centre of Nigeria Natural Medicine Development Agency. The animals were fed with improvised Ladoke feed (grower mash) and had free access to water and maintained under standard conditions of humidity, temperature and 12/12 ratio light and darkness cycle. The animals were allowed to acclimatize before the commencement of the study and standard protocol was drawn up in accordance with the Good Laboratory Practice (GLP). The principles of laboratory animal care were strictly adhered in this study.

Acute Toxicity Studies

Acute toxicity (LD_{50}) was estimated in mice following Lorke's method (1983).

Sub-chronic Toxicity Studies

Male and female rats were selected randomly and divided into four experimental groups of five rats each of both sexes. After fasting the animals overnight, the control group received 0.2ml saline solution daily. The remaining three groups were given 600, 1,200 and 1,800mg/kg of Detogen-B orally for 30 days. On the first day (Day 0) initial weights of the animals were recorded following the administration of the drug extract. The animals were weighed on a sensitive balance every five days from the start of the treatment to note any body weight variation.

At the end of the treatment period, the rats were sacrificed under light chloroform vapour and the blood collected via cardiac puncture in two tubes: one in EDTA for immediate analysis of haematological parameters, the second with heparin to separate plasma for biochemical estimation. The liver, heart, lung, pancreas and kidney were dissected out. Washed with saline solution and weighed (Ogbonna *et al.*, 2010). The collected blood was centrifuged within 20mins of the collection at 4000rpm for 10 min to obtain the plasma, which was analyzed for total cholesterol, total glycerin and HDL – cholesterol levels of precipitation and modified enzymatic procedures from Sigma Diagnostics (Wasan *et al.*, 2001). LDL – cholesterol levels were calculated using Friedwald equation (Crook, 2006). Plasma was analyzed for alanine amino transfers (ALT), aspartate aminotransferase (AST), and Creatinine by standard methods (Sushura *et al.*, 2006), Hematocrit was estimated using the method of (Ekaidem *et al.*,2006). Hemoglobin contents were determined using Cyanmethaemoglobin (Drabkin) method (Ekaidem *et al.*, 2006).

Statistical Analysis

The results were expressed as mean±standard error of the mean (SEM). The data were analyzed using one way analysis of variance (ANOVA) (Petersen, 1985 and Scheffe, 1959). 95 per cent level of significance ($p \leq 0.05$) was used in the statistical analysis.

3. RESULTS

Acute Toxicity Studies

There was no mortality recorded in the mice that received different doses of Det-B throughout the 72hr observational study even at the highest dose of 5000mg/kg.

Body Weight

From Day 0 – Day 30, (Table 16.1), there were changes in the body weight of all the rats in the treated groups throughout the study. Rats treated with doses 600mg/kg and 1,800mg/kg Det-B gained body weight between day 5 and 15, then a decline between day 15 and 25 and a slight gain in weight on day 30, while the control group experienced a progressive gained in weight between day 0 and day 30. Rats treated with 1,200mg/kg had a reduction in the body weight on Day 5, slight increase on day 10, slight decrease on the day 15 and increase in weight between day 20 and 30 when compared with the control groups.

Table 16.1: Mean Body Weight Changes in Rat Treated with Various Doses (600–1,800mg/kg) of Detogen-B for 30 Days Against the Control

Treatment	Day 0	Day 5	Day 10	Day 15	Day 20	Day 25	Day 30
Control	98±6.9	120±9.0	138±10.4	147±11.4	155±15.0	158±15.0	167±14.8
600mg/kg	103±8.1	185±17	193±19.6	185±18.2	177±20.2	177±22,6	189±21.8
1200mg/kg	174±17	154±15	156±13.7	147±17.4	153±11.8	152±12.7	161±21.1
1800mg/kg	124±14.7	161±23	166±21.1	170±21.1	156±16.9	161±17.9	167±20.0

n=5. value (Mean±SEM), *P≤0.05 Significant (one way ANOVA) versus control.

Organ Weight

In Table 16.2, the gain in weights of organs was significant (P<0.05) in the Lung, Heart, Liver and Kidney in the rats treated with doses 600mg/kg and 1,800mg/kg Det-B compared with the control group and the reduction in weights of these organs in the group administered 1200mg/kg dose of Det-B was significant compared to the control. However, changes observed in the weight of the pancreas compared with the control group was not significant.

Table 16.2: Mean Organ Weight Changes in Rat Treated with Various Doses (600–1,800mg/kg) of Detogen-B for 30 Days Against the Control

Treatment	Lung	Heart	Pancreas	Liver	kidney
Control	1.18±0.06	0.50±0.04	0.30±0.03	5.40±0.60	0.48±0.58
600mg/kg	1.08±0.1*	0.68±0.04*	0.50±0.06	7.00±0.63*	0.48±0.12*
1200mg/kg	1.12±0.07*	0.48±0.04*	0.26±0.04	4.80±0.48*	0.44±0.74*
1800mg/kg	1.82±1.3*	0.56±0.05*	0.72±0.32	6.30±0.35*	1.26±0.29*

n=5.value (Mean±SEM), *P≤0.05 Significant (one way anova) versus control.

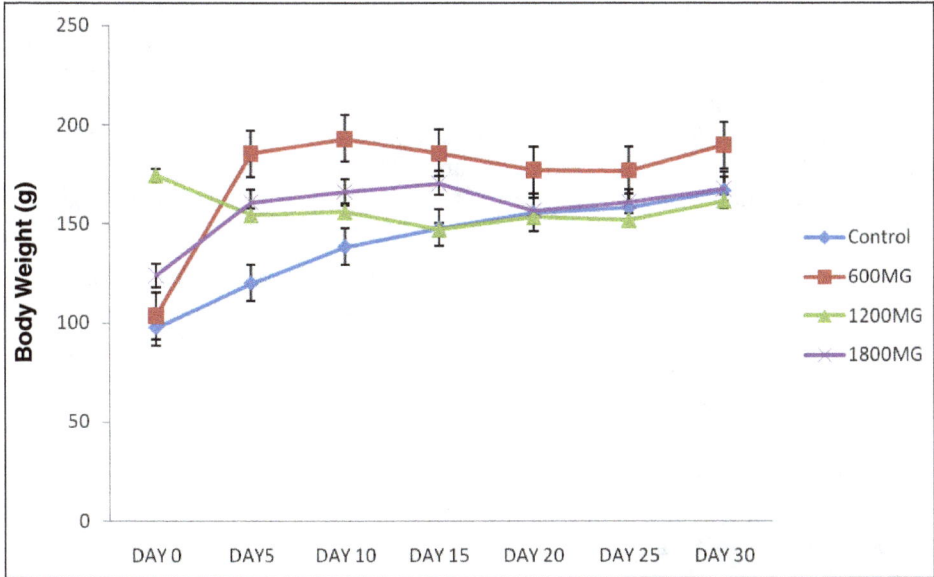

Figure 16.1: Mean Body Weight Changes in Rat Treated with Various Doses (600–1,800mg/kg) of Detogen-B for 30 Days Against the Control. n=5. Value (Mean±SEM), *P≤0. 05 Significant (one way anova) versus control.

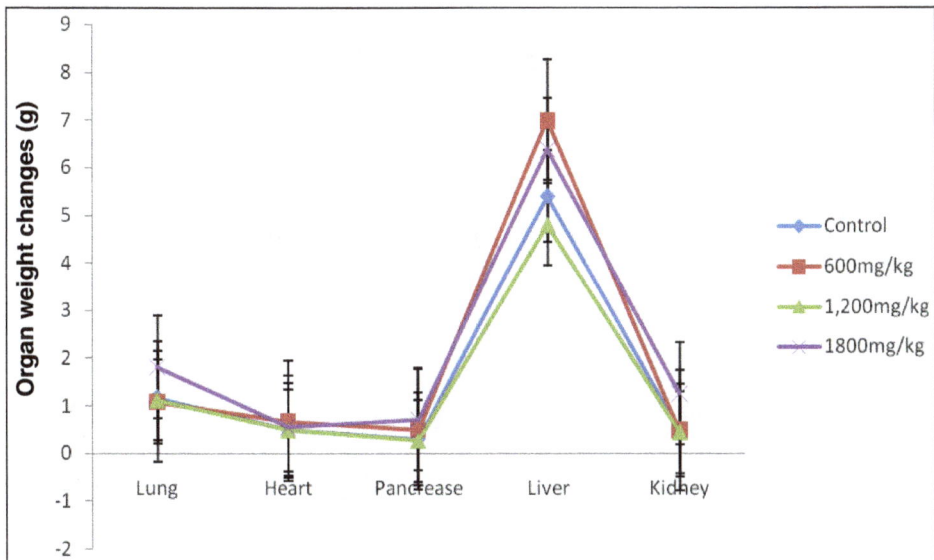

Figure 16.2: Mean Organ Weight Changes in Rat Treated with Various Doses (600–1,800mg/kg) of Det-B for 30 Days Against the Control. n=5. Value (Mean±SEM), *P≤0.05 Significant (one way anova) versus control.

Biochemical Profile

There was a significant (P<0.05) reduction in TC, HDL, LDL values in the rats treated with 600mg/kg – 1,800mg/kg dose of Detogen-B (Table 16.3), compared with the control groups. No significant reduction in the triglyceride (TG) compared to control while the reduction in total protein (TP) values was not significant between the rats that were treated with Detogen-B and and control group. The reduction in alanine amino transfearease (ALT) values compared with the control group were not significant while the increase in aspartate aminotransferase(AST) values compared with control groups at doses 1,200mg/kg and 1,800mg/kg was not significant. The increase in serum creatinine values at 1,800 mg/kg and decrease at 600mg/kg and 1,200mg/kg was not significant compared to the control.

Table 16.3: Mean Biochemical Profile in Rat Treated with Various Doses (600–1,800mg/kg) of Detogen-B for 30 Days Against the Control

Treatment	Control	600mg/kg	1,200mg/kg	1,800mg/kg
AST (u/l)	33.20±1.59	33.40±2.6	43.80±2.6	37.20±4.1
ALT (g/dl)	30.40±1.03	26.40±1.5	28.20.±1.7	2.40±2.4
TC (mmol/l)	164±10.8	110.±7.8*	129±8.4*	99±4.3*
HDL (mmol/l)	31.80±3.18	21.4±2.5*	22.8±1.96*	19.00±1.58*
LDL (mmol/l)	124±10.6	82.00±8.8*	100±10.5*	74.6±4.7*
TG (mmol/l)	40.0±2.9	32.20±1.9	33.4±3.8	29.6±1.6
TP (g/dl)	6.16±0.16	5.92±0.06	5.66±0.19	5.7±0.11
CREA (mg/dl)	0.82±0.03	0.68±0.37	0.74±0.24	0.92±0.37

n=5. Value (Mean±SEM), *P≤0. 05 Significant (one way anova) versus control.

Aspartate aminotransferase (AST), Alanine aminotransferase (ALT), total cholesterol (TC), high density lipoprotein (HDL), low density lipoprotein (LDL), triglyceride (TG), total protein (TP) and Creatinine.

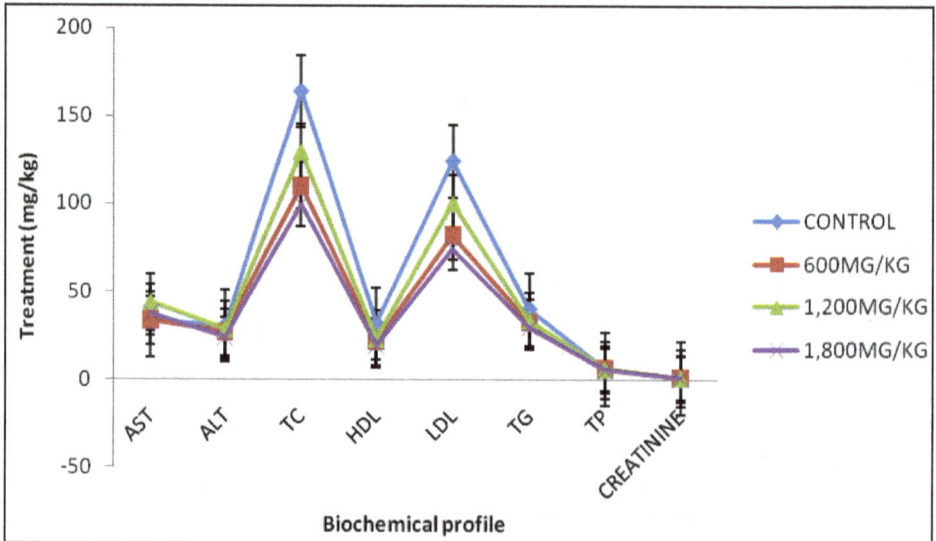

Figure 16.4: Mean Biochemical Profile in Rat Treated with Various Doses (600–1,800mg/kg) Det-B for 30 Days Against the Control.
n=5. Value (Mean±SEM), *P≤0. 05 Significant (one way anova) versus control.

Hematological Profile

In Table 16.4, there was a significant decrease in RBC of the rats administer 600 and 1200mg/kg doses of Det-B compared to control, while a significant increase in Red Blood Cells (RBC) count at 1800mg/kg of Detogen B compared to the control. There were no significant changes in White Blood Cell (WBC), hemoglobin (HGB), Haematocrit (HCT). Mean Corpuscular Volume (MCV), Mean Corpuscular Hemoglobin (MCH), Mean Corpuscular Hemoglobin Concentration (MCHC), Platelet (PLT) and Procalcitonin (PCT) Counts compared to the control.

Table 16.4: Mean Hematological Profile in Rat Treated with Various Doses (600–1,800mg/kg) of Detogen-B for 30 Days Against the Control

Treatment	Control	600mg/kg	1,200mg/kg	1,800mg/kg
WBC (10^3/l)	12.7±2.84	11.3±1.26	56.6±30.5	8.15±0.69
RBC (10^6/l)	5.95±0.38	5.44±0.32*	3.92±1.03*	6.23±0.10*
HGB (g/dl)	11.0±0.68	10.2±0.55	10.2±0.36	11.9±0.05
HCT (per cent)	33.7±2.37	31.2±1.88	26.2±5.68	35.3±0.05
MCV(fl)	56.6±0.94	57.4±0.87	64.1±5.63	56.7±0.90
MCH (pg)	18.5±0.37	18.8±0.23	70.9±34.7	15.9±1.60
MCHC (g/dl)	32.8±0.32	32.8±0.68	93.3±40	33.7±0.19
PLT(10^9)	570±116	688±59.3	845±259	628±165
PCT (per cent)	0.42±0.20	0.52±0.11	0.42±0.24	0.18±0.04

n=5 value (Mean±SEM), *P≤0. 05 Significant (one way anova) versus control.

WBC,RBC, HGB, HCT,MCV, MCH, MCHC, PLT-Platelet and PCT.

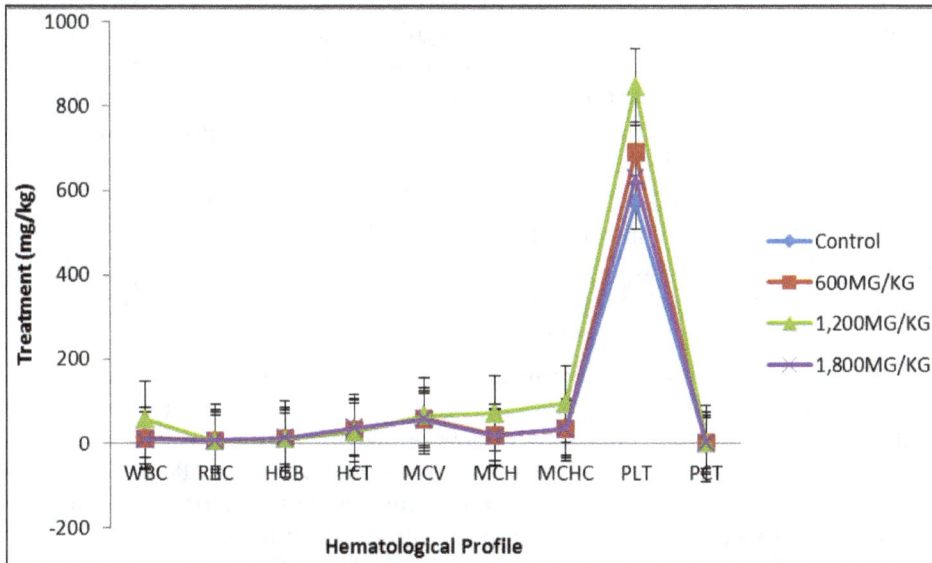

Figure 16.4: Mean Hematological Profile in Rat Treated with Various Doses (600–1,800mg/kg) of Det-B for 30 Days Against the Control.
n=5 value (Mean±SEM), *P≤0. 05 Significant (one way anova) versus control.

4. DISCUSSION

Herbal medicines have received greater attention as a medical alternative in clinical therapy in recent times leading to a subsequent increase in their demands (Sushruta *et al.*, 2006). In rural communities, exclusive use of herbal medicines, prepared and dispensed by herbalists without formal training for the management of various ailments is still a very common practice requiring that experimental screening method be established to ascertain the safety and efficacy of these herbal products. Detogen-B is one of such herbal formulation, formulated to treat various debilitating conditions.

Rats treated with doses 600mg/kg and 1,800mg/kg Det-B gained body weight between day 5 and 15, then a decline between day 15 and 25 and a slight gain in weight on day 30, while the control group experienced a progressive gained in weight between day 0 and day 30. Rats treated with 1,200mg/kg had a reduction in body weight on Day 5, slight increase on day 10, slight decrease on the day 15 and increase in weight between day 20 and 30, these changes were not significant compared with the control groups. This slight decrease could be due to breakdown of adipose tissue by the product.

In a similar fashion, there were significant $P<0.05$ changes in the weight of the lung, heart, liver and kidney but no significant changes in the weight of its pancreas when compared with the control groups. There was a significant ($P<0.05$) decrease in TC, HDL, LDL values in the rats treated with 600mg/kg – 1,800mg/kg dose of Detogen-B compared with the control groups. No significant decrease in triglyceride (TG) compared to control while the decrease in total protein (TP) values was not significant between the rats that were treated with Detogen-B and and control group. The decrease in alanine amino transfarease (ALT) values compared with the control group was not significant and the increase in aspartate aminotransferase(AST) values compared with the control group at doses 1,200mg/kg and 1,800mg/kg was also not significant. The increase in serum creatinine values at 1,800 mg/kg and decrease at 600mg/kg and 1,200mg/kg were not significant compared to the control.

No significant increase ($P>0.05$) in AST values in all the treated rats compared with the control groups, indicating that the formulation may not have any adverse effects on the cardiovascular functions. Also non significant reduction ($P>0.05$) in the ALT values in all treated rats compared with the control groups shows that the formulation has no deleterious effects on the liver functions.

In Figure 16.3, there were significant reduction ($P<0.05$) in the TC, HDL, LDL and TG values in all the treated rats compared with the control groups indicating that the formulation has beneficial effects on cardiovascular risk factors. Non significant reduction in the creatinine values at doses 600 and 1,200mg/kg in the treated rats respectively and a non significant increase in the value of creatinine at a dose 1,800mgkg compared with the control groups coupled with the decrease in the values of TP in all the treated groups compared with the control suggests that the formulation had no deleterious effects on the renal functions.

As it is presented in Figure 16.4, there were significant reduction (P<0.05) in the RBC values at doses of 600mg/kg and 1,200mg/kg in the treated group. Also, there was a significant increase in the treated groups in a dose of 1,800mg/kg compared with the control groups. Non significant decrease in all the treated groups compared with the control showed that the product has no hematinic effect. Changes in the values of WBC in all the treated groups were observed. But at dose 1,200mg/kg there was no significant increase in the value of WBC compared with the control groups.Also, there was an increase in the PLT values in all the treated groups compared with the control group.

5. CONCLUSIONS

The non-lethal effect of the formulation at 5000 mg/kg in mice suggest the product is not toxic, any compound or drug with an oral LD50 estimates greater than 1000 mg/kg body weight could be considered to be of low toxicity and safe (Clarke and Clarke 1967).

The slight decrease in body weights at some stages of the therapy could be due to breakdown of adipose tissue by the product. There was increase in weights of internal organs by the formulation, reduction in weights of internal organ are considered sensitive indices of toxicity after exposure to toxic substances (Rosa *et al.*, 2002 and Teo *et al.*, 2002).

Also the increase in weight of the liver without a significant increase in ALT means the product has no toxic effect on the liver. No significant increase in AST values in all the treated rats compared with the control groups indicate that the formulation may not have any adverse effects on the cardiovascular functions.

The reduction in creatinine with an increase in kidney weights suggests the product is not nephrotoxic. The reduction in TC, HDL, LDL and TG values in the experimental groups could suggest a beneficial effect on cardiovascular risk factors. The formulation reduces cholesterol level at low and high doses in animals, it may be beneficial on cardiovascular risk factor.

The product is thus relatively safe for human use. However, it is important to take key note of issues around GMP in the collection of raw materials and its subsequent manufacture into the herbal product.

Conflict of interest: There is no conflict of interest in this study

6. ACKNOWLEDGEMENT

The research team wishes to acknowledge the management and staff of the Nigeria Natural Medicine Development Agency, the management and staff of Ekenna Natures Lab, Mr. Kwaptoe Ts for their assistance in carrying out this study.

7. REFERENCES

1.   Clarke, M.L.and Clarke, *E.G.C.* 1967. Veterinary toxicology. London: Bailliere Tindall.

2. Crook, M.A. 2006. *Clinical chemistry and metabolic medicine,* 7th edn. London: Hodder Arnold.

3. Ekaidem I. S., Akpanabiatu M. I. And Uboh F. E., Eka O. U. 2006. Vitamin B supplementation: Effects on some biochemical and hematological indices of rat on phenytoin administration. *Biokemistri* 18(1): 31-37.

4. Fabio Firenzuoli and Luigi Gori (2007): Herbal Medicine Today: Clinical and Research Issues, Evidence-Based Complementary and Alternative Medicine, Volume 4 (2007), S1, Pages 37-40 http://dx.doi.org/10.1093/ecam/nem096 ECAM 2007;4 (S1).

5. Lorke D (1983). A new approach to practical acute toxicity testing. *Archives of Toxicology*, 54: 275-286.

6. Organization for Economic Cooperation and Development (OECD, 2013), Guideline for the Testing of Chemical TG No. 408. Repeated Dose 90-day Oral Toxicity Study in Rodents, 2013, http://www.oecd-ilibrary.org/ environment/test-no-408- repeated-dose-90-day-oral-toxicity-study-in-rodents 9789264070707-en.

7. Ogbonnia S. O., Mbka G. O., Igbokwe N. H., Anyika E. N., Alli P. and Nwakakwa N. (2010a): Antimicrobial evaluation, acute and subchronic toxicity studies of Leone Bitters, a Nigerian poly-herbal formulation, in rodents. *Agriculture and Biology Journal North America* 1 (3), 366-376.

8. Ogbonnia S.O., Mbaka G.O., Anyika E. N., Osegbo O. M. And Igbokwe N. H. (2010b): Evaluation of acute toxicity in mice and subchronic toxicity of hdroethanolic extract of Chromolaena odorata (L.) King and Robinson (Fam. Asteraceae) in rats. *Agriculture and Biology, Journal of North America*. 1(5), 1367-1376.

9. Ogbonnia S.O., Mbaka G.O., Adekunle A., Anyika E.O., Gbolade O.E. and Nwakakwa N. (2010c): Effect of poly-herbal formulation, Okudiabet on Alloxan – Induced diabetic rats. *Agriculture and Biology journal of North America* 1 (2), 139-145.

10. Petersen RG (1985). Design and Analysis of Experiments. Marcel Dekker, New York.

11. Raza M, Al-Shabanath OA, El-Hadiyah TM and Al-Majed AA (2002). Effect of prolonged vigabatrin treatment on hematological and biochemical parameters in plasma, liver and kidney of Swiss albino mice, *Scientia Pharmaceutica*, 2002, 70, 135-145.

12. Scheffé, H. (1959). The Analysis of Variance, New York: John Wiley and Sons, Inc.

13. Seeff L B (2007). "Herbal hepatotoxicity,"*Clinics in Liver Disease*, vol. 11, no. 3, pp. 577–596, 2007.

14. Tang, J L, B. Y. Liu, and K. W. Ma (2008), "Traditional Chinese medicine," *The Lancet*, vol. 372, no. 9654, pp. 1938–1940, 2008.

15. Tedong L., Dzeufiet P. D. D., Dimo T., Asongalem E. A., Sokeng S. N., Flejou J. F., Callard P. And Kamtchouing P. (2007): Acute and Subchronic toxicity of *Anacardium occidentale* Linn. (Anacardiaceae) leaves hexane extract in mice. *Afr Journal of Traditional and Alternative Medicine* 4 (2): 140-147.

16. Wasan K. M., Najafi S., Wong, J. And Kwong M. (2001): Assessing plasma lipid levels, body weight, and hepatic and renal toxicity following chronic oral administration of a water soluble phytostanol compound FMVP4, to gerbils. *J. Pharm. Sci.* (www.ualberta.ca/-csps) 4(3): 228-233.

Chapter 17

Acute Toxicity Studies of Ethanol Extracts of the Seeds of *Physostigma venenosum* (Balf.): Biochemical and Hematological Effects on Wister Albino Rats

M.O. Aihiokhai[1], J.O. Erhabor[2], M. Idu[2] and O. Timothy[2]

[1]*Department of Plant Biology and Biotechnology,*
Faculty of Life Sciences, University of Benin,
PMB 1154, Benin City, Nigeria
E-mail: markaihiokhai@yahoo.com
[2]*Phytomedicine Unit, Department of Plant Biology and Biotechnology,*
University of Benin, PMB 1154, Benin City, Nigeria
E-mail: erhaborjoseph@yahoo.com; mcdonald.idu@gmail.com;
odtimo@yahoo.com

ABSTRACT

Seeds of *Physostigma venenosum* (Balf.), used in traditional medicine for the treatment of glaucoma, Alzheimer's disease, delayed gastric emptying and acute tetanus, underwent an extraction process by way of maceration and evaporation in ethanol. The 16 Wister rats of either sex were weighed and placed into four groups of four animals per group. Graded doses of the extracts (5, 10 and 20 mg/kg) corresponding to groups 2, 3 and 4 were separately administered to the rats in each respective group by means of an orogastric tube. The control group representing group 1 was given oral distilled water of 10 ml/kg only. Hematological analysis revealed that the white blood cells (WBC) and platelets (PLT) counts were significantly higher than those of the rats in the control group while the packed cell volume (PCV) and red blood cell (RBC) counts dropped significantly in the experimental animals as compared to the control group ($P>0.05$). This significant decrease in PCV value may indicate destructive effects on the red blood cells. Also, the WBC count which showed a significantly high value at certain concentrations of the extract administered indicates that at a higher dose level, the extract may be hematologically toxic to rats. The serum total protein and bilirubin concentrations remained unaltered throughout the experimental period when compared with the control. However, the serum levels of alkaline phosphate (ALP) and alanine aminotransferase (ALT) were significantly higher than those of the control ($P>0.05$). This increase in ALP activity confirms a possible damage to the plasma membrane while an increase in ALT activity may be attributed to the release of enzymes from damaged organs which might have resulted from change in membrane permeability of the cells. The results of the effect of the extracts suggest selective toxicity. It is an indication that the plant is not completely safe as an oral remedy.

Keywords: *Physostigma venenosum, Ethanolic extracts, Medicinal plants, Acute toxicity, Biochemistry and hematology.*

1. INTRODUCTION

Plants have always been a common source of medicaments, either in the form of a traditional preparation or as pure active principles (Akerele, 1993). Generally, medicinal plants are considered to be affordable, reliable and constitute a cultural heritage to rural and most urban people (Idu, 2010). From a "Scientific" perspective, many herbal treatments are considered experimental (Berchtolf, 1989). The reality is however, that herbal medicine has a long and respected history (Katzung, 2009). Many familiar medications of the 20th century were developed from ancient healing traditions that treated health problems with specific plants (Saxton, 1971). Today, science has isolated the medicinal properties of a large number of plants and their healing components have been extracted and analyzed. However, substances derived from plants remain the basis of a large proportion of commercial medication used today for the treatment of heart disease, high blood pressure, pains, asthma and other ailments (Cunningham, 1998).

It has been stated by the World Health Organization (WHO) that the most critical assessment of herbal medicine is safety evaluation. Hence, standard toxicological protocols should be employed for acute, sub-chronic and chronic toxicity tests. Such data is mandatory for the registration of the product with National Health Authorities. It would also enhance the confidence of Health Professionals in the use of herbal medicines (WHO, 1991, 1992). The basic premise is that toxic effects caused by a drug are similar in man and other animals (Range *et al.*, 1995). If a chemical (or drug material) produces injury to a tissue, the capacity of the tissue to regenerate or recover will largely determine the reversibility of the effect (Curtis, 2001).

Walum (1998) described acute toxicity as the adverse effect of a substance that results either from a single exposure or in multiple exposures within a short space of time (usually less than 24 hours). It is widely considered unethical to use humans as test subjects for acute toxicity research (Somani, 1989). Gamaniel (2000) reported that some medicinal plants such as *Erythroxylon coca, Artemisia annua, Dichapetalum bateri, Abrus precatorius* and *Symphytum officinale* contain highly toxic principles. Their extracts are toxic to animals even at low doses. Although, it is generally agreed that medicinal plants are relatively safe, a detailed toxicological evaluation of all medicinal plant and their products are essential to avoid the various forms of "medicinal plant misadventuring" as highlighted by Andreas (2009).

The plant *Physostigma venenosum* belongs to the family Fabaceae. Its produces seeds that are commonly called Calabar beans. The plant is a large, herbaceous perennial plant that grows up to 16m high. It is a great climber with pinnately trifoliate leaves and pendulous racemes. The seeds are two or three together in dark brown pods of about 6 inches long and kidney-shaped. They are thick, roughish and a little polished. The beans, which are thick with a deep brown chocolate color ripens in all seasons. *P. venenosum*, is a native to an area of Africa around Nigeria in Calabar. It thrives on the banks of streams (Dacie *et al.*, 1977). The plant is widely known in Africa because the seed had been used as an ordeal poison to determine if a person was a witch or possessed by evil spirits (Duke, 1995). Today *P. venenosum*, has been implicated to produce alkaloids clinically used to contract the pupil, manage ocular pressure in glaucoma, reverse toxicity of certain other drugs, outstrip muscles of the intestines in chronic constipation, treat cholera and acute tetanus (Wickersham, 2003). In this study, the toxicological potential of the ethanol extract of the seeds of *Physostigma venenosum* was investigated for therapeutic assessment.

2. MATERIALS AND METHODS

Samples Preparation and Extraction Procedure

The seeds of *Physostigma venenosum* were obtained from a local market in Calabar metropolis, Cross River State of Nigeria and were identified by Prof. Macdonald Idu of the Department of Plant Biology and Biotechnology, University of Benin, Benin City, Nigeria. The seeds were purchased dried but were further subjected to drying in an oven at 60°C for complete dehydration. After dehydration, 200g of the dried samples were blended into fine powder using a mechanical grinder.

One hundred grams (100g) of the powdered sample were weighed into 200ml of absolute ethanol (98 per cent) for 48 hours is a tightly cored glass container. The solution was subsequently shaken and filtered using Whatman filter paper into a conical flask while the residue was discarded. The filtrate was evaporated to dryness using a rotary evaporator (Model type 349/2, Corning Limited). The yield of 9.1 per cent was obtained. The extract was then stored below ambient temperature until required.

Administration to Rats

Initial lethal dose (LD_{50}) studies were carried out to determine the maximum dose that did not produce any death in the rats. Sixteen rats weighing between 180g and 250g were randomly selected and kept four per cage. They were allowed to acclimatize for a period of fourteen days, during which time they were fed with commercial feed mash and water *ad libitum* before commencement of the experiment. After acclimatization, the rats were divided into three treatment groups and one control group of four rats each. Based on the LD_{50} studies, graded doses of the extract (5, 10 and 20 mg/kg) corresponding to groups 2, 3 and 4 were separately administered to the rats in each of the groups by means of an orogastric tube. The control group representing group 1 was given oral distilled water of 10ml/kg only.

Hematological and Biochemical Studies

At the end of 7 days of administration of doses, the rats were anaesthetized using chloroform, after which they were sacrificed and blood samples were collected from their abdominal aorta and kept in lithium heparin and ethylene diamine tetracetic acid (EDTA) treated bottles for biochemical and hematological assays. Blood samples were analyzed within 3 hours of collection for total red blood cells (RBC) count, white blood corpuscle (WBC) count, hemoglobin (Hgb) concentration, packed cell volume (PCV), platelet (PLT) count, lymphocyte (LY), monocyte (MO) and granulocyte (GR) counts using standard methods as prescribed by Dacie and Lewis (1977). Blood samples kept in the lithium heparin treated bottles were examined for biochemical parameters such as alanine aminotransferase (ALT), aspartate transferase (AST), alkaline phosphate (ALP), total protein, albumin, total and direct bilirubin using procedures described by Tietz (1995).

Statistical Analysis

Results were expressed as mean±standard error of mean (SEM) and the level of significance between means were computed by students t – test using SPSS 16.0 computer software package. The level of significance was determined at 0.05.

3. RESULTS

Table 17.1 shows the result of the hematological analysis of the blood samples of rats administered with different concentrations of the extracts. In general, WBC and PLT values were significantly higher, 16.63±7.16 and 67.90±1.54 than those of the control, 5.30±0.53 and 38.30±17.35, respectively. While the PCV values and RBC counts dropped significantly in the experimental animals as compared to the control ($P > 0.05$). For the biochemical parameters as shown in Table 17.2, the serum protein and bilirubin concentration remained unaltered throughout the experimental period when compared with the control. However, the serum levels of ALP and ALT were significantly higher than those of the control ($P > 0.05$) while the levels of AST and albumin decreased significantly at all dose levels investigated.

The observations of biochemical and hematological parameters have been employed in toxicological studies (Basu and Navukkaras, 2006). However, in this study, the result obtained from the hematological analysis indicated that the ethanol extract of *P. Venenosum* seed administered at a higher dose level may have toxic effect on the number of formed elements of blood given an indication of progressive hemolysis. Oduola *et al.* (2007) reported that an increase in PLT values may be an indication of improvement in the bone marrow function. Also, the significant decrease in PCV value (Table 17.1) in the experimental analysis may indicate destructive effects on the red blood cells. Table 17.2 shows the effects of *P. venenosum* seed extract on some biochemical indices. The increase in ALP activity

Table 17.1: Hematological Analysis of Blood Samples of Rats Administered with Different Concentrations of Ethanol Extract of Physostigma venenosum Seeds: Concentration of Extract Administered (mg/kg body weight)

Parameters	Control	5	10	20
WBC(x10³ Cells/mm³)	5.30±0.053	16.63±7.16	8.30±2.47	7.96±1.30
RBC (x 10⁶ Cells/mm³)	7.68±0.53	7.58±1.01	7.80±0.20	7.28±0.23
PLT (Cells/mm³)	38.30±17.35	67.90±1.54	49.00±27.83	57.61±14.54
LY (per cent)	89.30±0.86	65.48±11.99	56.93±6.53	90.30±1.87
MO (per cent)	3.40±0.61	6.03±1.48	8.63±2.32	3.48±0.59
GR (per cent)	7.30±0.25	28.50±11.39	34.43±4.29	6.23±1.75
Hgb (g/100ml)	12.90±0.16	14.88±0.09	13.68±0.39	11.28±0.72
PCV (per cent)	49.08±0.51	48.78±6.00	48.03±0.93	43.38±1.55

Mean±SEM (Standard Error of Mean), n = 4.

WBC: White blood cell; Hgb: Hemaglobin; RBC: Red blood cells count; GR: Granulocytes count; PCV: Pack cell volume; PLT: Platelets; LY: Lymphocytes; MO: Monocytes.

Table 17.2: Effect of *P. venenosum* on Biochemistry of Rat Serum: Concentration of Extract Administered (mg/kg body weight)

Parameters	Control	5	10	20
ALP (I/L--)	38.09±7.57	78.02±25.99	60.25±12.75	74.34±20.58
AST (/L)	26.67±5.75	23.67±1.89	15.00±1.23	18.3±2.68
ALT (/L)	8.33±2.39	2.67±7.26	14.67±3.47	14.67±1.84
Total protein (g/dl)	7.60±1.48	7.78±0.69	7.38±0.50	6.74±0.60
Albumin (g/dl)	13.68±0.85	13.29±1.66	11.53±5.20	10.79±0.70
Total bilirubin (mg/dl)	2.48±0.33	2.67±0.41	2.50±0.23	1.92±0.93
Direct bilirubin (mg/dl)	5.64±1.30	4.10±0.51	5.13±1.88	4.17±0.94

Mean±SEM (Standard Error of Mean).

confirms a possible damage to the plasma membrane, leading to a compromise of its integrity (Yakubu *et al.*, 2003). While an increase in ALT activity may be attributed to the release of enzymes from damaged organs which might have resulted from change in membrane permeability of the cells (Latha *et al.*, 1998). Ganong 2001) revealed that albumin, total bilirubin and globulin are mixtures of biomolecules that can be used to indicate the integrity of glomeruli and regulation of osmotic pressure respectively. They can also be used to access the synthetic ability of the liver as well as damage to the hepatocytes. Hence, the reduction in the serum albumin content observed in this study may be an indication of diminished synthetic function of the liver. Chowdhury (1989) reported that bilirubin is an important catabolic product of blood with biological and diagnostic values. The significant decrease in total bilirubin level recorded in this study may reduce the ease at which bilirubin becomes excreted in the bile.

It can be suggested from the result of the acute toxicity study that the ethanol extracts of *P. venenosum* seeds fall within the range of harmful substances. However, at a low dose level, it produces mild stimulation of inflammation and the immune system; which are protective responses. Therefore proper monitoring of the usage should be carried out before any drug formulation and administration is done.

4. REFERENCES

1. Akerele, O., 1993. Nature's medicinal bounty: Don't throw it away. *World Health Forum*, 14:390 – 395.

2. Andreas Luch, 2009. Molecular, clinical and environmental toxicity. *In*: Luch Andreas (Ed) Experimental supplement. Basel Press, New York, USA. 470p.

3. Basu, S.K and Navukkarasu, R., 2006. Acute toxicity and diuretic studies of aerial parts in rats. Filoterapia, 77:83 – 85.

4. Berchtof, N.C and Cotman, C.W., 1989. Evolution in conceptualization of Dementia and Alzheimer's disease: Greco-Roman period to the 1960s *Neurobiol. Aging*. 19(3): 173 – 189.

5. Chowdhury, J.R, Wolkoff, A.W, Arias, I.M., 1989. Disorders of bilirubin metabolism. *In*: The Metabolic Basis of Inherited Disease, 1(8): 1367 – 1408.

6. Cunningham, A.B., 1988. Medicinal Plants and sustainable trade. *In*: Medicinal Plant: A Global Heritage. Proceeding of the International Conference of Medicinal Plant for Survival. New Delhi: International Developing Research Centre. 121p.

7. Curtis, D.K., 2001. The Pharmacological Basis of Ecological and Treatment Poisoning Therapeutics. 10th Edn. McGraw Hill, New York, 321p

8. Dacie and S. Lewis, 1997. Practical Hematology. 11th Edn, Churchill Livingstone, USA, 668p.

9. Duke, J., 1995. Handbook of Medicinal Herbs. Boca Raton, Florida: CRC Press, Inc. 68p.

10. Gamaniel, K.S., 2000. Toxicity from Medicinal Plants and their Products. Nig. Journal of Natural Production and Medicine 4: 5 – 8.

11. Ganong, W.F., 2001. Review of Medical Physiology, 2nd Edn. Lange Medical Books, London, pp: 414 – 417.

12. Idu, M., 2010. Phytomedicine in Nigeria – Past, Present and Future 7th Professor James Ogonor Memorial Lecture. Women's Health and Action Research Centre, Benin City. 12p.

13. Katzung, G.G Masters, S.J, Trever, A., 2009. Basic and Clinical Pharmacology. McGraw Hill. 110p.

14. Latha, R.M Gentha, T. Varalakshmi, P. 1998. Effect of *Vernonia cinerea* Flower extract in adjuvant - induced arthritis, *General Pharmacology*, 31: 601 – 606.

15. Oduola T. Adeniyi, F. A Ogunyemi E.O Bello, I.S Idowu, T.O and Subair, H.G., 2007. Toxicity studies on an unripe *Carica papaya* aqueous extract: biochemical and hematological effects in Wister albino rats. *Journal of Medicinal Plant Research* 1(1): 01 – 004.

16. Range, H.P., M. Dale and J.M Riter, 1995. Pharmacology. 3rd Edn, Churchill Livingstone, USA, 800p.

17. Saxton, E., 1971. The Alkaloid. A Specialist Periodical Report. Vol.1. London. *The Chemical Society.* 3: 88 – 91.

18. Somani, M.A and Dube, S.N., 1989. Administration of Acute Oral Toxicity. *International Journal of Clinical Toxicology.* 21: 311 – 320.

19. Tietz, W. Norbert, 1995. Clinical Guide to Laboratory Tests. 2nd Edn. Philadelphia, PA., WB Saunders Company. 1096p.

20. Walum, E., 1988. Acute oral toxicity. *Evironmental Health Perspective.* 106 (3): 479 – 503.

21. WHO, 1991. Guidelines for the Assessment of Herbal Medicine. WHO/TRM/91.4.

22. WHO,1992. Quality Control Method for Medicinal Plant Material. WHO/PHARMA/92.559.

23. Wickersham, R.M., 2003. Drug Facts and Comparison. St Louis, MO: Wolters Kluwer Health Inc.

24. Yakubu, M.T., Bulbis, L.S., Lawal, M., Akanji, M.A. 2003. Effect of repeated administration of sildenafil citrate on selected enzymes activities of liver and kidney of male albino rats. *Nigeria Journal of Pure and Applied Science.* 18: 1395-14000.

Chapter 18

Complementary Medicines (Category D Medicines) Quality, Safety and Efficacy

Shameim Ahmed Adam

Deputy Chairman,
Complementary Medicine Committee, South Africa
E-mail: adam786@live.co.za

1. INTRODUCTION

The Complementary Medicines (CMs) that will be subject to these regulations are those associated with those disciplines regulated by the Allied Health Professions Council of South Africa (AHPCSA). These are commonly known as Homoeopathic medicines, Western Herbals, Traditional Chinese medicines, Ayurvedic medicines, Unani-Tibb and Aromatherapeutic medicines/oils. As per the Act, the term "practitioner" refers to a person registered as such under the Allied Health Professions Act, 1982 (Act No. 63 of 1982).

All actions arising from the application and use of the guidelines are aimed at benefitting the stakeholders involved in their use for prevention and treatment of disease(s). These stakeholders include commercial concerns, users, practitioners and the regulators.

It is thought that quality and safety is non-negotiable, whereas, depending upon the discipline, proof of absolute efficacy might prove challenging (for a variety of reasons). The approach of these guidelines is to enable the applicant to present, to the MCC, an application free of errors and easy to review. Each discipline will have its own set of requirements governed by its own references and pharmacopoeiae which are all subject to and compliant with the current science and knowledge of that particular discipline.

Medicines are not scheduled solely on the basis of toxicity. Although toxicity is one of the factors considered, and is itself a complex of factors; the decision to include a substance in a particular Schedule also takes into account many other criteria such as the purpose of use, potential for misuse, abuse, safety in use, the need for specialised (professional) knowledge in its prescription and the need for the substance.

Before submitting an application for registration of a complementary medicine, it is first necessary to establish that the product contains substances that are, in fact, complementary medicine substances. Essentially, if the substance is a designated active ingredient, as defined in the Regulations, it is a complementary medicine substance.

Good Manufacturing Practice (GMP) and Good Laboratory Practice (GLP)

All manufacturers of complementary medicines shall comply with all aspects of Good Manufacturing Practice as outlined in the latest version of the MCC.s **"Guide to Good Manufacturing Practice for Medicines in South Africa"** and Good Laboratory Practice by 2016. Also refer to WHO Guidelines on Good Agricultural and Collection Practices (GACP) for Medicinal Plants.

Good Dispensing Practice

All dispensing and compounding of medicines by Practitioners shall comply with the provisions of Act 101 of 1965 and all aspects of Good Dispensing Practice in accordance with the provisions of Act 53 of 1974.

Safety and Efficacy

Applications for the registration of complementary medicines must include appropriate data that demonstrate the safety and efficacy of the product as provided for in these guidelines.

Safety

Safety may be established by detailed reference to the published literature and/or the submission of original study data. Any complementary medicine that is of animal origin must comply with the requirements of the Animal Diseases Act, 1984 (Act 35 of 1984).

Description and Composition of the Product

A description of the finished product that includes the following information should be provided:

☆ Table of the ingredients in the product and their purpose in the formulation (*e.g.* active, disintegrant, antimicrobial preservative);

☆ Full/complete description of the dosage form, including any special character (*e.g.* modified release, film coated, uncoated); and

☆ The type of container and closure for the product, including the materials.

The table of ingredients should provide greater detail than simply the product formulation. It should include overages (additional quantities of ingredients, over the amounts nominated in the product's formulation, added during manufacture) if any.

Overview of Safety

☆ The overview of safety provides a concise critical assessment of the safety data, noting how the results may support and justify any restrictions placed on the product.

☆ The safety profile of the medicine may be motivated using relevant in vitro, in vivo evidence or clinical studies. The data should be outlined in a detailed, clear and objective manner. Tabulations of adverse events are often helpful.

☆ There should be a description of common and expected adverse events (both serious and non-serious). An accepted causality assignment determination protocol to show the relationship between the product and an event, or lack of relationship, should be provided.

Efficacy

The applicant must provide evidence (data) to support the product's efficacy for the proposed indication(s) and any claims that the applicant intends to make in the product labelling to determine whether the data supplied adequately support the requested indication(s)/claim(s) as provided for in these guidelines.

Criteria for Determining the Safety of Indications and Health Claims

The indications and health claims will be classified into two risk levels, namely High and Low risk indications or claims (Table 18.1).

☆ Health enhancement claims apply to enhancement of normal health. They do not relate to enhancement of health from a compromised state.

☆ All claims relating to symptoms must be accompanied by the advice "If symptoms persist consult your healthcare practitioner".

☆ In cultures where an oral tradition is clearly documented, evidence of use from an oral tradition would be considered acceptable provided the history of use is authenticated. Modern texts that accurately report or confirm the classical or traditional literature may be used to support claims. Traditional claims should refer to corresponding traditional descriptions of the condition(s).

☆ Terms used must be in accordance with the practice of the associated discipline registered with the AHPCSA.

Table 18.1

Risk Level	Type of Claim	Evidence Required to Support Claim
High risk	☆ Treats/cures/manages any disease/disorder. ☆ Prevention of any disease or disorder. ☆ Reduction of risk of a disease/disorder. ☆ Relief of symptoms of a named disease or disorder. ☆ Treatment of proven vitamin or mineral deficiency diseases.	☆ Clinical data to be evaluated. AND ☆ Two of the following four sources that demonstrates adequate support for the indications claimed: 1. Recognised Pharmacopoeia; 2. Recognised Monograph; 3. Three independent written histories of use in the classical or traditional medical literature, or ; 4. Citations from other *in vivo*, *in vitro* studies, case reports or others.
Low risk	☆ General health enhancement without any reference to specific diseases or conditions ☆ Health maintenance, including nutritional support. ☆ Relief of minor symptoms (not related to a disease or disorder)	☆ Clinical data to be evaluated AND/OR: ☆ Two of the following four sources that demonstrates adequate support for the indications claimed: 1. Recognised Pharmacopoeia; 2. Recognised Monograph; 3. Three independent written histories of use in the classical or traditional medical literature, or 4. Citations from other *in vivo*, *in vitro* studies, case reports or others.

The Naming of Complementary Medicines and Substances

Chemical Substance Name

The approved name *i.e.* International Non-Proprietary Name (INN) or chemical name of substances used as inactive ingredients in topical products must be stated. In the absence of such name being available, a chemical description or characterisation of the substance should be given.

The approved name (INN) or chemical name of mineral, metal or chemical substances or prepared mineral substances used in Homoeopathic, Traditional Chinese, Ayurvedic or Unani Tibetan Medicines must be stated.

Efficacy Criteria

The criteria to be considered in the evaluation of efficacy for all complementary medicines may include established traditional use, pre-clinical data and evidence from clinical trials in animals and human beings as well as those references specified below appropriate for the risk level of associated claim.

Generally acceptable evidence in support of efficacy includes:

(i) Appropriately designed clinical trials using the product for which an application is being made,

(ii) Appropriately designed qualitative and observational studies preferably using South African-validated instruments/methods,

(iii) Published systematic reviews such as in the Cochrane database,

(iv) Published clinical trials,

(v) Published case reports,

(vi) Evidence-based databases (*e.g.* Natural Medicines Comprehensive Database, Natural Standards Database),

(vii) Accepted Herbal monographs or pharmacopoeiae,

(viii) Monographs from any other source equivalent in standard to any of the above.

(ix) In the case of homoeopathic medicines, justification of the use of the medicines from the relevant Materia Medica or Repertory listing

Herbal Name

For purposes of the registration procedure, herbal names are stated in the Latin binomial format, which should include the genus, species, subspecies, variety, subvariety, form, subform or chemotype and author where appropriate. Reference should be made to the internationally accepted name for the plant, fungus or alga by referring to the following databases where appropriate (in order of priority):

1. The Plant List (Available at: http://www.theplantlist.org)
2. The Index Fungorum (Available at: http://www.indexfungorum.org)
3. The International Plant Names Index (Available at: http://www.ipni.org)
4. OR other recognised major flora

Examples of correct herbal names include:

☆ *Olea europaea* subsp. *africana* (Mill.) P.S. Green

☆ *Crataegus curvisepala* Lindm.

☆ *Thymus zygis* subsp. *gracilis* (Boiss.) R.Morales ct. thymol

Herbal Ingredient

The Latin binomial name (as above), the part and the preparation (including solvents and ratio if applicable) are used to fully name a herbal ingredient.

Herbal Substance

For purposes of labelling, a simple Latin binomial or pharmacopoeial names of Herbal ingredients that are fully characterised in a monograph of an accepted Pharmacopoeia may be used provided it is clear to the consumer exactly which herb is being used.

Herbal Component Name (HCN)

HCNs are names for classes of constituents that are found in herbal ingredients. The need for a HCN most often arises when a herbal extract is standardised to a particular class of constituents or where particular classes of constituents are restricted (*e.g.* hydroxyanthracene derivatives). Where a herbal extract is standardised to a single constituent, the single constituent should have a chemical name. The HCN is not a stand-alone name and should be used only when expressing a herbal substance.

Registration of Medicines

Nature identical oils are synthetic aromatic compounds which are made in the laboratory and have fewer synthetic compounds than 100 per cent synthetic oils. Nature identical essential oils are NOT suited for aromatherapy or any therapeutic applications. Nature identical Oils cannot be used therapeutically as complete substitutes for the naturally occurring aromatic materials.

Common names, Materia Medica Name, Traditional Chinese Pin Yin, Traditional Sanskrit and other Traditional Unani Tibb Names may be used in addition to the approved names. The Pin Yin name of the plant may also be used in addition to the English names of the plant parts in the case of Traditional Chinese medicines.

Combination Products

☆ A combination product means a single product that contains:

(a) a mixture of substances of various discipline specific origin or philosophy, or

(b) a mixture of at least one substance of discipline specific origin and other allowable substances which make no therapeutic claim.

In the Case of Combination Products

☆ Applicants will need to demonstrate explicit, cogent philosophies of use amongst all ingredients or will be referred for Category A registration;

☆ The registration–sub-category will be "Combination Product" and the discipline(s) it relates to;

☆ Where vitamins, minerals or other substances of food origin are included in a combination product and where such items fall below prescribed maximum food levels and provided that no medicinal claim is made, CM registration will be permitted, and

☆ where classified foods further purport to make medical claims or are above prescribed maximum food levels, these products will be referred for Category A registration.

☆ Accepted References

☆ Accepted references (in addition to any further specified accepted references for each discipline) should be consulted for purposes of motivating that the product or substances used originate from the discipline indicated.

Quality

Refer also to Pharmaceutical and Analytical Guideline:

☆ Information on the quality of a complementary medicine substance is required to characterise the substance for the purpose of developing a compositional guideline. Information that should be provided includes the substance name, composition, structure (chemical and/or morphological where possible) and general properties; manufacturing details, including process and controls; substance characteristics, including impurities and incidental constituents; specifications and details of analytical test methods, with method validation data; stability data; and a proposed compositional guideline.

☆ Some complementary medicines are comprised of relatively simple ingredients (*e.g.* single herb, mineral salts) and, unless the medicine contains multiple active ingredients, the quality parameters applying to such products are essentially the same as for pharmaceutical medicines.

☆ However, complementary medicines that contain complex ingredients that are difficult to characterise and/or certain combinations of multiple active ingredients require special consideration.

Manufacture of the Active Ingredient

☆ The manufacture of the active ingredient must be described.

☆ State the part of the plant or animal used and its form, *i.e.* whether it is a fresh or dried material, together with details of any processing it undergoes before use in the manufacture of the product. Where appropriate it may be necessary to state the country or region of origin of the ingredient, or give other details such as time of harvesting and stage of growth, which are pertinent to the quality of the ingredient.

Decision Tree: Category D Medicines

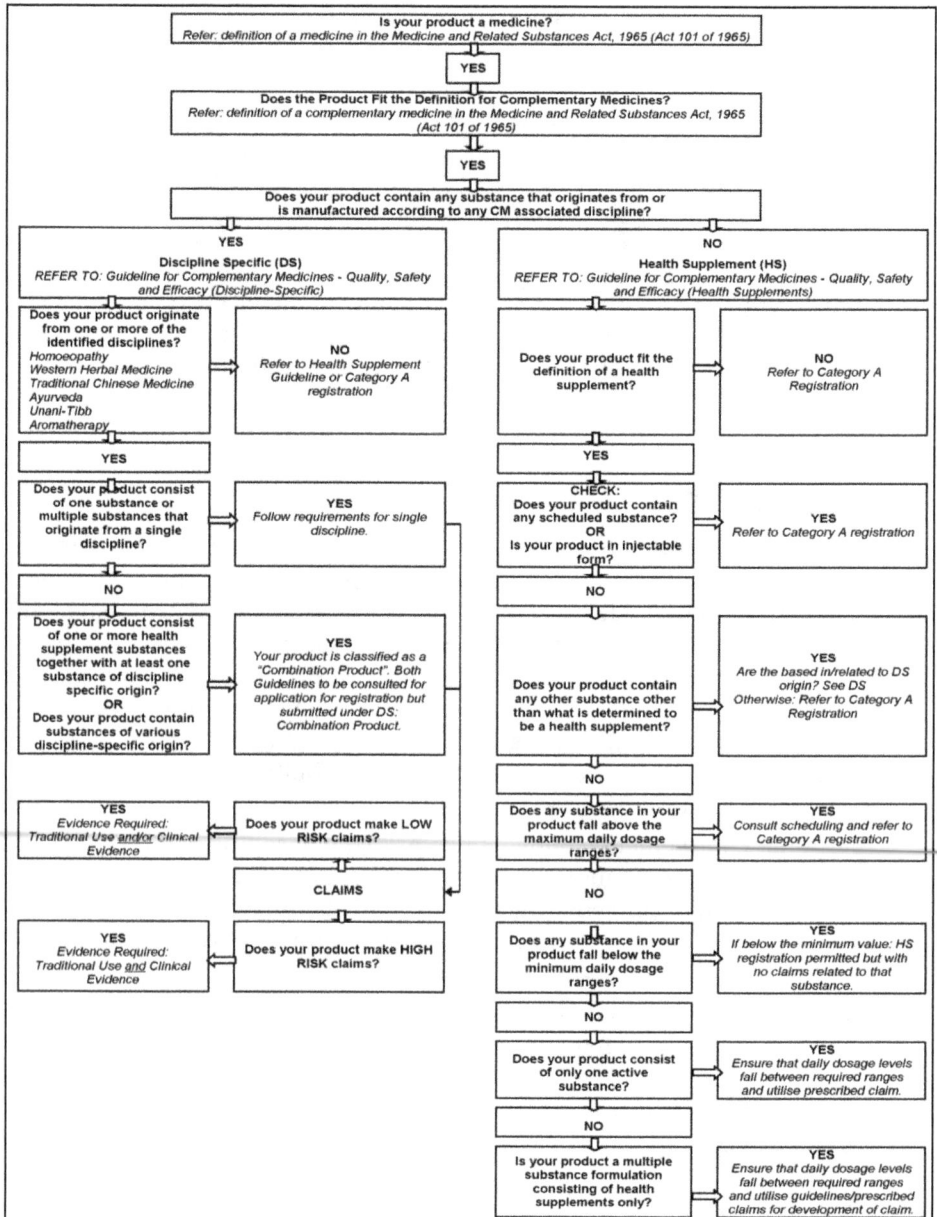

Is your product a medicine?
Refer: definition of a medicine in the Medicine and Related Substances Act, 1965 (Act 101 of 1965)

YES

Does the Product Fit the Definition for Complementary Medicines?
Refer: definition of a complementary medicine in the Medicine and Related Substances Act, 1965 (Act 101 of 1965)

YES

Does your product contain any substance that originates from or is manufactured according to any CM associated discipline?

YES → **Discipline Specific (DS)**
REFER TO: Guideline for Complementary Medicines - Quality, Safety and Efficacy (Discipline-Specific)

NO → **Health Supplement (HS)**
REFER TO: Guideline for Complementary Medicines - Quality, Safety and Efficacy (Health Supplements)

Does your product originate from one or more of the identified disciplines?
Homoeopathy
Western Herbal Medicine
Traditional Chinese Medicine
Ayurveda
Unani-Tibb
Aromatherapy
→ **NO** *Refer to Health Supplement Guideline or Category A registration*

YES

Does your product consist of one substance or multiple substances that originate from a single discipline?
→ **YES** *Follow requirements for single discipline.*

NO

Does your product consist of one or more health supplement substances together with at least one substance of discipline specific origin? OR Does your product contain substances of various discipline-specific origin?
→ **YES** *Your product is classified as a "Combination Product". Both Guidelines to be consulted for application for registration but submitted under DS: Combination Product.*

YES *Evidence Required: Traditional Use and/or Clinical Evidence*
← **Does your product make LOW RISK claims?**

CLAIMS

YES *Evidence Required: Traditional Use and Clinical Evidence*
← **Does your product make HIGH RISK claims?**

Does your product fit the definition of a health supplement?
→ **NO** *Refer to Category A Registration*

YES

CHECK: Does your product contain any scheduled substance? OR Is your product in injectable form?
→ **YES** *Refer to Category A registration*

NO

Does your product contain any other substance other than what is determined to be a health supplement?
→ **YES** *Are the based in/related to DS origin? See DS Otherwise: Refer to Category A Registration*

NO

Does any substance in your product fall above the maximum daily dosage ranges?
→ **YES** *Consult scheduling and refer to Category A registration*

NO

Does any substance in your product fall below the minimum daily dosage ranges?
→ **YES** *If below the minimum value: HS registration permitted but with no claims related to that substance.*

NO

Does your product consist of only one active substance?
→ **YES** *Ensure that daily dosage levels fall between required ranges and utilise prescribed claim.*

NO

Is your product a multiple substance formulation consisting of health supplements only?
→ **YES** *Ensure that daily dosage levels fall between required ranges and utilise guidelines/prescribed claims for development of claim.*

"The reasonable man adapts himself to the world; the unreasonable man persists in trying to adapt the world to himself. Therefore all progress depends on the unreasonable man."

– **George Bernard Shaw**

☆ If the herb is processed to produce a galenical form, the extraction and any concentration processes should be described or a reference cited, indicating whether the extract or additives, such as calcium phosphate in dry extracts, are present in the final product formulation.

☆ In the case of "low dose "starting substances these must in all cases be manufactured according to suitable pharmacopoeiae to ensure reproducible quality.

Scheduling

All medicines are subject to a scheduling process on the basis of the active ingredients they contain. The overall aim of this process is to classify substances into Schedules linked to degrees of prescriptive control to be exercised with respect to their sale to, and use by, the general public. These controls are based on the safety profile of a substance as well as the therapeutic indications for its use and may result in a substance being placed in one or more of the eight possible Schedules.

A substance may be rescheduled to a higher Schedule should it be found to be less safe than originally believed, or when reports of abuse are received. Likewise, if, following experience gained during use, it can be demonstrated that the medicine is safe for use with self-diagnosis, reclassification may be considered by removing the prescription requirements to allow for sale or supply without supervision of a healthcare practitioner.

Within the Complementary arena, we feel that there should only be a few Schedules pertaining to the dispensing or prescribing of CMs, and it would be difficult to populate all the Schedules pertaining to Allopathic medicines, given the nature of the substances that we use, and more importantly, the traditional manner these substances were administered directly to patients by traditional practitioners.

Chapter 19

Neurobehavioural Toxicity Study of a Hydro-Ethanolic Extract of *Boophone disticha* in Sprague Dawley Rats

Temba Ganga, Louis L. Gadaga and Dexter Tagwireyi

Drug and Toxicology Information Service (DaTIS),
School of Pharmacy, College of Health Sciences,
University of Zimbabwe,
P.O. Box A178, Avondale, Harare, Zimbabwe

ABSTRACT

Extracts of *Boophone disticha* (Amaryllidaceae) have been widely used in Zimbabwe and neighbouring countries for the management of a wide range of ailments, including psychotropic and inflammatory conditions.

The present work seeks to describe neurobehavioural toxicity of an extract of *B. disticha* with Sprague Dawley rats in 28-days sub-acute test. Sprague Dawley rats aged 4-5 weeks were allocated to four treatment groups (n=6); control and three experimental group (100, 200 and 400mg/kg p.o. Boophone extract). Functional Observational Battery (FOB) and motor activity testing were carried on day 1, 14 and 28 of the study. Signs of toxicity included stupor followed by tremor of effort affecting the whole body. With more severe cases the rat would progress to tremors at rest with the head sloping to one side. These responses were most pronounced at higher doses, with fatalities in the 800mg/kg resulting in this group being abandoned for humane reasons. Retropulsion was observed in some rats after repeated dosing. At low single doses, however, it appears that *Boophone disticha* has stimulant properties. The toxic responses lasted approximately 30 minutes.

Keywords: Boophone disticha, Subacute, Toxicity.

Abbreviations

FOB: Functional observational battery

RAPD: Randomly amplified polymorphic DNA

1. INTRODUCTION

The use of plants or plant extracts to treat diseases is a therapeutic modality, which has stood the test of time (Gilani and Rahman 2005; Sariæ-Kundaliæ *et al.,* 2010). In some developing countries, up to 80 per cent of the populace relies on traditional practitioners and herbal plants to meet their primary health care needs (WHO 2008). Moreover, recently there has been a cultural renaissance globally towards more natural methods of healing (Barnes 2003; WHO 1999), with an increasing proportion of people in western settings turning to alternative and complimentary medicines (Makunga *et al.,* 2008).

Southern Africa has a wide diversity of plants which have claimed medicinal properties and these have been documented in ethnopharmacological publications, highlighting their healing value and bioactivity (van Wyk *et al.,* 1997; Watt and Breyer-Brandwijk 1962). However despite the existence of such information, most of these plants have not been the subject of systematic toxicological tests to evaluate their safety (Fennell *et al.,* 2004; McGaw and Eloff 2005; Botha and Penrith 2008). In many cases, assumptions about the safety of these medicinal plants is made based on their long term traditional use (Elgorashi *et al.,* 2003; McGaw and Eloff 2005; Verschaeve and van Staden 2008).

However, reports published in the medical literature have shown that southern African traditional medicines (most of which are derived from plants) are a significant and important cause of hospital admissions in these African countries (Nhachi and Kasilo 1992; Tagwireyi *et al.,* 2002; van Wyk *et al.,* 1997). This has led to an increase in advocacy for research to elucidate the toxic components and to describe the toxicological effects of these so called 'safe' traditional medicines (Tagwireyi *et al.,* 2002).

Boophone disticha (Amaryllidaceae) is a bulbous plant which has been widely used in Southern African traditional medicine for the management of a wide range of ailments, including psychotropic and inflammatory conditions (Gelfand *et al.,* 1985, Sorbiecki 2002, van Wyk and Gericke 2000). In a recent review, this plant was described as the most important medicinal plant in the Amaryllidaceae family (Nair *et al.,* 2014). Apart from its medicinal uses, bulb infusions of this *B. disticha* are also used culturally by some Southern African tribes to induce hallucinations for divinatory purposes and for initiation ceremonies for young men (Gelfand *et al.,* 1985, Sorbiercki 2002). In addition, bulb infusions of the plant are used as a recreational 'drug' for purported euphoric action in Zimbabwe (Acuda and Eide, 1994; Gelfand *et al.,* 1985). In spite of the widespread use of extracts of *B. disticha* in Southern Africa, there is currently no published literature related to toxicological effects resulting from repeated use of extracts from this plant.

Thus, given the increasing amount of research interest on *B. disticha* as a potential source in drug discovery (Nair *et al.*, 2014, Van Wyk and Gericke 2000; Sorbiercki 2002), as well as the continued chronic use for cultural and recreational activities outlined above, there is a need to investigate the toxic effects of *B. disticha* after repeated daily dosing. This information would be useful in describing management protocols in cases of poisoning, as well as providing toxicity data relevant to the plant should it be developed into a medicine. Furthermore, this information would assist in filling the information gap with regards to toxicological properties of Southern African medicinal plants. Therefore, in the present work, we describe the neurotoxicological effects of a crude extract of the bulb of *B. disticha* after a 28 day repeated single dose oral administrations to Sprague Dawley rats. This work is a follow up study on earlier work done on the acute toxicity of the same plant extract which showed adverse effects directed primarily at the central nervous system (Gadaga *et al.*, 2011).

2. MATERIALS AND METHODS

The experimental protocols, care and handling of animals used in this study were in accordance with international guidelines (European Community guidelines, EEC Directive of 1986; 86/609/EEC) on the use and care of laboratory animals and were approved by the Division of Veterinary Services, Zimbabwe (Pote *et al.*, 2013).

Plant Materials, Extraction and Qualitative Analysis of Alkaloids

Bulbs of *B. disticha* were harvested, authenticated, dried and the aqueous ethanolic (70 per cent v/v) extract was prepared as described previously (Gadaga *et al.*, 2011; Pote *et al.*, 2013). Chromatographic analysis (Zulu *et al.*, 2011) confirmed the presence of isoquinoline alkaloids reported in previous studies (Adewusi *et al.*, 2012; Cheesman *et al.*, 2012; Hauth and Stachaffer, 1961; Neergard *et al.*, 2009; Sandager *et al.*, 2005; Steenkamp, 2005).

Animals and Housing Conditions

Sprague Dawley rats (4-5 weeks old) of both sexes (4 females and 20 males) purchased from the Animal Unit of the University of Zimbabwe were used for the experiments. These were housed in wire mesh cages with a maximum of 6 animals per cage and fed on standard mouse pellets purchased from National Foods (Pvt) Limited, Harare, Zimbabwe. Food and tap water were available *ad libitum.* The bedding of sawdust was changed at least three times a week. At the end of testing, animals were humanely sacrificed by chloroform asphyxiation. Animals that became moribund during experiments and did not recover after a period of 24 hours were humanely sacrificed by chloroform asphyxiation. The animal holding facility was kept under a normal day/night which is almost a 12 hour light/dark cycle (summer, Sub-Saharan Africa). There was equal distribution of male and female rats across all groups. Relative humidity was 50±10 per cent and the room temperature ranged between 20 and 25°C.

a) Observations

All the experiments were carried out between 08:00 and 16:00. The rats were carried to the test room in their home cage, and they were left undisturbed for about 30 minutes. Rats were allocated to 4 groups (n=6); control (water), three experimental groups (100, 200 and 400mg/kg *Boophone* extract). The rats were dosed daily by oral gavage using an intubation tube for 28 days. Animals that died during the test period were necropsied. The functional observational battery and motor activity testing were carried on days 1, 14 and 28 of the study.

b) Neurobehavioral Toxicity and Motor Activity Assessment

The functional observational battery (FOB) was used to evaluate neurobehavioural toxicity and physiological changes. The experimental protocol for the FOB was based on procedural details and scoring criteria for FOB as described by McDaniel and Moser (1993). On the test days, rats were transported to an observation room and allowed at least one hour to acclimate before testing began. Animals were then observed using the FOB 30 minutes after dosing. Soon after the FOB motor activity was evaluated using the Rota-rod apparatus as described by Franco and colleagues (2005) with each rat being given three successive trials and the longest period it remained on the rod taken as its score.

c) Gross Necropsy and Histopathology

On the final day (Day 28) of experiments, all the rats that survived were weighed and sacrificed by chloroform asphyxiation. The rats were immediately dissected and blood was collected from the heart using a syringe and needle. The testes, kidneys, spleen, stomach, small intestines, large intestines, liver, heart, lungs, and brain were harvested in that order. Organs were weighed and immediately preserved in formalin. Blood samples were all centrifuged within 30 minutes at 3000rpm for 5minutes and refrigerated at zero degrees Celsius overnight before being stored at -80 °C for future tests.

Statistical Analysis

Data collected with the FOB was analysed by STATVIEW 5.1. statistical package. The statistical analyses were performed by non-parametric Kruskal-Wallis to evaluate significant differences between the groups. Differences were considered significant at $p < 0.05$.

3. RESULTS

General Observations

At commencement of the study there were 28 rats and only 16 rats survived to the final day. Onset of toxic responses were from 5 minutes to 15 minutes post dosing and lasted between 35 and 60 minutes, followed by a rapid recovery. It was also observed that the latency to toxic responses decreased with increasing dose and with repeated dosing. Signs of toxicity included stupor followed by whole body tremor. In the severe cases, the animal would progress from tremors at rest with the head sloping to one side to explosive tonic-clonic jerks and eventually death in the majority of the cases. This occurred most with the high doses, particularly with the 800 mg/kg dose after single dosing. This led to the abandonment of this dose for the repeated dosing study. Death usually occurred within 10 minutes of the onset of severe toxic responses. Lower doses were associated with mild responses, including piloerection, palpebral closure, drunken gait, walking on tiptoe, and hunched body position. Retropulsion was observed in some rats after repeated dosing.

Body and Organ Weight

Rats were weighed on a weekly basis. At the commencement of the study, mean body weights of rats in all the groups was below 100 g. However after a week the average mass for the group receiving 400mg/kg of extract had increased to about 130g (Table 19.1). This was significantly more than its previous average and that of the other groups on this day (p<0.05). By day 14, average mass for 100 mg/kg had increased from 100g to110g, but without significant statistical difference from the rest (p>0.05). On the whole, animals receiving the plant extract showed a higher growth rate than the control group (Table 19.1).

Table 19.1: Evaluation of Relative Organ Weight to Body Weight

	Boophone Disticha mg/kg/day			
	Control	*100*	*200*	*400*
Body weight				
Initial	88.60±6.14	74.50±8.01	81.50±11.78	98.00±10.39
Day 7	95.20±5.75	99.25±11.54	87.25±13.20	129.0±8.72
Day 14	103.0±5.53	113.3±13.55	97.50±11.24	131.33±4.67
Day 21	104.0±4.46	112.8±12.80	94.75±7.79	121.33±3.53
Final	103.2±4.29	108.8±13.26	108.8±13.26	114.0±1.53
Weight Gain (per cent)	17.66±5.036	45.39±3.182	26.96±19.77	19.06±12.9
Relative Organ weight (per cent body weight)				
Spleen	0.239±0.048	0.203±0.023	0.247±0.023	0.198±0.017
Kidneys	0.474±0.043	0.443±0.068	0.538±0.049	0.451±0.061
Testes	1.107±0.109	0.680±0.000	0.739±0.130	0.706±0.000
Stomach	1.453±0.126	1.555±0.305	4.153±1.097*	3.010±0.758*
Large intestines	4.379±0.305	3.810±0.689	5.628±0.673*	4.758±0.670*
Small intestines	3.542±0.270	3.869±0.661	6.232±0.655*	7.081±1.357*
Liver	3.378±0.178	3.274±0.393	5.234±0.588*	4.595±0.544*
Heart	0.411±0.013	0.482±0.062	0.506±0.045	0.401±0.038
Lungs	0.829±0.074	0.754±0.099	0.798±0.056	0.667±0.046
Brain	1.511±0.135	1.544±0.269	1.514±0.100	1.277±0.120

Data expressed as mean±S.E.M; Significant difference: *p<0.05 versus control group.

Significant differences in organ weights of the stomach, large intestines, small intestines, and the liver were noted for animals receiving the plant extracts (200 mg/kg and 400mg/kg) when compared to animals in the vehicle control group (Table 19.1). The weight of the testes in animals receiving the plant extracts were generally less than for animals receiving water in a dose related manner with animals receiving higher doses having a smaller average mass. However, this trend was not statistically significant (p>0.05). Animals receiving the plant extracts had larger heart masses and smaller lung masses compared to animals receiving water, however these differences were not statistically significant (p>0.05).

Neurobehavioural Observations

a) Activity Observations

i) Posture

Generally posture abnormality significantly increased with dose for the experimental groups compared to the control (p<0.05) and this trend was consistent throughout study duration. The scores were also higher for all experimental groups over the 28 days when compared to the vehicle control (Figure 19.1).

ii) Arousal

On day 1, the 100 mg/kg group had the highest score for arousal, with 60 per cent (3) of the rats being slightly excited. However, this trend decreased with repeated dosing at days 14 and 28 (Figure 19.1). Rats in the control group were alert throughout the study period. Rats in the 400mg/kg group had the least score for arousal throughout the whole study period, with p<0.05 for all test days. By day 14 levels of unprovoked activity for the 200 mg/kg and 100 mg/kg groups had dropped to below that of the control (Figure 19.1).

iii) Involuntary Motor Movements

The control group had no involuntary movements through out the study period. The group receiving 100 mg/kg of the plant extract had slight tremors on day 1 which worsened slightly over time (Figure 19.1). A similar trend was also observed with groups receiving the 200 mg/kg and the 400 mg/kg doses in a dose-response related fashion with animals in the 400mg/kg dose group exhibiting the most severe involuntary movements. In addition, the involuntary movements worsened with time (Figure 19.1). In addition to the movements mentioned above, animals in the 400 mg/kg group also exhibited explosive convulsions which were absent from animals in the other groups. The forelimbs were mainly affected and were observed to have varying degrees of paralysis.

iv) Ease of Removal from Home Cage

Generally the rats became easier to remove from their home cage with repeated dosing (Figure 19.1). On Days 1 and 14, rats in the 100mg/kg group were the most difficult to remove from their home cages when compared to those of the other groups. Rats in the 200mg/kg group were the easiest to handle on days 14 and 28 when compared to the other groups (p<0.05) (Figure 19.1). By Day 28, all groups receiving the plant extract were almost stuporous, showing no resistance at all to being handled.

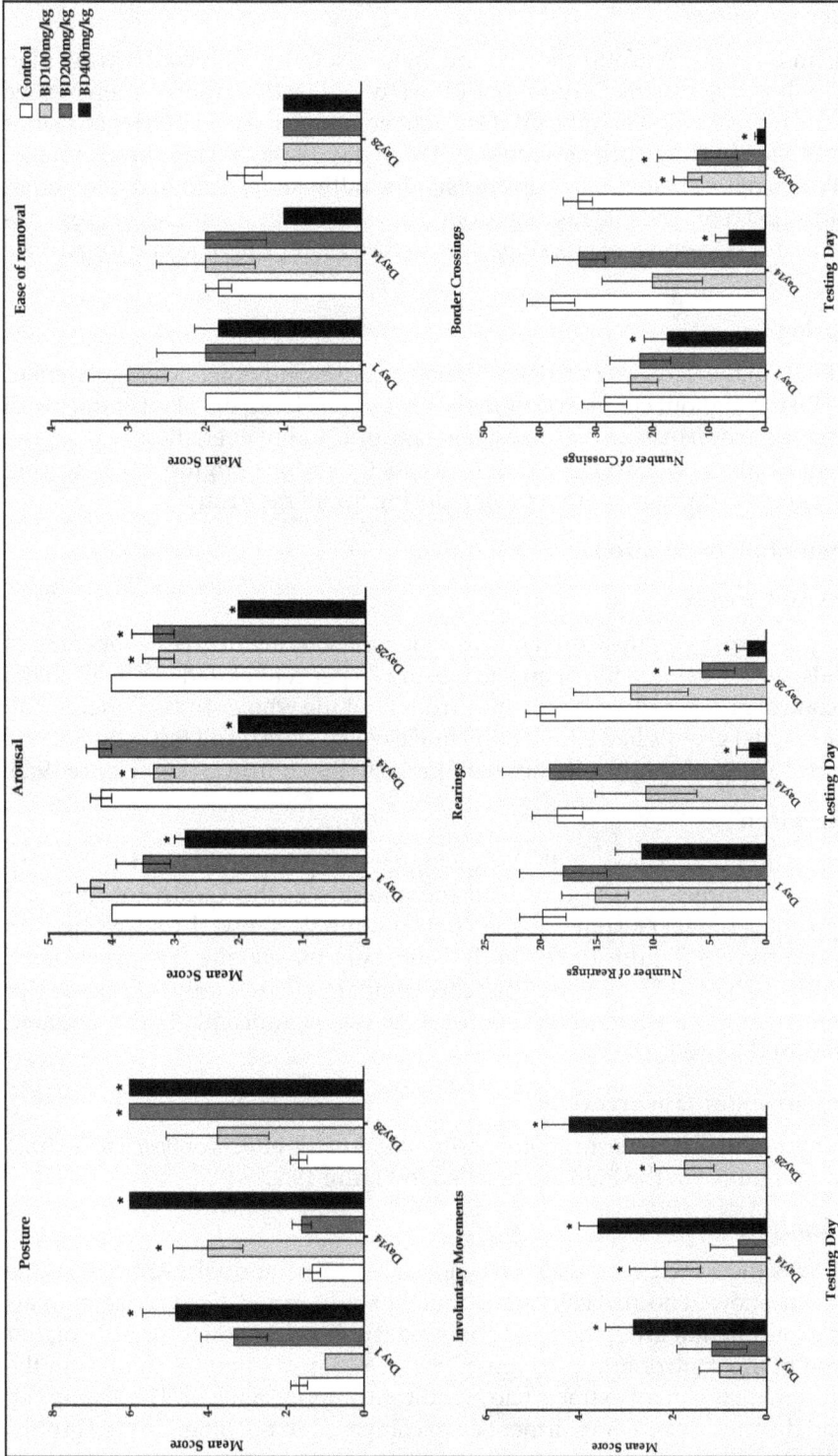

Figure 19.1: Effects of *Boophone* Extract Repeated Oral Intake on Activity Endpoints in the FOB. Significant difference: *p<0.05 versus control.

v) Border Crossings

Rats in the control group showed the highest activity with consistently high number of border crossings, throughout the study. For the groups receiving the plant extract, there was a significant dose dependent decrease in border crossings which was also time dependent (Figure 19.1). On day 14 the 400 mg/kg group had average crossings of below 7, which was significantlly lower than that of all other test groups (p<0.05). The 400 mg/kg group showed the least number of crossings for all three days. Number of crossings decreased with repeated dosing for all dose groups.

vi) Rearing

Animals in the vehicle control group had a consistently high average number of rears (20) throughout compared to all the groups receiving the plant extract with the 400mg/kg group having the lowest (average of 13). However, there was a dose-dependent decrease in rearing on day 14 across the groups (Figure 19.1). Rearing showed a general decrease with repeated dosing for all the groups.

b) Autonomic Observations

i) Palpebral Closure

This parameter was measured by observing the varying degrees of opening of the eyelids, with 1 being wide open and 3 being 75 per cent closed. Animals in the vehicle control group had a score of one throughout the whole study. Generally all the experimental groups had a high palpebral closure score on all test days and the score increased significantly with time compared to the control group (Figure 19.2).

ii) Piloerection

Piloerection was scored, with 0 showing lack of piloerection and 1 showing presence of piloerection. On day 1 all the animals in the Control group had no piloerection however some of the control animals showed piloerection on days 14 and day 28 (Figure 19.2). On day one, only the 200 mg/kg group had a significantly high (p<0.05) piloerection score compared to the control. From day 14 all experimental groups had piloerection, which was significantly high compared to the control (Figure 19.2).

c) Neuromuscular Observations

Neuromuscular and sensorimotor observations of the effects of *Boophone* extract repeated oral intake in the FOB are detailed in Figure 19.2.

i) Gait Analysis

Gait was measured on a scale of 1 (normal) to 5 (abnormal). Animals in the control group showed normal gait throughout the study period. On day one, animals in all the experimental group had gait abnormality, which was significantly higher than those in the control groups (Figure 19.2). On days 14 and 28 animals in the groups receiving the plant extracts had significantly abnormal gait. The 400mg/kg group had the most severe impairment even compared to the other dosage groups (Figure 19.2).

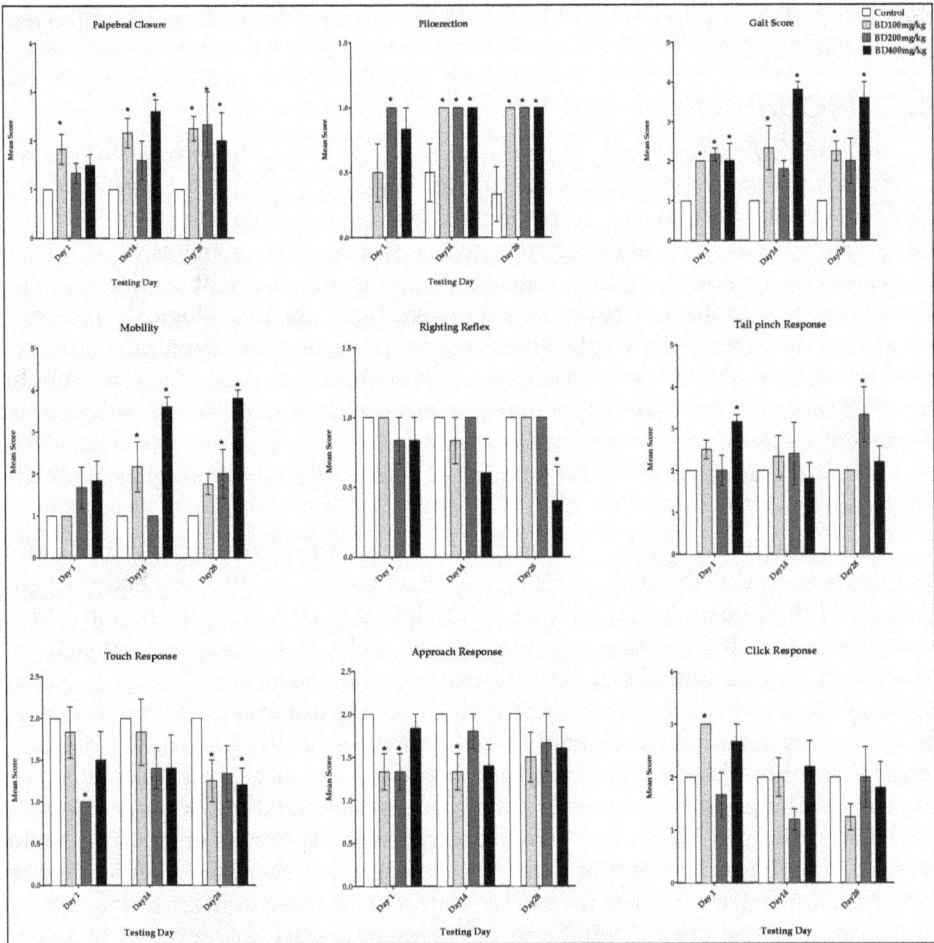

Figure 19.2: Effects of *Boophone* Extract Repeated Oral Intake on Neuromuscular and Sensorimotor Endpoints in the FOB. Significant difference: *p<0.05 versus control.

ii) Mobility

Mobility followed an almost similar trend to gait analysis. Initially all experimental groups showed slight impairment which was not significant compared to the control. However after repeated dosing (day 14 and day 28) there was significant impairment in the mobility (p<0.05) with the animals receiving the 400 mg/kg dose being the most seriously affected. This trend was dose dependent with the exception of the 200 mg/kg dose on day 14 (Figure 19.2).

iii) Righting Reflex

This parameter was ranked 0 (absence of reflex) to 1 (presence of reflex). On day 1 all animals from all the groups had the righting reflex intact, however on days 14 and 28 some rats in the 400mg/kg group had lost this reflex. The rest of the rats in other dosing groups maintained the righting reflex throughout the test period (Figure 19.2).

4. DISCUSSION

Although *Boophone disticha* has been used traditionally to treat long term conditions like depression and anxiety (van Wyk*et al.*, 1997, 2002; van Wyk and Gericke, 2000), there are no documented reports about its long term adverse effects after both acute intake or repeated exposure. Since some adverse effects can be delayed, subacute and chronic assessments are pertinent to determine its chronic toxicity. In this paper we sought to investigate the toxicological effects of a hydroethanolic extract of this plant by using standard neurobehavioural toxicology methods. Although the study was designed as a subacute toxicity study, we sought to delve more on the neurotoxicological effects of the plant. This was based on documented literature in both humans (Gelfand *et al.*, 1985; van Wyk *et al.*, 1997) and animals (Gadaga *et al.*, 2011) pointing to the fact that *B. disticha* has alkaloids which are neurotoxic.

In this study it was noted that animals receiving the plant extract gained more weight compared to those in the vehicle control. We believe that this body weight gain could have been attributed increases in levels of circulating serotonin which is known to lead to increased appetite (Purves *et al.*, 2004). To support this, there is evidence from a number of other researchers that extracts of *Boophone disticha* may lead to increases in serotonergic activity (Mutseura *et al.*, 2013; Perderson *et al.*, 2008; Risa *et al.*; Sandager *et al.*, 2005). However, in the present study, daily food consumption was not measured and hence we can only speculate on this. The groups receiving the higher doses of the plant extract (200 mg/kg and 400 mg/kg) had significantly larger livers when compared to the control group. This could have been perhaps the result of a direct toxicological effect of the liver leading to inflammation or perhaps some form of hypertrophy resulting from the liver adapting to the toxicological insult brought on by the plant extract. Concerning the latter, Dominic and colleagues (1993) found hypertrophy to be an adaptive response of the liver to an orally administered toxicant. However, since no specific studies were done to confirm the above, this would be difficult to confirm. It is more likely that the larger liver weights were more because the animals in these groups also had general body weight gain. Nevertheless detailed hepatic assessments are necessary for the exact mechanism of this effect to be determined. There were also differences in the weights of the various organs across the different groups however since there were not statistically significant, these are likely to have been as a result of chance. Notwithstanding the above, there is still a need for further histopathology analysis after repeated dosing to investigate whether some the plant extract causes toxic effects at that level.

Literature reporting on cases of acute poisoning in humans (Gelfand *et al.*, 1985; Laing 1979), as well as acute toxicity studies in animals (Gadaga *et al.*, 2011) indicates that poisoning from *Boophone disticha* results in neurotoxicological effects. In the present study, we also showed that the toxicological effects of this plant are directed towards the central nervous system. The severity in abnormality of posture and head orientation with dose and with repeated dosing may be attributed to a possible effect of the extract or its metabolites on the vestibular system which is responsible for head orientation and posture during movement. However, there is no literature to suggest exactly which phytochemical components could be responsible for this action. The observed dose dependent increase in involuntary motor movements, might point to a possible effect on motor co-ordination due to an action on the basal ganglia which is responsible for gating the proper initiation of movements that is, suppressing unwanted movements and priming upper motor neuron circuits for initiation of movement (Purves *et al*.2004). However, more neuromuscular investigations are required to affirm this postulation.

Boophone disticha has been used traditionally for its sedative effects (Gelfand *et al*.1985). Eyelids were drooping for all the experimental groups, indicative of its CNS depressive effects (Purves *et al., 2004*). This result is complemented by homecage observations that showed that rats became easier to remove from home cages and to manipulate after repeated dosing and administration of higher doses. This may also be attributed to effects on central transmission since it is documented that serotonin, as a central transmitter, is partly responsible for wakefulness (Purves *et al., 2004*; Katzung *et al., 2009*) and evidence from Sandager *et al.* (2005) suggests that *Boophone disticha* has Selective Serotonin Reuptake Inhibitory activity, thus the postulated drowsiness. Drowsiness may also have possibly resulted from effects on the parasympathetic transmission (Katzung *et al., 2009*) of which some alkaloids in extracts of *Boophone disticha* have been reported as having anticholinesterase activity (Risa *et al., 2004*). However the lack of lacrimation and salivation in any of the experimental groups could perhaps mean that either, the anticholinesterase activity is localised in the central nervous system or that it does not exist at all. We are inclined to believe the former since extracts of this plant have been used in South African traditional medicine for the management of age related dementia's and Alzheimer's Disease (Nair and..., 2014). It would be useful to carry out some follow up study examining the acetylcholinesterase activity in the brains of animals administered high doses of the plant extracts.

Arousal, which showed the level of unprovoked activity, was higher on the first few days for the lower dose groups with bursts of movement as compared to the control (Figure 19.1). This may suggest that at lower doses *Boophone disticha* has central nervous system stimulatory effects. However over repeated dosing, the animals became less active suggesting that the alkaloids in the extract accumulate on repeated dosing and are highly lipophilic and thus at cause central nervous system at high doses.

After repeated dosing retropulsion was observed and this may suggest presence of hallucinations as a result of the extract, a finding that would corroborate with the use of *Boophone disticha* by traditional healers to get into a trance (Gelfand *et al.,* 1985). According to Gadaga and colleagues (2011), this retropulsion may suggest involvement of the dopaminergic and or the serotonergic pathways. In line with the above, Mutseura and colleagues (2013) showed that toxicity from the plant extracts resulted from an overdrive in the serotonergic system.

The decrease in border crossings and rearing (Figure 19.1) with repeated dosing points to reduced exploratory activity may be due to decreased levels of consciousness, or poor motor coordination which may be a result of toxicity on the neural structures (local spinal cord and brain stem circuits, descending modulatory pathways, or basal ganglia) involved in control of movement. This may also result from interference with central dopaminergic transmission (Purves *et al.*, 2004). Some symptoms (tremor of effort) were comparable to those for Parkinsonism, hence the inference to the dopaminergic system, and central cholinergic transmission. Thus, either the alkaloids themselves or their primary metabolites can cross the blood brain barrier and retain pharmacologic activity (Gadaga *et al.*, 2011). Rearing is an activity that requires significant motor coordination again pointing to a possible effect on the sensory-motor coordination pathway and/or the vestibular pathway.

Gait analysis showed a tendency (day 1 and day 14) to move on tiptoes with a scurrying motion and with the backs in a hunched posture. This observation infers a possible effect on muscle contractility and excitability, either directly or as a result of interference with transmission to the muscles as was postulated above. Beyond day 14 the rats were moving with forelimbs dragging, in crouched positions or with the body dragging against the surface. This again shows a possible effect on the motor complex, especially the lower motor neuron section as damage to this region is usually associated with paralysis, muscle atrophy (as observed by the decreasing body weight of the experimental groups), decreased superficial reflexes and reduced tone (Purves *et al.*, 2004). On the ranked gait abnormality there was a worsening of this parameter with repeated dosing as well as with increasing dosage. These increases in abnormality may infer a build-up of alkaloids in the system possibly as a result of the detoxification processes being overcome by the frequency of dosing or amounts administered respectively.

Righting reflex was lost in the 400 mg/kg group by day 14, while all other groups retained this reflex. These rats were stuporous and thus were disqualified from this test. However, even severely intoxicated rats from other groups retained the reflex suggesting that *Boophone disticha* may have no effect on reflex transmission that is the spinal cord and brain stem circuitry.

5. CONCLUSIONS

Boophone disticha has constituents that are either highly toxic or have their primary metabolites being toxic especially at higher doses. At low single doses, however, it appears that *Boophone disticha* has stimulant properties. The toxic responses lasted approximately 30 minutes. Death due to *Boophone disticha* poisoning can occur as early as 10 minutes post dosing or can take up to 12 hours. The main toxic effects of *Boophone disticha* after repeated administration are directed towards the central nervous system and include hallucinations, disturbances in gait as well as convulsions at high doses. These neurotoxicological effects seem to increase with repeated doses and as the dose is increased.

6. REFERENCES

1. Acuda, S.W. and Eide, A.H. 1994. Epidemiological study of drug abuse in rural secondary schools in Zimbabwe. *Central African Journal of Medicine* **8**, 207-212.

2. Botha, C.J. and Penrith, M.L. 2008. Poisonous plants of veterinary and human importance in Southern Africa. *Journal of Ethnopharmacology* **119**, 513-537.

3. Elgorashi, E.E., Taylor J.L.S., Verschaeve, L., Maes, A., van Staden, J. and De Kimpe, N. 2003. Screening for medicinal plants used in South African traditional medicine for genotoxic effects. *Toxicology Letters* **143**, 195-207.

4. Fennell, C.W., Lindsey, K.L., McGaw, L.J., Sparg, S.G., Stafford, G.I., Elgorashi, E.E., Grace, OM., van Staden, J. 2004. Assessing medicinal plants for efficacy and safety; *Pharmacological screening and toxicology in Journal of Ethnopharmacology* **94**, 205-217.

5. Gadaga, L.L., Tagwireyi, D., Dzangare, J., Nhachi, C.F.B. 2011. Acute Oral Toxicity and Neurobehavioural Toxicological Effects of a Hydroethanolic Extract of Boophone disticha in Rats. *Human and Experimental Toxicology* **30**, 972-980.

6. Gelfand, M., Mavi, S., Drummond, R.B., Ndemera, B. 1985. The Traditional Medical Practitioner in Zimbabwe. Mambo Press Gweru. pp. 296.

7. Gilani, A.H. and Rahman, A.U. 2005. Trends in ethnopharmacology. *Journal of Ethnopharmacology* **100**, 43-49.

8. Katzung, B.G., Masters, B. and Trevor, A.J. 2009. *Basic and clinical Pharmacology*, 11th edition. Tata McGraw Hill Education.

9. Makunga, N.P., Philander, L.E. and Smith, M. 2008. Current perspectives on an emerging formal natural products sector in South Africa. Journal of Ethnopharmacology **119**, 365–375.

10. McDaniel, K.L. and Moser, V.C. 1993. Utility of a neurobehavioural screening battery for differentiating the effects of two pyrethroids, Permentrin and cypermethrin. *Neurotoxicology and Teratology* **18**, 929–938.

11. McGaw, L.J. and Elloff, J.N. 2005. Screening of 16 poisonous for antibacterial, antihelmintic and cytotoxic activity *in vitro*. *South African Journal of Botany* **71**, 302-306.

12. Nhachi, C.F.B. and Kasilo O.M. 1992. The pattern of poisoning in urban Zimbabwe. *Journal of Applied Toxicology* **12**, 435-438.

13. Organisation for Economic Co-operation and Development [OECD]. 1997. Test guideline 424: Neurotoxicity study in rodents.

14. Purves, D., Augustine, G.J., Fitzpatrick, D., Hall, W.C., Lamantia, A.S., Mcnamaras, J.O., Williams, M. 2004. Lower Motor Neuron Circuits and Motor Control In: Neuroscience. Third edition Sinauer Associates. Sunderland, Massachusetts U.S.A. pp. 371-441.

15. Reddy, C.S. and Hayes, A.W. 2001. Acute toxicity and eye irritancy. In Hayes A.W. (ed) Principles and Methods of Toxicology Fourth Edition Taylor and Francis, Philadelphia pp. 853-906.

16. Saric-Kundalic, B., Dobes, C., Klatte-Asselmeyer, V. and Saukel, J. 2010. Ethnobotanical study on medicine use of wild and cultivated plants in middle, south and west Bosnia and Herzegovina. *Journal of Ethnopharmacology* **131**, 33–55.

17. Sandager, M., Nielsen, N.D., Stafford, G.I., Van Stadden, J. and Jagger, A.K. 2005. Alkaloids from *Boophone disticha* with affinity for the serotonin transporte in the rat brain. *Journal of Ethnopharmacology* **98**, 367-370.

18. Sorbiercki, J.F. 2002. A preliminary inventory of plants used for psychoactive purposes in South African healing traditions. *Transactions of the Royal Society of South Africa* **57**, 1-24.

19. Tagwireyi, D., Ball, D.E. and Nhachi, C.F.B. 2002. Traditional medicine poisoning in Zimbabwe;: clinical presentation and management in adults. *Human and Experimental Toxicology* **21**, 579-586.

20. van Wyk, B.E. and Gericke, N. 2000. People's plants; a guide to useful plants in South Africa.Briza publications, South Africa.

21. van Wyk, B.E., van Heerden, F.R. and van Oudtshoorn, B. 2002. Poisonous Plants of South Africa. Briza Publications, Pretoria., pp. 60.

22. van Wyk, B.E., van Oudtshoorn, B., Gericke, N. 1997. *Medicinal Plants of South Africa*, first editionBriza Publications, Arcadia, South Africa, pp. 60–61.

23. Verschaeve, L. and van Staden, J. 2008. Mutagenic and antimutagenic properties of extracts from South African Traditional medicinal plants. *Journal of Ethnopharmacology* 119, 575-587.

24. Watt, J.M. and Breyer-Brandwijk, M.G. 1962. The medical and poisonous plants of Southern Africa. 2nd edition. Livingstone pp. 23-35.

25. Weiss, B. and Cory-Slechta, D.A. 2001. Assessment of Behavioral Toxicity. In Hayes A.W. (ed) *Principles and Methods of Toxicology* Fourth Edition. Taylor and Francis, Philadelphia pp. 1451-1520.

26. World Health Organization (WHO). 1999. Monographs on Selected Medicinal plants. Vol 1.

IV. Herbal Drug Research in Developing Countries

Chapter 20

Traditional Medicine in Cambodia

Bota Chengli

Technical Officer,
National Center for Traditional Medicine, Cambodia
E-mail: chenglibota@yahoo.com

Key words: *Traditional medicine, Integration, Traditional healer/health practitioner, Thnam Boran Khmer, Traditional healer association, Municipal health department, University of Health Science, Phytochemistry, Flore photographique.*

As from ancient to modern times, almost every society and culture of human kind have been using plants as ingredients to produce medicine. Traditional medicine is currently being used more widely and is increasingly considered as an important part of health care systems, as well as playing an increasing role in the various economies as a whole. Prosperous Khmer traditional medicine, which predated the Angkorian era, is currently playing a highly popular role in health care, particularly in rural areas. Khmer traditional medicine depends on medicinal plants and other natural products that are locally available and are part of indigenous knowledge after having gone through usage for many generations. The application of traditional medicine with equity, safety and efficacy remains a vital contribution to equitable health care at all levels and to sustaining economic and social development.

Traditional medicine policy of the Kingdom of Cambodia recognizes and determines the role of traditional medicine in the national health care system and provides a basis for promoting and developing traditional medicine to ensure support for and response to medical needs and to become a form of affordable, safe and effective healthcare. The national traditional medicine strategic plan 2012-2020 had been developed and launched, in which 5 key strategic objectives are highlighted: integration of traditional medicine into the national health system, promotion of the rational use of traditional medicine, promotion of access to safe and effective traditional medicine, promotion of protection and sustainable use of traditional medicine (TM), and strengthening of national and international cooperation in generating and sharing traditional medicine knowledge and skills.

The Government of the Kingdom of Cambodia, especially the municipal and provincial health departments led by the Ministry of Health, have taken into account the need in promoting, encouraging and gathering a database of traditional medicine stores in the municipalities and provincial cities in order to register their names/ stores in their respective locations so as to easily monitor and enable access; this would enhance control of their respective quality of service and product delivery. 92 stores, both traditional Chinese medicine and traditional Khmer medicine, locally called "Thnam Boran Khmer", have currently been registered in Phnom Penh; however, the majority of the stores are still non-registered due to the lack of knowledge, information, and capacity.

Currently, Cambodia has no pharmaceutical factory facilities available to solely produce medicinal plant-based products. Traditional medicine products presently on the market are being imported from neighbouring Asian countries such as Thailand, Vietnam, Korea, India and mainly from China. Local traditional medicine products are usually sold as pre-packaged/packaged and in different dosage forms: solid, powder, pills, balm, herbal tea, decoction, sometimes fresh herb/s, and fruit extract. The medicinal plant- based products that are being imported require prior registration by the Drug Department and further authorization by the Ministry of Health before being circulated on the Cambodian market.

The National Center for Traditional Medicine (NCTM) plans to start the integration of traditional medicine into the health care system as a pilot project in one health center at the outskirts of Phnom Penh municipality as of this year. It is in the process of discussing these matters with the Ministry of Health: screening and encouraging traditional health practitioners, learning ways to provide maximum quality and effective service and products on primary health care (PHC), and promoting access of this service to the general population.

The Traditional Healers have formed a few Traditional Healer Associations in Cambodia to support their careers, to enable easy and rapid access to news and appropriate health related knowledge from the Ministry of Health via NCTM, and are sometimes involved in NCTM activities.

As at 2013, thus far 345 traditional health practitioners had been trained by the NCTM in collaboration with the Cambodian Traditional Medicine Organization (CATMO) and the Cambodian Traditional Healer Association (CATHA) with support from the Nippon foundation. Two 3-day training programs on involving traditional health practitioners (THP) in primary health care have been provided to the provincial traditional health practitioners with the support of WHO in the latter part of last year. It is important to engage with the traditional health practitioners, as they are one of the first contact persons in the community when members of that community seek help with a health problem or concern. Referral of their clients to the nearest health center may be facilitated in this way for more serious cases. Moreover, it is anticipated that there will be more TM related training/refresher training programs aimed at the traditional healers and workshops aimed at provincial health providers to be conducted in the future in order to enhance the quality of TM services.

The NCTM has published in the local language 5 volumes of medicinal plants in Cambodia, your medicine in your garden booklet, your medicine around your house booklet, the translation from records on palm leaves, (ancient form of books) about the use of medicinal plants for treatment of illnesses from Bali into Khmer, and a certain number of formulas gathering from "Kru Khmer" (traditional healers) for the treatment of illnesses.

A botanical garden has been organized to retain local medicinal plants of approximately 300 species for demonstration to the public and particularly to teach students from local pharmaceutical faculties on the recognition of medicinal plants in the country.

Currently, Cambodia has not established a school or laboratory control specific to traditional medicine. However, it has the potential to do so in the future in order to build and strengthen local capacity on traditional medicine practices and products, to bring about the successful and effective integration of traditional medicine into the health care system in the future thus assisting in reducing poverty of the population in the rural areas.

The University of Health Science (UHS) with the collaboration from Pierre Fabre laboratory based in France established a well-equipped laboratory mostly for conducting phytochemistry research including qualitative and quantitative analysis. The UHS has annually received grants from major donors such as Pierre Fabre, the World Bank and the French government via Bio Asia's project to conduct the valorization of TM, quality control and the preservation of specimens of medicinal plants. The screening of different extracts derived from different parts of Cambodian medicinal plants, based on traditional use, has been conducted in the UHS for anti-bacterial activity, anti-oxidant activity etc. It has been identified that some extracts potentially have anti-bacterial, antioxidant and anti-cancer activities. In collaboration with NCTM, the first 7 monographs of selected medicinal plants are planned to be finalized by the end of year 2015 and the Cambodian traditional medicine pharmacopoeia will potentially be established within this future plan. Thus far, 1300 specimens of plants have been preserved at the UHS, of which only 50 per cent have been identified. The "Flore photographique du Cambodge" book have been published. Researchers at the UHS presently publish 2-3 articles relating to TM research internationally per year, which are peer reviewed. The UHS currently has the potential to look for funding to support a clinical trial pilot project on TM products used in primary health care.

The National Center for Traditional Medicine in Cambodia has strongly cooperated, continues to collaborate with, and strives to look for more support from local and international organizations, including Asian and other countries to improve traditional medicine in the country as a whole and in particular to develop and build staff capacity through ongoing training, research techniques and methods to ensure safety, efficacy and effectiveness of the use and the production of traditional medicine. The modernization of traditional medicine is also considered to align with different forms and presentations of products, research and services used in the region.

The NCTM nowadays faces a lack of competent staff in dealing with traditional medicine improvement, laws and regulations, technical and financial support to conduct activities, and tools and equipment necessary in establishing its own laboratory facilities.

Finally, this training workshop is a great opportunity for the NCTM to increase the capacity of its employees in order to learn and obtain updated knowledge related to the research approaches on drug discovery and the innovation strategy derived from this workshop and the interaction with other fellow NAM S&T members.

Chapter 21

Accumulation and Production of Total Indole Alkaloids, Vinblastine and Vincristine from Egyptian *Catharanthus roseus* (L.) G. Don. Calli Cultures by using Levels of Cytokinins (Kin)

M. Abd-El-Kareem Fathalla

Plant Production Department, City for Scientific Research and Technology Applications, Borg El-Arab, Alexandria, Egypt
E-mail: mafaboaish @yahoo.com

ABSTRACT

Catharanthus roseus is still the only source for the powerful antitumor drugs vinblastine and vincristine. Effects of cytokinins **(Kin)** on enhancement the accumulation rate of total indole alkaloids; antineoplastic agents (vinblastine and vincristine) in calli cultures of Egyptian *Catharanthus roseus* (L.) G. (Don) were studied. Cytokinins **(Kin)** was used at levels of 0.1,0.2,0.5,1.0,2.0 and 3.0 mg/l added to modified alkaloid production medium (MAPM). The obtained results showed that: All tested concentrations of Kin significantly increased the different measured growth parameters of shoot and root derived calli, however, Moreover, the percentage of total indole-alkaloids was significantly increased. The highest values of the accumulation rate of total indole- alkaloids, further, vinblastine and vincristine were recorded at the rate of 3.0 mg/l/Kin in shoot and root derived calli, respectively.

Keywords: *Catharanthus roseus, Calli cultures, Cytokinins (Kin), Total indole alkaloids, Vinblastine and/or vincristine.*

1. INTRODUCTION

Catharanthus roseus has warranted significant study due to its production of two valuable alkaloids used in chemotherapy. These indole-alkaloids, vincristine and vinblastine, are produced at extremely low levels within the plants and remain resistant to feasible chemical synthesis due to their complex structures (Verpoorte *et al.*, 1997, Hughes and Shanks, 2002). *C. roseus* (Madagascar periwinkle) is the natural source for about 130 terpenoid indole alkaloids (TIA) including the well-known anticancer agent's vinblastine and vincristine. These indole-alkaloids are spindle toxins that have proven effective in chemotherapy treatments for leukemia and Hodgkin's disease (lymph node and spleen cancer). In addition, it is an important medicinal plant, belongs to the family Apocynaceae and cultivated mainly for its alkaloids (Datta *et al.*, 1980). Pink flowered cultivar gives higher yield of foliage; roots and total alkaloids as compared with white flowered cultivar. In addition to anticancer activity, the total alkaloids and chloroform fraction of crude drug showed significant sustained hypotensive action and selective transquilizing properties (CSIR *et al.*, 1992). Moreover, the ethanolic extract of leaf has significant hypoglycemic activities, nrendering it as a good anti-diabetic drug (Kar *et al.*, 2003). The current work aimed to study the induction and accumulation of the active constituents (vincristine and/or vinblastin) in different types of *C. roseus* calli cultures in response to cytokinins (Kin).

2. MATERIALS AND METHODS

Plant Materials

Seeds of Egyptian *Catharanthus roseus* (L.) Don.c.v. alba were kindly obtained from Institute of Horticulture Research, Agricultural Research Centre, Giza, Egypt. Seeds were surface sterilized under aseptic conditions of laminar flow hood, using 70 per cent EtOH for 30 Sec, and then transferred to a solution of 50 per cent Clorox (containing 5.25 per cent NaOCl and drops of twin 20) for 15 min. Then they were aseptically germinated on basal MS (Murashige and Skoog, 1962) medium for 4 weeks and used as plant materials.

Calli Production

Three aseptically segment of shoots and roots were excised from *C. roseus* sterilized plantlets and placed in 200 ml jars containing 40 ml of B5 (Gamborg and Shyluk, 1981) solid medium containing 1mg/l each of 2,4-D and BA (The best medium for callus production according to Taha *et al.* (2008). Sub-culturing had been done every 4 weeks on MS-medium containing 1mg/l BA.

Supplementation of B5 Medium with Cytokinins (Kin)

The effect of addition of cytokinins **(Kin)** in formula of Kin to MAPM medium on achievement of shoot and root calli growth parameters as well as on accumulation of total indole alkaloids was determined. The tested levels of Kin 0.1, 0.2, 0.5, 1.0, 2.0 and 3.0 mg/l gm/l; were used. The control treatment was MAPM medium without kin. Thus, there were 6 separated experiments considering the effect of cytokinins **(Kin).** At different concentrations, and each one contained 6 treatments.

Determination of Total Indole Alkaloids

Preparation and Extraction of Crude Indole-Alkaloid

Crude alkaloid extract was obtained from each kind of dry plant materials by a modification of the method mentioned by Hirata *et al.* (1987). Where two grams as a fine powder of dry plant material was extracted 4 times with ethanol (10 ml) for 15 minutes in a sonic bath at 60 °C. After filtered, the extracted solution was evaporated at 60°C under reduced pressure in a rotary evaporated and portioned between 5 per cent sulphuric acid and diethyl either (1:1). The organic phase was washed 3 times with 5 per cent sulfuric acid (1:1). The combined acid solution was adjusted to pH 10 with 10N sodium hydroxide and extracted 3 times with chloroform (1:1). The combined chloroform phase was evaporated under reduced pressure to give the crude alkaloid extract.

Quantitative Determination of Total Indole Alkaloid and Antineoplastic Agents (mg/g Dw) Using HPLC Technique

Quantitative analysis of alkaloid contents was carried out using high – performance liquid chromatography (HPLC) according to the described method by Zhao *et al.* (2000). Alkaloids enriched medium were directly extracted into ethyl acetate after adjusting the medium to pH 10. HPLC conditions: Shimadsu LC -4A instruments with CR3A integrator and SPD -4A UV-detector (280nm), Nucleosil 5 C18 column (250 x 4.6mm, 5µ m) ; Samples were eluted with a mobile phase of methanol/acetonitrile/0.025 mol l-1 ammonium acetate/triethylamine (15:40:45:0.1) at a flow rate of 1 ml/min. Alkaloids were identified by TLC, co- elution in HPLC and HPLC Diode Array Detection of spectra and compared with standard substances.

Statistical Analysis

The statistical analysis was carried out according to the described method by Snedecor and Cochran (1980) using least significant difference (L.S.D.) at 0.05 level of probability.

3. RESULTS AND DISCUSSION

Effect of (Kin) on Qualitative and Quantitative Determination of Total Indole Alkaloids

Total Indole Alkaloid Contents

Table 21.1 shows the total alkaloid contents expressed as mg per jar in both cases of shoot and root derived calli. Statistical analysis indicated that the differences among the different treatments were highly significant in all cases. It was clear that all tested Kin concentrations significantly increased the average of total alkaloids per jar for both shoot and root derived calli either before or after subculture as shown also in Figure 21.1a.

Regarding the shoot derived calli, the highest average of total alkaloids per jar was observed at the Kin concentration of 3.0 mg/l (17.839 and 18.589 mg/jar before and after subculture; respectively).The treatment of 3.0 mg/l Kin concentration

Table 21.1: Effect of Different Concentrations of Kin on Total Indole Alkaloid Contents (mg/g dry weight) of Shoot and Root Derived Calli of Catharanthus roseus L.G.Don, Before and After Subculture[1]

SHOOT

Kin Conc. mg/l	Dry Weight (g/Jar)		Total Alkaloid (mg/jar)		Total Alkaloid (mg/g DW)		Total Alkaloid (per cent)	
	Before	After	Before	After	Before	After	Before	After
control	0.4966c	0.5259c	6.703f	7.341f	13.498f	13.958f	1.350 (6.8)f	1.396 (6.8)f
0.1	0.5227c	0.5286c	10.447e	10.802e	19.987e	20.435e	1.999 (8.13)e	2.044 (8.13)e
0.2	0.5587b	0.5574b	12.081cd	12.563d	21.624d	22.538d	2.162 (8.53)d	2.254 (8.72)d
0.5	0.4963cd	0.5124c	11.794d	12.268d	23.763c	23.942c	2.376 (8.91)c	2.394 (8.91)c
1.0	0.4814d	0.5347bc	12.359c	13.920c	25.672b	26.034b	2.567 (9.28)b	2.603 (9.28)b
2.0	0.5703b	0.5624b	15.387b	15.367b	26.981a	27.324a	2.698 (9.46)a	2.732 (9.46)a
3.0	0.6374a	0.6631a	17.839a	18.589a	27.987a	28.034a	2.799 (9.63)a	2.803 (9.63)a

ROOT

Kin Conc. mg/l	Dry Weight (g/Jar)		Total Alkaloid (mg/jar)		Total Alkaloid (mg/g DW)		Total Alkaloid (per cent)	
	Before	After	Before	After	Before	After	Before	After
control	0.4400d	0.4596d	5.530e	5.964e	12.567f	12.976g	1.257 (6.55)e	1.298 (6.55)f
0.1	0.5465c	0.5458b	9.146cd	9.269d	16.735e	16.982f	1.674 (7.49)d	1.698 (7.49)e
0.2	0.5758b	0.5846a	10.039b	10.512c	17.435d	17.982e	1.744 (7.49)d	1.798 (7.71)d
0.5	0.5326c	0.5205c	9.816bc	9.694d	18.431c	18.625d	1.843 (7.71)c	1.863 (7.92)c
1.0	0.4691d	0.4775d	8.907d	9.226d	18.987c	19.321c	1.899 (7.92)b	1.932 (7.92)c
2.0	0.6112a	0.6090a	12.025a	12.177a	19.674b	19.995b	1.967 (8.13)a	1.999 (8.13)b
3.0	0.5640bc	0.5537b	11.473a	11.386b	20.342a	20.563a	2.034 (8.13)a	2.056 (8.33)a

1) Figures followed by the same letters are not significantly different using L.S.D. test at 0.05 level of probability.

Each value is the average of eight replicates.

Parenthetical values are angles corresponding to the mean of percentages and were used in ANOVA and for comparison.

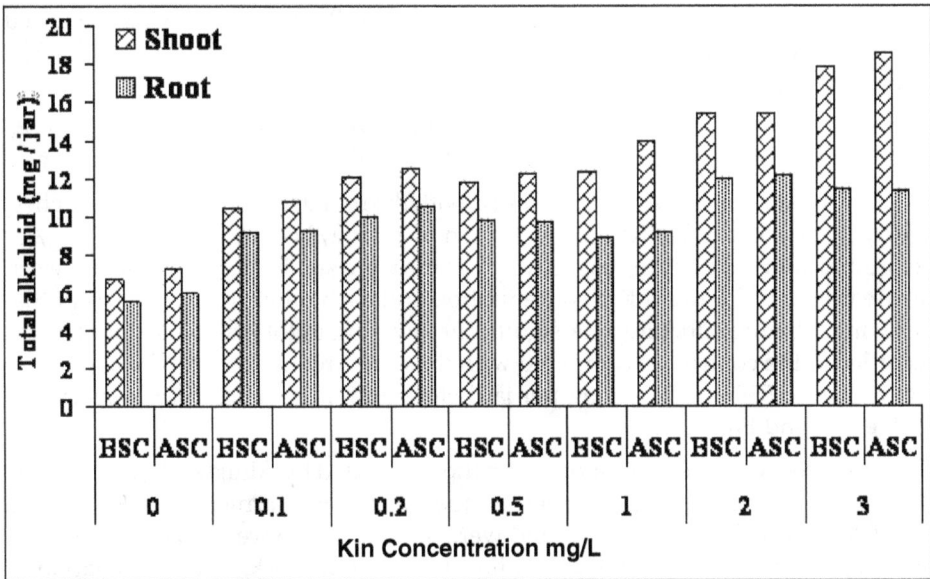

Figure 21.1a): Effect of Different Concentrations of Kin on Total Alkaloids
per Jar Before and After Subculture (BSC and ASC) of Shoot and
Root Callus Tissues of *Catharanthus roseus*.

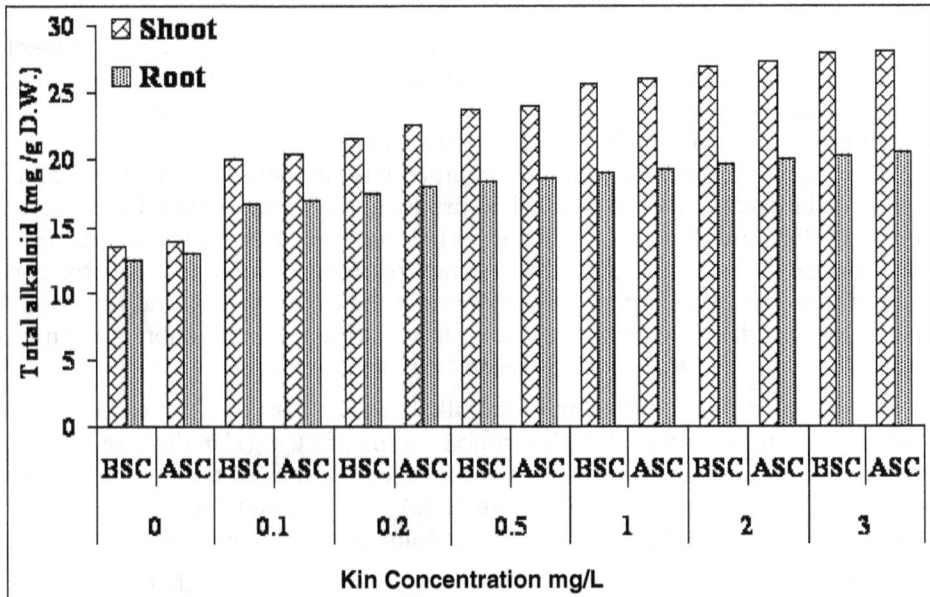

Figure 21.1b: Effect of Different Concentrations of Kin on Total Alkaloids
per gram Dry Weight Before and After Subculture (BSC and ASC) of Shoot and
Root Callus Tissues of *Catharanthus roseus*.

significantly overcame all other treatments. Generally, with increasing the tested Kin concentration from 0.1 to 3.0 mg/l, the total alkaloid content increased. The lowest average per jar was seen.

At the control treatment and was equal to 6.703 and 7.341 mg per jar before and after subculture; respectively (Table 21.1 and Figure 21.1a).

Nearly, similar results were recorded with the root derived calli, but the highest average was recorded at Kin concentration of 2.0 mg/l (12.025 and 12.177 mg/jar before and after subculture; respectively). There were significant difference between the treatment of 2.0 mg/l Kin concentration and each other treatment except that of 3.0 mg/l before subculture. A significant lowering in total alkaloids per jar was recorded at the control, as compared with the other treatments and the average of control was 5.530 and 5.964 mg per jar before and after subculture; respectively (Table 21.1 and Figure 21.1a).

These results were in harmony with those reported by Ahuja *et al.* (1982); Kodja *et al.* (1989); Decendit *et al.* (1992); Ferderique *et al.* (1996); Garnier *et al.* (1996); Yahia *et al.* (1998) and Zhao *et al.* (2001) and were not in accordance with those reported by Rudge and Fowler (1989). Regarding the current work, The combination of IAA (0.1752 mg/l) and tested Kin (from 0.1 to 3.0 mg/l) gave the greatest amount of alkaloid production and the highest degree of growth of calli, also increased the dry weight and percentage of dry matter, while the control gave the lowest of the mentioned parameters. According to the results presented in Table 21.1b, it could be concluded that the higher parameters mentioned before were usually accompanied by higher alkaloid production. Since the growth rate and the increase value can represent the degree of callus growth and callus growth could represent the dry weight and cellular/tissue activation, the higher degree of growth and callus weights may be responsible for more stable and higher amounts of alkaloid biosynthesis and accumulation in callus tissues. In fact, plant growth regulators (at tested combination) and other medium components influenced the degree of callus growth and increased the indole alkaloid production of callus tissues. Lindsey and Yeoman (1983) and Zhao *et al.* (2001) reported that some differentiated structures were formed in callus tissues. These structures functionally transport oxygen and nutrients across the interior and exterior of the callus tissues, and are closely related to secondary metabolism since certain level of differentiation is required for production and storage of some secondary metabolites.

After making subculture, the total alkaloids per jar were increased at the different treatments except at the Kin concentration of 2.0 mg/l in the case of shoot formed calli and at those of Kin concentrations of 0.5, and 3.0 mg/l in the case of root formed calli (Table 21.1 and Figure 21.1a). Similar results were recorded by Stafford *et al.* (1985), Scragg *et al.* (1989) and Rijhwani and Shanks (2000).

Regarding the total alkaloids expressed as mg per g dry weight of calli, there were highly significant differences among the different treatments either before or after subculture for calli derived from shoots or roots. In all cases the tested Kin rates were significantly able to produce higher amounts of total alkaloids compared with the control and these amounts increased with increasing the tested Kin concentration (Table 21.1 and Figure 21.1b).

For the shoot derived calli, the highest amount of total alkaloids was recorded at the Kin concentration of 3.0 mg/l (27.987 and 28.034 mg/g dry weight with a percentage of 2.798 and 2.803 per cent before and after subculture; respectively). There was a significant differences between each of the treatments of 2.0 and 3.0 mg/l Kin concentrations on one side and each of the other treatments including the control, on the other side. The control treatment significantly had the lowest amount of alkaloids comparing with the tested Kin concentrations (13.498 and 13.957 mg/g dry weight with a percentage of 1.350 and 1.396 per cent before and after subculture; respectively) as listed in Table 21.1 and shown in Figure 21.1b.

For the root derived calli, The Kin concentration of 3.0 mg/l had the highest amount of total alkaloids (20.342 and 20.563 mg/g dry weight with the percentage of 2.034 and 2.056 per cent before and after subculture; respectively) and significantly differed with the other treatments. On the contrary, the lowest amount of total alkaloids was detected at the control and significantly differed with the other treatments (12.567 and 12.976 mg/g dry weight with the percentage of 1.257 and 1.298 per cent before and after subculture; respectively) as shown in Table 21.1 and Figure 21.1b.

These mentioned results were supported with those mentioned by Ahuja *et al.* (1982); Kodja *et al.* (1989); Decendit *et al.* (1992); Ferderique *et al.* (1996); Garnier *et al.* (1996); Yahia *et al.* (1998) and Zhao *et al.* (2001). The declaration was reported before with the total alkaloids produced in each jar.

After making subculture the amount of total alkaloids increased at each treatment in both cases of shoot and root derived calli which was on line with Stafford *et al.* (1985); Scragg *et al.* (1989) and Rijhwani and Shanks (2000).

The results of the percentage of total alkaloids were similar to those of the total alkaloid per jar and the total alkaloid expressed as mg per g dry weight (Table 21.1).

Some differences were observed between the results of the shoot formed calli and those of the root formed calli which was in accordance with those reported by Bereznegovskaga *et al.* (1977); Lapinjoki *et al.* (1987) and Valenzuela *et al.* (1998). These differences may due to the different response of each explant to the used medium and Kin concentrations or to the environmental conditions of cultures.

Quantitative Determination of Antineoplastic Agents using HPLC Technique

Table 21.1 shows the total alkaloids contents and vincristine levels extracted from shoot and root derived calli grown on alkaloid production medium supplemented with 3.0 mg/l Kin, before and after subculture. The highest total alkaloids content was recorded at this Kin concentration.

Regarding the shoot derived calli, the highest amount of total alkaloids content was 27.987 and 28.034 mg per gram dry weight before and after subculture; respectively. These amounts of total alkaloids were higher (nearly 2-fold) than that recorded in the original mother plants. Concerning this result, several workers reported that cytokinins can stimulate the production of total alkaloids in callus cultures of *C. roseus* (Miura *et al.*, 1978; Ahuja *et al.*, 1982; Smith *et al.*, 1987a; Kodja *et al.*, 1989; Laszlo and Berci, 1990; Decendit *et al.*, 1992; Ferderique *et al.*, 1996; Garnier *et al.*, 1996; Abd- elouahab *et al.*, 1998 and Yahia *et al.*, 1998). Yahia *et al.* (1998) concluded that cytokinins can up- regulate alkaloid production in *C. roseus* cultures through independent pathways when added exogenously to the cultures. On the contrary of the mentioned observation, Miura *et al.* (1988) stated that alkaloids in callus cultures of 222 were lower than those in the parent plant.

Vincristine levels in the shoot derived calli were lower than those in the original mother plants and were 100.860 and 100.917µg per gram dry weight before and after subculture; respectively and with a percentage of 0.36 per cent relative to the total alkaloids content (Table 21.1 and Figures 21.3 and 21.4). Lapinjoki *et al.* (1987) suggested that alkaloid spectra of shoot cultures of *C. roseus* and those of intact plants were similar to each other. Hirata *et al.* (1991) pointed out that the production of some alkaloids was stimulated, while that of others was decreased in callus cultures of *C. roseus*.

The shoot derived calli that had two passages on alkaloid production medium (subculture) were indirectly proliferated to rooted plantlets. The analysis of these rooted plantlets showed that the total alkaloids content was 1.8- fold of that of the original mother plant and it was 25.875 mg per gram dry weight of the whole plantlets. In this concern, Lapinjoki *et al.* (1987) reported that alkaloid spectra of shoot cultures of *C. roseus* were similar to those of aerial parts of mother plants. Regarding vincristine concentration of in vitro shoot callus formed rooted plantlets, it was 251.134 µg per gram dry weight with a percentage of 0.97 per cent relative to the total alkaloids content (Table 21.1 and Figure 21.5). It was lower than that of the original plant materials.

Vinblastine alkaloid was not detected in the shoot callus or in the in vitro derived plantlets similar to that recorded in the mother plants.

Regarding the root derived calli grown on alkaloid production medium supplemented with 3.0 mg/l Kin, it was found that these calli produced total alkaloids contents of 20.342 and 20.563 mg per gram dry weight before and after subculture; respectively, and these amounts were more than that recorded for the mother plants by 5.612 and 5.833 mg; respectively (nearly 1.4 -fold) as shown in Table 21.1. Vincristine levels of root derived calli were more lower than that of the mother plants (0.458 and 0.497µg per gram dry weight with percentages of 0.002 per cent relative to the total alkaloids contents before and after subculture; respectively) as listed in Table 21.1 and shown in Figures 21.6 and 21.7. These results were supported by Hirata *et al.* (1991) where they reported that in *C. roseus* cultures some alkaloids increased and others decreased. With regard to this point Lapinjoki *et al.* (1987) reported that the alkaloid spectra of the root cultures were different from those of intact plants of *C. roseus*.

Quantitative Determination of Antineoplastic Agents Using HPLC Technique

Determination of indole alkaloid in whole cultures was carried out using HPLC technique. As shown in chromatogram of Figure 21.2 Vincristine was detected (retention time = 10.86 min.). The analysis indicated that vinblastine was not detected and this may be related to the used cultivar. Table 21.2 show that total alkaloids

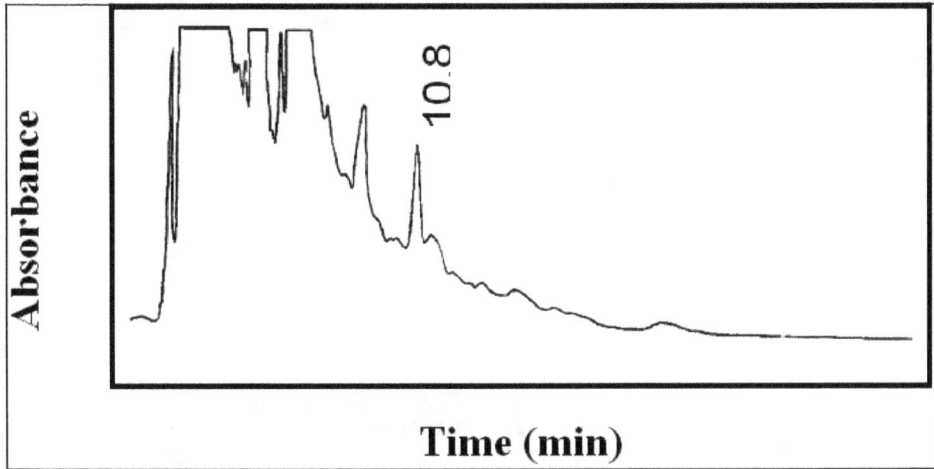

Figure 21.2: Chromatogram of Alkaloid Extracted from Whole Original Mother Plant of *C. roseus*, where Vincristine was Detected at 10.86.

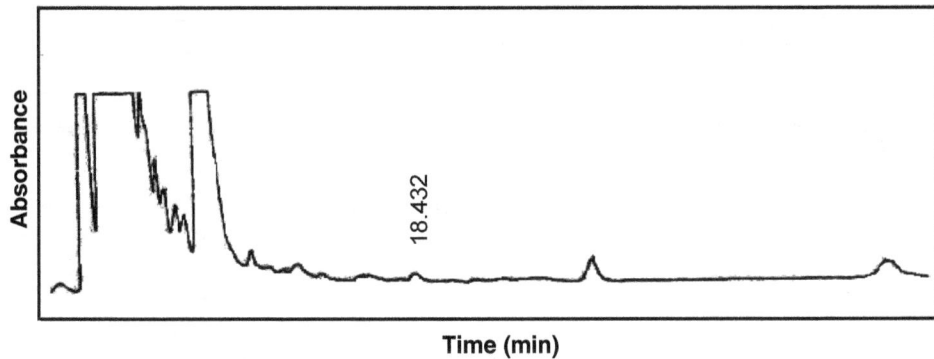

Figure 21.3: Chromatogram of Alkaloid Extract from Shoot Derived Calli of *C. roseus*. Grown on APM Medium Supplemented with 3.0 mg/l of Kin, Before Subculture. Where vincristine was detected at 18.432.

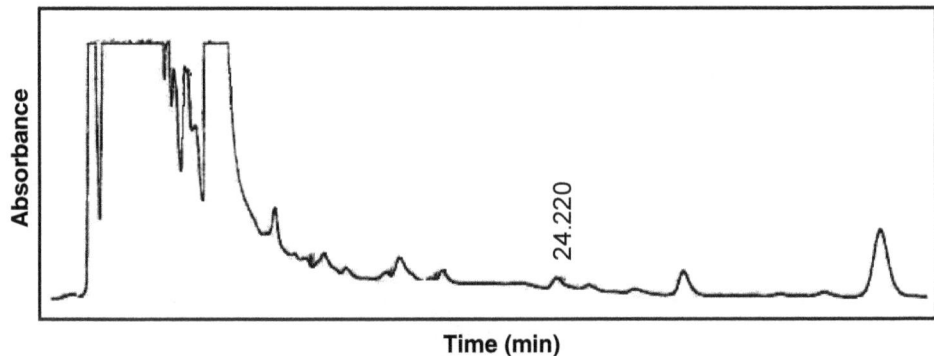

Figure 21.4: Chromatogram of Alkaloid Extract from Shoot Derived Calli of *C. roseus*. Grown on APM Medium Supplemented with 3.0 mg/l of Kin, Before Subculture. Where vincristine was detected at 24.220.

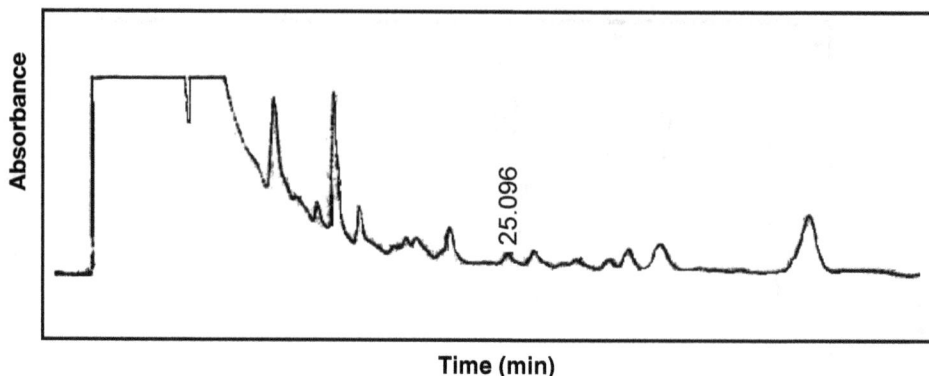

Time (min)

Figure 21.5: Chromatogram of Alkaloid Extract from Indirect Regenerated Plants from Shoot Derived Calli of *C. roseus*. Grown on APM Medium Supplemented with 3.0 mg/l of Kin, Before Subculture. Where vincristine was detected at 25.096.

content 14.730 mg/g DW was recorded with whole tissues, while vincristine content recorded 404 µg/g DW with a percentage of 2.743 per cent relative to the total alkaloids. Vincristine was isolated from the crude indole alkaloid extracted by preparative TLC technique.

Total Indole Alkaloids and Antineoplastic Agent Contents Cultured on B5 Medium

As shown in Table 21.2 shoot derived calli contained a total indole alkaloid 15.530 mg/g DW and those overcame the original plant with 0.8 mg/g DW. The concentration of vincristine was 128.672 µg/g DW with a percentage of 0.829 per cent relative to the total alkaloids (Table 21.2). It was clear that vincristine was decreased comparing with the original material. This decrease was supported by the obtained results of Morris *et al.* (1988). Absence of vinblastine in this study was similar to that mentioned by Tikhomiroff and Jolicoeur (2002) on *C. roseus* and in

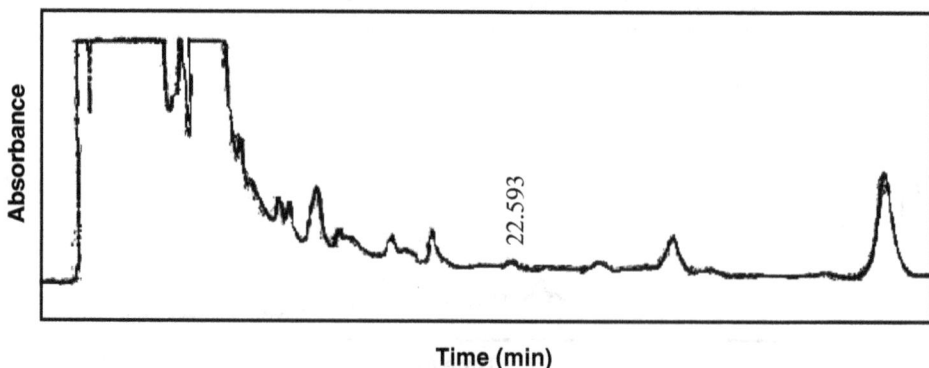

Time (min)

Figure 21.6: Chromatogram of Alkaloid Extract from Root Derived Calli of *C. roseus*. Grown on APM Medium Supplemented with 3.0 mg/l of Kin, Before Subculture. Where vincristine was detected at 22.593.

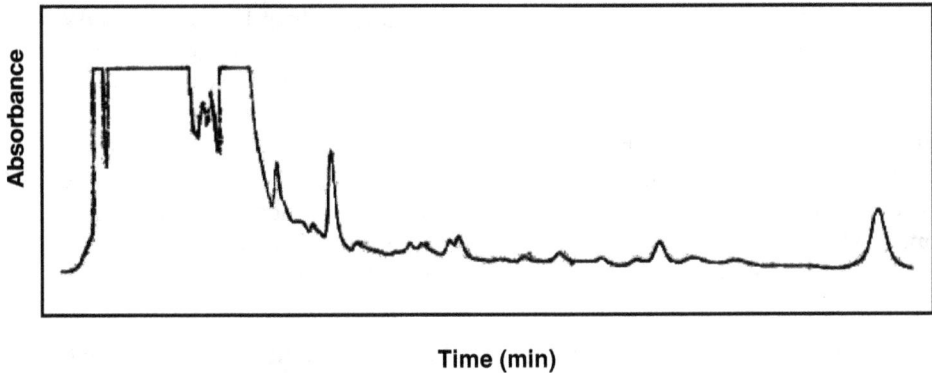

Time (min)

Figure 21.7: Chromatogram of Alkaloid Extract from Root Derived Calli of *C. roseus*. Grown on APM Medium Supplemented with 3.0 mg/l of Kin, Before Subculture. Where vincristine was detected at 23.964

contrast to the results reported by Pardsani *et al.* (1979), Mandel and Maheshwani (1987), Hirata *et al.* (1989), Parr *et al.* (1989), Renault *et al.* (1999) and Uniyal *et al.* (2001) may be this related to the used genotype. Regarding root derived calli, the total indole alkaloids content was 12.765 mg/g DW, which less than those of shoot derived calli and origin materials, (Table 21.2). Vinblastine was not detected, but vincristine was found at 117.524 µg/g DW with a percentage of 0.921 per cent relative to the total indole alkaloids content (Table 21.2). These results were similar to those mentioned by Bereznegovskaga *et al.* (1977), Lapinjoki *et al.* (1987) and Fatah Allah (1997) on *C. roseus*. The difference between shoot derived calli and root derived ones may be due to the biochemical potential of each calli type under the culture conditions used in the present work.

In the same field, Moiura *et al.* (1988) found that the yield of alkaloids in the leaf tissues was higher than that in the root calli cultures of *C. roseus* which similar to the present results.

Determination of Total Indole Alkaloids and Antineoplastic Agent Contents Cultured on MAP Medium

The results of shoot derived calli indicated that the total indole alkaloids contents were 13.498 and 13.959 mg/g DW before and after subculture on MAP

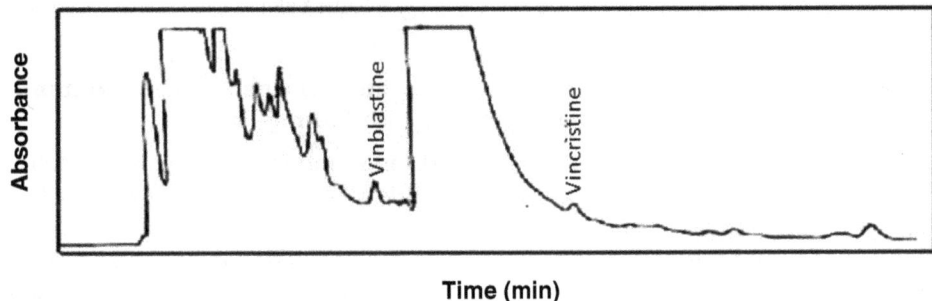

Time (min)

Figure 21.8: Standard Curve of Vincristine and Vinblastine.

medium. Concerning, vincristine was detected at the rate of 101.970 µg/gDW with a percentage of 0.755 per cent relative to the total indole alkaloids content. After subculture vincristine was detected at the rate of 101.580 µg/g DW with a percentage of 0.728 per cent relative to the total indole alkaloid contents. As shown in Table 21.2 all these values were less than those detected in the original materials. Regarding root derived calli, the total alkaloids contents were 12.567and 12.976 mg/g DW before and after subculture on MAPM, respectively. Vincristine rates were 0.346 and 0.342 µg/g DW before and after subculture; respectively with percentages of 0.003 per cent relative to the total alkaloids content either before or after subculture (Table 21.2).

Table 21.2: Determination of Total Alkaloids and Vincristine Contents and Vincristine Relative Percentage using HPLC for Original Intact Plants, different Types of Calli Cultures of *Catharanthus roseus*

Analyzed Material	Original Plants and Different Calli		
	Total Alkaloids (Mg/g d w)	Vincrstine (µg/g d w)	Vincrstine (per cent)
Original intact plants	14.730	404.000	2.74
Shoot callus grown on B5 medium	15.530	128.672	0.83
Root callus grown on B5 medium	12.765	117.524	0.92
Shoot callus grown on APM before sub.	13.498	101.970	0.76
Shoot callus grown on APM after sub.	13.959	101.580	0.73
Root callus grown on APM before sub.	12.567	0.346	0.003
Root callus grown on APM after sub.	12.976	0.342	0.003
Shoot callus grown on APM with 3.0 mg/l Kin before sub.	27.987	100.860	0.36
Shoot callus grown on APM with 3.0 mg/l Kin after sub.	28.034	100.917	0.36
Root callus grown on APM with 3.0 mg/l Kin before sub.	20.342	0.458	0.002
Root callus grown on APM with 3.0 mg/l Kin after sub.	20.563	0.497	0.002

Kin Concentration mg/L

In both shoot and root derived calli, vinblastine was not detected. All results indicated that both callus types contained lower amounts of the total alkaloids and vincristine corresponding to those detected in the original plants (Table 21.2). These results were in harmony with those reported by Moiura *et al.* (1988) and Toivonen *et al.* (1989).

Determination of Total Indole Alkaloids and Antineoplastic Contents Cultured on MAP Medium Supplemented with 3.0 mg/l of Kin

Table 21.2 show the total indole alkaloids and vincristine contents, extracted from the shoot or root derived calli before and after sub culturing onto MAPM supplemented with *3.0 mg/l* of Kin. Regarding the average of total alkaloid expressed as mg/g DW, and it found that the highest averages 27.987 and 28.034 mg/g DW were recorded before and after subculture, respectively. The reduction of total alkaloids was increased with increasing of Kin concentrations. In the case of root derived calli, the highest average of total indole alkaloid recorded with Kin of *3.0 mg/l*. HPLC analysis showed that vincristine contents 100.89 and 100.917 µg/g DW with the same percentage of 0.005 relative to the total alkaloids content were recorded with shoot calli formed.

Before and after subculture. However, vincristine contents 0.458 and 0.497 µg/g dry weight with the percentages of 0.00255 and 0.00316 relative to the total alkaloids content were recorded with formed callus by roots, before and after subculture; respectively.

These results mentioned before were in contrast with those obtained by Tallevi and Cosma (1988) who reported that the reduction of alkaloids accumulation was associated with the high amounts of heavy metal and which had an inhibitory effect on metabolic processes. It was clear that the Kin element caused a stress for the callus when added to the media, however, enhanced the callus to produce higher total alkaloid contents. The Kin supplementation to the medium was a good stress factor for indole alkaloid accumulation in spite of the depression of weight. The total alkaloid content in callus grown on MAPM supplemented with *3.0* mg/l. of Kin was higher than those in the original plants, but vincristine was lower.

4. REFERENCES

1. Ahuja, A.; R. Parsad and S. Grewal, 1982. *Catharanthus roseus* L.G.Don. Tissue culture: growth differentiation and secondary metabolism. *Indian Journal of Pharmaceutical Sciences*. 44(4): 78-81.

2. Bereznegovskaya, L.N., N.A. Trofimova and T.I. Andreeva, 1977. Characteristics of *Catharanthus roseus* tissue culture. *Rastitel'nye resursy*. 13(3): 455-460.

3. Datta, S.C., 1980. Cultivation of *Catharanthus roseus* in West Bengal, in: C.K.Atal, B.M. Kapur, (Eds.), Cultivation and Utilization of Medicinal Plants, CSIR, Jammu Tawi, and pp: 279-283.

4. Decendit, A.; D. Liu; I.Ouelhazi; P.Doireau; J.M.Merillon and M.Rideau, 1992. Cytokinin enhanced accumulation of indole alkaloids in *Catharanthus roseus* cell cultures: The factors affecting the cytokinin response. *Plant Cell Reports* 11 (8): 400- 403.

5. Fatah- Allah, M.A., 1996. In vitro studies on *Catharanthus roseus* (L.) G.Don. *M.Sc. Thesis*, Faculty of Agriculture - Cairo University.

6. Garnier, F.; S.Carpin; P.Labelb; C. Joel; H.Said and Marc R., 1996. Effect of cytokinin on alkaloid accumulation in periwinkle callus cultures transformed with a light-inducible ipt gene. *Plant Sciences*.120:47-55.

7. Hirata, K., A. Yamanaka, N. kurano, K. Msyamoti and y. Miura, 1987. Production of indole alkaloids in multiple shoot culture of *Catharanthus roseus* L. G. Don. *Agricultural and Biological Chemistry*, 51(5): 1311-1317.

8. Hirata, K., M. Kabayashi, K. Miyamoto, T. Hoshi, M. Okazaki and Y. Miura, 1989. "Quantitative determination of vinblastine in tissue cultures of *C. roseus* by radioimmunoassay. *Planta medica*, 55(3): 262-264.

9. Hirata, K.; M.Horiuchi; T.Ando; K. Miyamato and Y.Miura, 1990. Vindoline and catharanthine production in multiple shoot cultures of *Catharanthus roseus*. *Journal of Fermentation and Bioengineering*. 70: 193-195.

10. Hirata, K.; M. Horiuchi; T. Ando; M.Asada; K. Miyamato and Y.Miura, 1991. Effect of near-ultraviolet light on alkaloid production in multiple shoots cultures of *C. roseus*. *Planta Medica*. 57 (5): 499-500.

11. Kar, A., B.K. Choudhary, N.G. Bandyopadhyay, 2003. Comparative evaluation of hypoglycemic activity of some Indian medicinal plants in alloxan diabetic rats, *Journal of Ethnopharmacology*., 84: 105-108.

12. Kodja, J.; D. Liu; J.M. Merillon; F. Andreu ; M. Rideau and J.C. Chenieux, 1989. Stimulation of indole alkaloid accumulation in cell suspensions of *Catharanthus roseus*. By cytokinins. Comptes *Rendus* de l' *Académie* des *Sciences, Sciences* de La *Vie*, In: *B*. Loft (Ed.), Physiology of the amphibia, vol. II. *309–521*. *Comparative Biochemistry and Physiology*, 127:453–467.

13. Lapinjoki, S., H. Verajankorva, J. Heisknen, A. Huhtikangas and M. Lounsmaa, 1987. "Immunoanalytical methods for screening vindoline from cell culture". *Planta medica*, 53(6): 565-567.

14. Laszlo, M. and I. Berci, 1990. The effect of cultural conditions on the alkaloid production of *Catharanthus roseus* L.G.Don. Tissue cultures. *Napjaink Biotechnologiaja* 26: 30-34.

15. Lindsey,K. and M.M. Yeoman,1983. The relationship between growth rate, differentiation and alkaloid accumulation in cell cultures. *Journal of Experimental Botany*, 34: 1055-1065.

16. Mandal, S. and M.L. Maheshwari, 1987. "HPLC determination of vindolin, catharanthine, vincaleukoblastine and vincristine in periwinkle leaf". *Indian. Journal of Pharmaceutical. Sciences*. 49: 205-209.

17. Mollard, A.; G. Hustache and F. Barnoud, 1974. Some aspects of the action of indolyl- acetic acid on the tissue of *Rosa glauca* cultured in vitro. *Phytomorphology* 23(1-2): 99-104.

18. Morris, P., K. Rudge, R. Cresswell and M.W. Fowler, 1988. Regulation of product synthesis in cell cultures of *Catharanthus roseus*.V.Long - term maintenance of cells on a production medium. *Plant Cell, Tissue and Organ Culture*, 17: 79-90.

19. Miura, Y.; K.Hirata, and N.Kurano, 1987. Isolation of vinblastine in callus with differentiation roots of *Catharanthus roseus*. *Agriculture Biological Chemistry*. 51 (2): 611-614.

20. Murashige, T. and F. Skoog, 1962. A revised medium for rapid growth and bioassays with tobacco cultures. Physiol. Plant., 15: 473-478. *Journal of Applied Sciences Research*, 7(4): 542-549, 2011 549.

21. Pardsani, K.M., S. Singh and J.P.S. Sarin, 1979. Chromatogrphic estimation of vincaleukoblastine in *C. roseus* L. *Indian Journal of Pharmaceutical Sciences*, 41: 207-208.

22. Parr, A-J., A.C.J. Peerless, J.D. Hamill, N.J. Walton, R.J. Robins and M.J.C. Rhodes, 1988. "Alkaloids production by transformed root cultures of *C. roseus*". *Plant Cell Reports*, 7(5): 309-312.

23. Renault, J.H., J.M. Nuzillard, G. Le Crouerour and P.Z. Thepeiner, 1999. Isolation of indol alkaloids from *C. roseus* by centrifuged partition chromatography in the pH zone refining mode. *Journal Chromatography*, A. 849(2): 421-431.

24. Rudge, K. and M.W. Fowler, 1989. The effects of plant growth regulators on the production of secondary metabolites in plant cell callus and suspension cultures. *Plant Growth Regulator Society of America*. 13: 42-49.

25. Rijhwani, S.K. and J.V. Shanks, 2000. Effect of elicitor dosage and exposure time on biosynthesis of indole alkaloids by *Catharanthus roseus* hairy root cultures. *Biotechnology Progress*. 16: 442-449.

26. Scragg, A.H., 1999. Alkaloids accumulation in *Catharanthus roseus* suspension cultures. *Methods Molecular Biology*, 111:393-402.

27. Snedecor, G.W and W.G. Cochran, 1980. *Statistical Methods* Seventh Ed. The Iowa State University Press, Ames, Iowa.

28. Taha, H.S., M.K. El-Bahr and M.M. El-Nasr, 2008. *In vitro* studies on Egyptian *Catharanthus roseus* (L.) G.Don. I-Calli production, direct shootlets regeneration and alkaloids determination. *Journal of Applied Sciences Research*, 4(8): 1017-1022.

29. Tallevi, S.G. and F.D. Cosma, 1988. Stimulation of indole alkaloid content in Vanadium – treated *Catharanthus roseus* suspension cultures. *Planta Medica*, 2: 149-152.

30. Tikhomiroff, C. and M. Jolicoeur, 2002. Screening of *Catharanthus roseus* Secondary metabolites by highperformance liquid chromatography. *Journal of Chromatography*, A, 955: 87-93.

31. Toivonen, L., J. Balsevich and W.G.W. Kurz, 1989. Indole alkaloid production by hairy root cultures of *Catharanthus roseus*. *Plant Cell Tissue and Organ Culture*. 18(1): 79-93. (Hort., Abst., 1990, *60*:9218.).

32. Uniyal, G.C., S. Bala, A.K. Mathur and R.N. Kulkarni, 2001. Symmetry C18 column: a better choice for the analysis of indole alkaloids of *Catharanthus roseus*. *Phytochemical Analysis*, 12: 206-210.

33. Valenzuela, O.A.; R.M. Galaz-Avalos; Y. Minero-Garcia and V.M.Loyola-Vargas, 1998. Effect of differentiation on the regulation of indole alkaloids production in Catharanthus *roseus* hairy roots. *Plant Cell Reports*.18 (1-2): 99-104.

34. Verpoorte, R, R. Van der Heijden and P.R.H. Moreno, 1997. Biosynthesis of terpenoid indole alkaloids in *Catharanthus roseus* cells. In Cordell G.A. (Edited). The alkaloids chemistry and pharmacology (Academic Press, San Diego, London Boston New York Chapter 3) 49: 221–299.

35. Yahia, A.; C. Kevers; T.Gaspar; J.C. Che-Nieux; M. Rideau and J. Creche, 1998. Cytokinins and ethylene stimulate indole alkaloid accumulation in cell suspension cultures of *Catharanthus roseus* by two distinct mechanisms. *Plant Science*. 133: 9–15.

36. Zhao, J., W. Zhu, H. Hu, J. Zhao, W. Zhu and Q. Hu, 2000. Enhanced ajmalicine production in *C. roseus* cell cultures by combined elicitor treatment: from shake-flask to 20–l airlift bioreactor. *Biotechnology Letters,* 22(6): 509-514.

37. Zhao, J; W. Zhu; Q. Hu and X. He, 2001. Enhanced indole alkaloid production in suspension compact callus clusters of *Catharanthus roseus*: Impacts of plant growth regulators and sucrose. *Biotechnology Letters,* 33: 33-41.

Chapter 22

Herbal Medicine in Egypt between Ancient and Modern Civilization

Nora ElSayed Ameen Megahed

Head,
Dietary Supplement Registration Department,
Central Administration of Pharmaceutical Affairs (CAPA),
Ministry of Health, Cairo, Egypt
E-mail: norasayed79@hotmail.com

ABSTRACT

The Ancient Egyptians were quite advanced in their diagnoses and treatments of various illnesses. Their advancements in ancient medical techniques were quite extraordinary, considering the lack of "modern" facilities, sterilization, sanitation, and researching capabilities. The remedies used by Ancient Egyptian physicians came mostly from nature, more especially medicinal herbs. We can surely say that Ancient Egyptians put the bases for natural healings. Most of the complementary medicine modalities originated from ancient Egyptians; one of these modalities being herbal medicine. The plant medicines mentioned in the Ebers Papyrus for instance include opium, cannabis, myrrh, frankincense, fennel, cassia, senna, thyme, henna, juniper, aloe, linseed and castor oil - though some of the translations are less than certain. Nowadays Herbs in Egypt are regulated as dietary supplements, taken orally to fortify food in order to improve bodily functions and support human health. They are not considered alternative to medicine or food and are not used alone for the treatment or diagnosis or prevention of diseases. Medicinal herbs are divided into three categories in accordance to specifications of the World Health Organization 2006 as follows: Medicinal herbs with established monographs, Traditional or folk herbs, and New Medicinal herbs.

Keywords: *Ancient Egyptian, Medicinal herbs, Remedies, Herbal medicine, Ebers Papyrus, Dietary supplements.*

1. INTRODUCTION

If you had to be ill in ancient times, the best place to do so would probably have been Egypt. The Ancient Egyptians were quite advanced in their diagnoses and treatments of various illnesses.Their advancements in ancient medical techniques were quite extraordinary, considering the lack of "modern" facilities, sterilization, sanitation, and researching capabilities.

Along with their strong faith in their gods, the Ancient Egyptians used their knowledge of the human anatomy and the natural world around them to treat a number of ailments and disorders effectively. Their knowledge and research is impressive still today, and their work paved the way for the study of modern medicine. The remedies used by Ancient Egyptian physicians came mostly from nature, especially medicinal herbs.

We can surely say that Ancient Egyptians laid the groundwork for natural healing. Still, there are lots of secrets about the life of Pharaohs and their respective civilizations that require more research in trying to solve these puzzles and secrets that will eventually help in encouraging the return to nature and natural healing.

Most of the complementary medicine modalities originated from ancient Egyptians; one of which is herbal medicine.

According to Herodotus there was a high degree of specialization among physicians. The Egyptians were advanced medical practitioners for their time. They were masters of human anatomy and healing mostly due to the extensive mummification ceremonies.

Ancient Egyptians were as equally familiar with pharmacy as they were with medicine. They were also familiar with drug preparation from plants and herbs such as cumin, fennel, coriander, caraway, aloe, basil, safflower, glue, pomegranates, castor and linseed oil. Other drugs were made of mineral substances such as copper salts, plain salt and lead. Eggs, liver, hairs, milk, animal horns and fat, honey and wax were also used in drug preparation.

Ebers Papyrus (approximately 1500 BCE), one of the two oldest maintained medical documents, contains a most extensive record of Egyptian medical history. The papyrus also has many prescriptions showing the treatment of various disorders using animal, plant and mineral toxins still being used today.

Nowadays Herbs in Egypt are regulated as dietary supplements(according to the definition of the FDA(taken orally to fortify food in order to improve body functions and support human health. They are not considered to be alternative to medicine or food and are not used solely for the treatment or diagnosis or prevention of diseases.

Medicinal herbs are divided into three categories in accordance to specifications of the World Health Organization 2006 as follows:

1. Medicinal herbs with established monographs.
2. Traditional or folk herbs.
3. New Medicinal herbs.

2. MATERIALS AND METHODS

Ancient Egyptians used simple recipes for medications, which were commonly added to common food in the form of spices and drinks. Egyptian foods and herbs spread to the rest of the world since antiquity, and became part of universally known culinary ingredients. Egyptian herbs and medications can be considered as "healthy foods", though they are not effective medicines for acute diseases, and they are safe to take and eat.

While many ailments would have been difficult or impossible to treat, the Egyptians were able to treat many less serious conditions through the use of natural remedies (N. H. Aboelsoud, 2010).

If you had to be ill in ancient times, the best place to do so would probably have been Egypt. The Ancient Egyptians were quite advanced in their diagnoses and treatments of various illnesses.Their advancements in ancient medical techniques were quite extraordinary, considering the lack of "modern" facilities, sterilization, sanitation, and researching capabilities.

Along with their strong faith in their gods, the Ancient Egyptians used their knowledge of the human anatomy and the natural world around them to treat a number of ailments and disorders effectively. Their knowledge and research is impressive still today, and their work paved the way for the study of modern medicine. The remedies used by Ancient Egyptian physicians came mostly from nature, especially medicinal herbs.

We can surely say that Ancient Egyptians laid the groundwork for natural healing. Still, there are lots of secrets about the life of Pharoahs and their respective civilizations that require more research in trying to solve these puzzles and secrets that will eventually help in encouraging the return to nature and natural healing.

Most of the complementary medicine modalities originated from ancient Egyptians; one of which is herbal medicine.

According to Herodotus there was a high degree of specialization among physicians. The Egyptians were advanced medical practitioners for their time. They were masters of human anatomy and healing mostly due to the extensive mummification ceremonies.

Ancient Egyptians were as equally familiar with pharmacy as they were with medicine. They were also familiar with drug preparation from plants and herbs such as oils from castor (Figure 22.1) and linseed (Figure 22.2) and cumin (Figure 22.3), fennel, coriander (Figures 22.4 and 22.5), caraway, aloe, basil (Figure 22.6), safflower, glue, pomegranate. Other drugs were made of mineral substances such as copper salts, plain salt and lead. Eggs, liver, hairs, milk, animal horns and fat, honey and wax were also used in drug preparation.

Herbs played a major part in Egyptian medicine. The plant medicines mentioned in the Ebers Papyrus for instance include opium, cannabis, myrrh, frankincense, fennel, cassia, senna, thyme, henna, juniper, aloe, linseed and castor oil - though some of the translations are less than certain. Cloves of garlic have been found in Egyptian burial sites, including the tomb of Tutankhamen and in

Figure 22.1: Castor

Figure 22.2: Linseed

Figure 22.3: Cumin

Figure 22.4: Dry Coriander Fruits

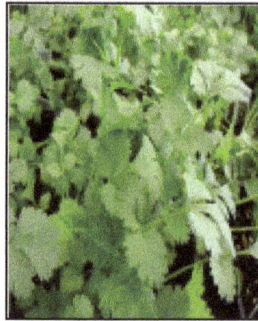

Figure 22.5: Green Coriander Leaves

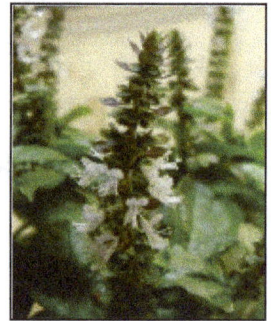

Figure 22.6: Basil

the sacred underground temple of the bulls at Saqqara. Many herbs were steeped in wine, which was then drunk as an oral medicine.

Egyptians thought garlic (Figure 22.7) and onions (Figure 22.8) aided endurance, and consumed large quantities of them. Raw garlic was routinely given to asthmatics and to those suffering with broncho-pulmonary complaints. Onions helped against problems of the digestive system.

Medical prescriptions were written with high skill. A prescription usually began with a description of the medicine, *e.g.*, "Medicine to discharge blood out of wounds", followed by ingredients and measures used in addition to method of preparation and usage whether in tablet form, ointment or by inhaling.

The Ebers Papyrus (approximately 1500 BCE), one of the two oldest maintained medical documents, preserved for us a most extensive record of Egyptian medical

Figure 22.7: Garlic

Figure 22.8: Onions

history. In it, the Egyptians show a degree of knowledge of the workings of the human body, its structure, the work of the heart and blood vessels. This oldest and well preserved medical document from ancient Egyptian records dated from approximately 1500 BC contains 110 pages on anatomy and physiology, toxicology, spells, and treatment written on papyrus. This papyrus also has many prescriptions showing the treatment of various disorders by animal, plant and mineral toxins still being used today.

Figure 22.9: Ebers Papyrus.

Nowadays herbs in Egypt are regulated as dietary supplements (according to the definition of the FDA).

A dietary supplement product is in a pharmaceutical dosage form taken orally to fortify the human food in order to improve body functions and support human health and it is not an alternative to medicine or food and is not used alone for the treatment or diagnosis or prevention of diseases.

Dietary supplements may include one or any combination of the following:

☆ Vitamins [should comply with Recommended Daily Allowance (RDA)]

☆ Minerals [should comply with Recommended Daily Allowance (RDA)]

☆ Amino acids

☆ Fatty acids

☆ Medicinal herbs, their extracts and oils

☆ Probiotics for GIT disorders

Medicinal herbs are divided into three categories in accordance with specifications of the World Health Organization 2006 as follows:

1. Medicinal herbs with established monographs

2. Traditional or folk herbs

3. New Medicinal herbs

Requirements for Registration of Medicinal Herbs with Established Monographs:

1. Identification of the used raw herb or herbal extract by stating the Latin name and the part of the herb, supporting the submitted data by documented studies of safety and efficacy (scientific reference or monograph).

2. Specifications of the used raw herb or herbal extract, including a chromatographic fingerprint, the method of analysis and the physical and chemical specifications (standardization).

Requirements for Registration of Traditional or Folk Herbs

These are medicinal herbs (individually or in combination) used inside the Arab Republic of Egypt for a period of not less than 15 years supported by documented studies of safety and effectiveness.

Their registration requires the following:

1. Identification of the used raw herb or herbal extract by stating the Latin name and the part of the herb, supporting the submitted data by documented studies of safety and efficacy.

2. Specifications of the used raw herb or herbal extract, including a chromatographic fingerprint, the method of analysis and the physical and chemical specifications (standardization).

Requirements for Registration of a New Medicinal Herb

These are medicinal herbs (individually or in combination) as follows:

☆ A herb not previously used inside the Arab Republic of Egypt, or, alternatively, previously used for a short period of time (less than 15 years), by a limited number of individuals.

☆ A herb that is added to a mixture of herbs without previous usage documented in that well-documented existing herbal mixture, (that is, making a new formulation by adding a new herb to an already existing herbal mixture).

This category of herbs requires the following:

A reference or monograph to ensure safety and efficacy or

1. All toxicity tests to ensure safety.
2. Pre-clinical and clinical studies to ensure efficacy and safety.

3. RESULTS AND DISCUSSION:

Herbal medicines have been used since ancient times as medicines for the treatment of a wide range of diseases. Herbs have been used in a wide variety of ways. These include the fresh or dried herbal plant parts including the leaves, stems, roots, flowers, seeds, or fruits. A few papyri have survived, from which we can learn about Egyptian medicine such as The Ebers Papyrus on ophthalmology, diseases of the digestive system, the head, the skin and specific maladies. This is a compilation of earlier works as it contains a large number of prescriptions and recipes.

Medical prescriptions were written by Ancient Egyptian physicians with a high level of skill. A prescription usually began with a description of the medicine, *e.g.*, "Medicine to discharge blood out of wounds", followed by ingredients and measures used in addition to method of preparation and usage whether in tablet form, ointment or by inhaling.

Most of the complementary medicine modalities originated from ancient Egyptians; one of these modalities being herbal medicine.

Nowadays herbs in Egypt are regulated as dietary supplements taken orally to fortify food in order to improve bodily functions and support human health. They are not considered alternative to medicine or food and are not used solely for the treatment or diagnosis or prevention of diseases. Medicinal herbs are divided into three categories in accordance to specifications of the World Health Organization 2006 as follows: Medicinal herbs with established monographs, Traditional or folk herbs, and New Medicinal herbs.

4. CONCLUSIONS

Civilization in Ancient Egypt was not only about the pyramids and tombs, but it also involved all aspects of human life. Health and wellbeing was one of the most cared for arts by the Pharaohs. Both physicians and magicians participated in the field of medical care. Most of the complementary medicine modalities originated from ancient Egyptians. One of these modalities being herbal medicine, and this was recorded in their medical papyri.

The Ancient Egyptians were quite advanced in their diagnoses and treatments of various illnesses. Ancient Egyptians were as equally familiar with pharmacy as they were with medicine. They were also familiar with drug preparation from plants and herbs.

The majority of present day Egyptians has great confidence in the role of herbs in treatment and alleviation of diseases purchasing them from attar shops that sell all types of plants either non-medicinal or medicinal; or they may get pharmaceutical products which contain herbal preparations as they are licensed by The Ministry of Health as dietary supplements.

5. REFRENECES

1. Comparison between Egyptian and Chinese Herbal Medicines. http://www.aldokkan.com/science/herbal_remedies.htm

2. Ebers papyrus http://www.toxipedia.org.

3. Guidelines on Registration of Dietary Supplements in Egyptian Ministry of Health http://www.eda.mohp.gov.eg

4. N. H. Aboelsoud - Herbal medicine in ancient Egypt. *Journal of Medicinal Plants Research* Vol. 4(2), pp. 082-086, 18 January, 2010.

Chapter 23

Identification of Herbal Drugs

Kreft Samo

University of Ljubljana,
Faculty of Pharmacy, Askerceva 7,
1000 Ljubljana, Slovenia
E-mail: samo.kreft@ffa.uni-lj.si

ABSTRACT

The correct identification of the plant species present in a herbal drug is of great importance both in research and manufacturing. Misidentified plants can cause the lack of therapeutic activity or can be even toxic. The methods for identification of herbal material can be divided into four groups: 1) Macroscopic and microscopic examination; 2) DNA analysis; 3) Chromatographic phytochemical investigation and 4) Spectroscopic analysis. The advantages and disadvantages of each method will be discussed in this paper and our own experiences will be presented.

Keywords: Identification, Microscopy, DNA, Chromatography, Spectroscopy.

1. INTRODUCTION

Herbal medicinal products have been used for treatment since early human history. In recent years, the public interest in natural products has increased significantly. One ofthe reasons is the belief in the better safety of the herbal preparations compared to synthetic drugs. The quality assurance of these products is thereforevery important. The use of herbal product of poor quality can result in the absence of the therapeutic effect or even worse in poisoning, allergic reactions and even death (Butt *et al.*, 1996; Ernst, 1998, Sperl *et al.*, 1995). The main quality concerns are contaminations (with microorganisms, heavy metals, pesticides, etc.), or misidentification of the plant species (Elvin-Levis, 2001; Calixto, 2000).

The identification of the herbal drug is an extremely important control step for safety and efficacy reasons. Herbal drugs can be adulterated with inert plant material, which lowers the efficacy of such preparations. Herbal drugs can also be mistaken for poisonous herbal drugs, whichis even more dangerous. Proper identification of herbal samples is therefore of crucial importance in every step of production of herbal medicinal product and also in research activities.

In recent years our laboratory has tested several approaches for identification of plants. Here we review our experiences and the experiences of other research groups.

The identification (and confirmation of identity) can in principle be achieved by several methods, which can be divided into four groups:

1. *Macroscopic and microscopic examination (EDQM, 2014)*: This is a time consuming procedure and requires specific knowledge on the botanical anatomy, histology and cell biology. It cannot be automated and in some cases, when dealing with a heterogeneous crushed sample, it isnot successful. This method can only be used when the botanical structures are preserved, and it cannot be used with herbal extracts of essential oils. The advantage of this method is that it does not require specialised equipment.

2. *DNA analysis*: This is a time consuming method, whichrequires specialised laboratory equipment and it is also relatively expensive(Slanc *et al.*, 2006). It can only be used for samples which contain DNA, therefore in most cases it cannot be used for extracts.

3. *Chromatographic phytochemical investigation*: By identifying and quantifying some of the main constituents (markers). This procedure is also time-consuming, requiring specialized equipment and the use of organic solvents. A database of typical markers and their concentrations must be made for each plant species and in some cases even for each source of the plant.

4. *Spectroscopic analysis*:   Infrared(IR) spectroscopy has proven to be effective in the classification of plant material and in the detection of adulteration (Rohman *et al.*, 2010; Quiñones-Islas *et al.*, 2013; Cozzolino, 2013; Downey, 1998). Infrared spectroscopy is fast, easy to implement, requires a very low quantity of the sample, does not use organic solvents and can be implemented automatically. Again a database of typical IR spectra must be made for each plant species and possibly for each source of the plant.

In this article we will present the advantages and disadvantages of different methods in more detail.

2. RESULTS AND DISCUSSION

Infrared Spectroscopy

In our research(Kokalj *et al.*, 2011a; Kokalj *et al.*, 2011b; Kokalj *et al.*, 2010, Kokalj *et al.*, 2014, Strgulc Krajsek *et al.*, 2008)it has been shown that in some samples, MID infrared spectroscopy is more appropriate when compared to NIR. MID ATR and MID KBr methods showed to be the most suitable [12].MID and NIR infrared spectroscopy have successfully been used for qualitative and quantitative analysis of tea (Camelia sinensis), however studies on herbal teas are lacking in the field of infrared spectroscopy (Budinova *et al.*, 1998; Zhang *et al.*, 2004; Chen *et al.*, 2006).

In order to apply IR to plant samples it is necessary to combine it with chemometrical methods. We found that different statistical approaches for analysing the IR spectra of leaves can be employed successfully to discriminate amongst six species of genus *Epilobium* and *Hypericum* (Kokalj *et al.*, 2011a; Kokalj *et al.*, 2011b; Kokalj *et al.*, 2010). Infrared spectra are very large and complex sets of data; therefore the mathematical and statistical analysis is complex. However once the optimal model for given data is established, the use is simple and can be automatized (Bunaciu *et al.*, 2011; Gad *et al.*, 2011).

The analysis is composed of three main steps: spectra collection, spectra pre-treatment and chemometrical analysis. Each of these steps needs to be optimized in every specific case since the properties of the samples used and the application needs can be very different. In different modes of collecting the spectra (transmission, reflectance) it is important to take in account from which part of the sample the spectra are collected, the surface or the inside (Kokalj *et al.*, 2011a). With different pre-treatments(smoothing, derivatisation) of the spectra important information can be emphasized or, if an inappropriate technique is used, important information may be lost (Kokalj *et al.*, 2011b). Since the infrared spectra are a complex set of data it is also important to carefully choose the mathematical and statistical methods for their interpretation (Kokalj *et al.*, 2010). There is no universal rule;therefore optimal conditions have to be found for each specific case.

DNA Analysis

The research performed in our laboratory (Slanc *et al.*, 2006) has established two methods to identify the constituent drugs in a sedative tea made of Valeriana officinalis radix, Humulus lupulus strobuli, Melissa officinalis folium and Mentha piperita folium. Thefirst method involved amplification of the internal transcribed spacers (ITS) region of nuclear ribosomal DNA and restriction analysis with selected restriction endonucleases Nae I, PshA I and Xcm I. The second method involved real-time PCR, using primers specifically designed for each individual herbal drug. Real-time PCR proved to be a method for identifying individual herbal drugs in a tea mixture with a single DNA extraction in a single PCR run, since its limit of detection is lower than that for restriction analysis.

In our recent research (Ravnikar *et al.*, 2015) we screened the extracts of fungal endophytes for antimicrobial activity and taxonomically identified the active strains. 73 strains of endophytic fungi were isolated from plant samples, mainly from needles of conifers. Extracts of cultured fungi were tested for antimicrobial activity using a microdilution assay. Genomic DNA from the five most active fungal strains was isolated and species-specific DNA regions (ITS regions) were amplified, sequenced and compared to the GenBank database. Active endophytic strains were identified as two strains of *Lophodermium pinastri*, two strains of *Lophodermium seditiosum* and one of *Phoma herbarum*.

Chromatographic Phytochemical Investigation

We performed (Obradovic *et al.*, 2007) a HPLC and capillary electrophoretic analysis of pulverized plant samples, for the authentication of medicinal plant species. In the conventional procedure, individual compounds were identified on the chromatograms and their contents were used for the authentication ofplants. We proposed a new approach in analysing chromatograms, in which chromatograms were split into sections, as described by four variables (number of peaks in the section, average retention time of peaks in the section, total area of peaks in the section and average area of peaks in the section); and these variables were then used in the statistical analysis. The method was especially useful when the peaks on the chromatogram were not well separated and it was not easy to link individual peaks on one chromatogram with corresponding peaks on other chromatograms. In comparison with the standard procedure, our approach in analysing chromatographic data of willow-herb (*Epilobium* and *Chamaenerion* spp.) extracts was more objective, gave better results and easier to perform.

In other research (Janeš *et al.*, 2012) we used the classical approach and identified phytochemical markers that discriminate Tartary buckwheat (*Fagopyrum tataricum*) and common buckwheat (Fagopyrum esculentum). The second aim of this study was to identify and quantify individual compounds responsible for the tartary buckwheat aroma. Volatiles were extracted with simultaneous extraction and distillation methods using the Likens-Nickerson apparatus and analysed by GC-MS. A total of 48 compounds were quantified and their odour activity values (OAV) were calculated. OAV of 26 compounds were higher than 10; therefore, they significantly contribute to the overall tartary buckwheat aroma. The most important difference of tartary buckwheat from common buckwheat is the absence of salicylaldehyde and presence of naphthalene. Using these markers, the products from different species of buckwheat can be distinguished by analysis of aroma compounds.

3. CONCLUSIONS

Each of the described methods for identification of plant species in herbal drug samples has some advantages and disadvantages. The corresponding method has to be chosen on a case by case basis.

4. REFERENCES

1. Budínová G, Vláèil D, Mestek O, Volka K. Application of infrared spectroscopy to the assessment of authenticity of tea. *Talanta* 1998; 47: 255-260.

2. Bunaciu A: A., Aboul-Enenin H. Y. and Fleschin S. (2011). *Applied Spectroscopy Reviews* 46, 251-260.

3. Butt PPH, Tomlinson B, Cheung KO, Yong SP, Szeto ML, Lee CK. Adulterants of herbal products can cause poisoning. *BMJ* 1996; 313: 117.2.

4. Calixto JB. Efficacy, safety, quality control, marketing and regulatory guidelines for herbal medicines (phytotherapeutic agents). *Brazilian Journal of Medical and Biological Research* 2000; 33: 179-189.

5. Chen Q, Zhao J, Zhang H, Wang X. Feasibility study on qualitative and quantitative analysis in tea by near infrared spectroscopy with multivariate calibration. *Analytica Chimica Acta* 2006; 572: 77-84.

6. Cozzolino A. An overview of the use of infrared spectroscopy and chemometrics in authenticity and traceability of cereals. *Food Research International* 2013, *article in press*

7. Downey G. Food and ingredient authentication by mid infrared spectroscopy and chemometrics. *Trends in Analytical Chemistry* 1998; 17(7): 418-424.

8. EDQM, European Directorate for the Quality of Medicines and Health Care. European pharmacopoeia. (8th ed). Strasbourg: Council of Europe; 2014.

9. Elvin-Levis M. Should we be concerned about herbal remedies. *Journal of Ethnopharmacology* 2001; 75: 141-164.

10. Ernst E. Harmless Herbs? A Review of the Recent Literature. *The American Journal of Medicine* 1998; 104: 170-178.

11. Gad H. A., El-Ahmady S. H., Abou-Shoer M. and Al-Azizi M. M. (2011). *Phytochemical Analysis* 24, 1-24.

12. Janeš D, Prosen H, Kreft S. Identification and quantification of aroma compounds of tartary buckwheat (*Fagopyrum tataricum* Gaertn.) and some of its milling fractions. *J Food Sci.* 2012; 77(7): C746-51.

13. Kokalj M, Kolar J, Trafela T, Kreft S. Differences Among Epilobium and Hypericum Species Revealed by four IR Spectroscopy modes: Transmission, KBr Tablet, Diffuse Reflectance and ATR. *Phytochemical Analysis* 2011a; 22: 541-546.

14. Kokalj M, Rihtariè M, Kreft S. Commonly applied smoothing of IR spectra showed inappropriate for the identification of plant leaf samples. *Chemometrics and Intelligent Laboratory Systems* 2011b; 108: 154-161.

15. Kokalj M, Štih K, Kreft S. Herbal tea identification using mid-infrared spectroscopy. *Planta Med.* 2014; 80(12): 1023-1028.

16. Kokalj M, Strgulc Krajšek S, Omahen Bratuša J, Kreft S. Comparison and improvement of commonly applied statistical approaches for identification of plant species from IR spectra. *Journal of Chemometrics* 2010; 23: 611-616.

17. Obradovic M, Krajsek SS, Dermastia M, Kreft S. A new method for the authentication of plant samples by analyzing fingerprint chromatograms. *Phytochem Anal.* 2007 Mar-Apr; 18(2):123-32.

18. Quiñones-Islas N, Meza-Márquez OG, Osorio-Revilla G, Gallardo-Velazquez T. Detection of adulterants in avocado oil by Mid-FTIR spectroscopy and Multivariate analysis. *Food Research International* 2013; 51: 148-154.

19. Ravnikar, M., Terèelj, M., Janeš, D., Štrukelj, B., and Kreft, S. (2015). Antibacterial activity of endophytic fungi isolated from conifer needles. *African Journal of Biotechnology*, 14(10), 867-871.

20. Rohman A, Che Man YB. Fourier transform infrared (FTIR) spectroscopy for analysis of extra virgin olive oil adulterated with palm oil. *Food Research International* 2010; 43: 886-892.

21. Slanc P, Ravnikar M, Strukelj B. Identification of individual herbal drugs in tea mixtures using restriction analysis of ITS DNA and real-time PCR. *Pharmazie* 2006; 61(11): 912-5.

22. Sperl W, Stuppner H, Gassner I, Judmaier W, Dietze O, Vogel W. Reversible hepatic veno-oclusive disease in an infant after consumption of pyrrolizidine-containing herbal tea. *European Journal of Pediatrics* 1995; 154: 112-116.

23. Strgulc Krajsek S, Buh P, Zega A, Kreft S. Identification of herbarium whole-leaf samples of Epilobium species by ATR-IR spectroscopy. *Chem Biodivers.* 2008; 5(2):310-317.

24. Zhang MH, Luypaert J, Fernández Pierna JA, Xu QS, Massart DL. Determination of total antioxidant capacity in green tea by near-infrared spectroscopy and multivariate calibration. *Talanta* 2004; 62: 25-35.

Chapter 24

Herbals from the High Mountains in the East Mediterranean

Münir Ozturk[1], Volkan Altay[2] and Tuba Mert Gönenç[3]

[1]*Botany Department, Science Faculty,*
Ege University, Bornova, Izmir, Turkey
E-mail: munirozturk@gmail.com
[2]*Biology Department, Faculty of Science and Arts,*
Mustafa Kemal University, Antakya, Turkey
E-mail: volkanaltay34@gmail.com
[3]*Department of Pharmacognosy, Faculty of Pharmacy,*
Ege University, Bornova, Izmir Turkey
E-mail: tuba.mert@ege.edu.tr

ABSTRACT

In this presentation attempt has been made to present an overview of the MAPS from the East Mediterranean (Turkey, Cyprus, Syria, Lebanon, Palaestine and Jordan) covering altitudes 1000 m above sea level. Folk medicine is still commonly practiced in the region, and preservation of ethnopharmacological practices is high. In all 183 wild plant taxa from high altitudes belonging to 59 families and 123 genera are used in traditional medicine. The richest families are; Lamiaceae (37), Asteraceae (19) and Rosaceae (18). The richest genera are *Sideritis* (7 taxa), *Origanum* and *Teucrium* (5 taxa each), *Juniperus, Quercus, Hypericum, Mentha,* and *Salvia* (4 taxa each). Local people generally use herbals as leaves (21.85 per cent), flowers (13.31 per cent), fruits (9.54 per cent), shoots (7.53 per cent) and roots (7.28 per cent). These are used in the form of infusion (32.89 per cent), decoction (25.43 per cent), boiled (11.40 per cent) and poultice (6.14 per cent), and are generally used for the treatment of urinary system (18.49 per cent), stomach (10.79 per cent), respiratory (8.32 per cent) and skin disorders (7.40 per cent).

Keywords: East Mediterranean, High altitudes, Herbals, Ecology.

1. INTRODUCTION

The mountain ecosystems are complex systems influenced by varying environmental factors; in particular topographical conditions like elevation, directional extensions of the mountains, ruggedness, slope, temperatures, precipitation, plant distributions and soil conditions play a significant role in these systems, leading to the formation of differing ecological habitats (Atalay, 2004, 2008). There is also formation of microclimatic areas, therefore plant distribution at different altitudes shows variations (Atalay, 2006; Atalay *et al.*, 2014; Ozturk *et al.*, 2015). The elevation factor in mountainous terrain in general is one of the most important ecological factors used in the differentiaiton of the ecological systems, because altitude has a more dominating impact on the climate, plant physiology and soil dynamics than other habitat factors (Tecimen *et al.*, 2012).

These ecosystems; especially alpine and subalpine areas; play important roles with different functions (Ozturk *et al.*, 1991; Atay *et al.*, 2009; Sarý, 2010; Guleryuz *et al.*, 2010). Mountainous areas are isolated environments causing changes in the plant diversity and also an increase in the number of locally distributed plant taxa (Duran, 2013). The deep valleys here are a common site as the natural distribution areas of the endemic and relict taxa. The reason for high endemism ratios in such areas are because the plant taxa can not find the opportunity to spread or there is progressive decrease in the distribution area of the previously widely distributed plants or a withdrawal from the restricted areas in the mountainous regions (Ozturk *et al.*, 1991). In addition, there is no or very limited social and cultural activity in the mountainous areas, which prevents degradation as compared to the lower elevations.

The mountain ecosystems constitute approximately 3 percent of the world terrestrial surface but host approximately 4 percent of the world flora (Heywood, 1995a). Surprisingly a fixed number of plant taxa varying between 200-300 in the local vegetation of mountains (excluding isolated volcanic peaks) in different parts of the world has been reported (Körner, 1995). Numerous studies have been undertaken on the useful plants of such areas. A significant part of the existing flora in the high altitudes of the Himalayas is reported to be used in the traditional folk medicine including the treatment of diseases by Tibetan doctors, who use nearly 61 percent of the alpine plants for this purpose (Salick *et al.*, 2006, 2009).

Many studies have been carried out globally related to the biodiversity, flora, vegetation, ecology and ethnobotany of different mountain ecosystems, mainly alpine and subalpine areas (Theurillat and Guisan, 2001; Kala and Mathur, 2002; Körner and Spehn, 2002; Grabherr *et al.*, 2003; Körner 2003; Kala, 2002, 2005, 2008; Bahn and Körner, 2003; Klanderud and Birks, 2003; Moiseev and Shiyatov, 2003; Ozenda and Borel, 2003; Pickering and Armstrong, 2003; Price and Neville, 2003; Shrestha and Dhillion, 2003; Vare *et al.*, 2003; Anderson *et al.*, 2005; Casazza *et al.*, 2005; Walther *et al.*, 2005; Erschbamer *et al.*, 2006; Pauli *et al.*, 2007; Kadereit *et al.*, 2008; Grabherr, 2009; Arya, 2010; Gairola *et al.*, 2010; Kala and Ratajc, 2012).

Floristic and ecological studies on the plants occurring at high altitudes in the Eastern Mediterranean are limited, notable among these being; Ozturk *et al.* (1990), Gemici *et al.* (1994), Kürschner *et al.* (1998) and Parolly (2015).

The high altitude mountainous habitats in the East Mediterranean are rich in medicinal and aromatic plants (MAPS). These plants produce an array of chemicals that are known as secondary metabolites; many have been utilized by human beings for various purposes, especially for making herbal medicines. In view of the fact that plant biodiversity is disappearing at an alarming rate, there is a new sense of urgency behind the search for plants that cure. Future climate change scenarios pose additional threats for these plants at higher altitudes. The characteristic plant diversity in the mountains strongly depends on elevation.

According to the WHO reports, 80 percent of the world's population relies chiefly on traditional medicine, a major part of which involves the use of plant extracts or their active ingredients. Determination and conservation of the economic potential of these resources is of special importance for sustainable development. In this review an attempt has been made to enlighten medicinal and aromatic plant diversity (MAPS) from elevations of 1000 meters and above in the East Mediterranean, together with their use in traditional folk medicine.

2. MATERIALS AND METHODS

The data on the diversity of medicinal and aromatic plants from the high altitudes was collected from different published sources for Syria, Lebanon, Palaestine and Jordan. The information for Turkey and Cyprus was gathered by using a questionnaire distributed among the local inhabitants of the region in the Turkish Mediterranean, as well as from the published data. The list of plant taxa was organised alphabetically according to the families and genera. The list also shows the distribution of the taxon in the East Mediterranean countries, their altitudes, chorotypes, and life forms in a coded form. The parts used, preparations, and ways for treatment have also been included.

Study Area

The East Mediterranean (EMED) includes Turkey, Cyprus, Syria, Lebanon, Palestine, Jordan and Israel (Figure 24.1). This part lies within the transition zone between the Saharo-Arabian desert biome and temperate climates (Heywood, 1995b, 2003a, b). It is also called the Old World region where olive trees grow, and it abounds in forest, woodlands and scrub vegetation with different ecoregions (Öztürk *et al.*, 2008). EM possesses a rich diversity of plants (Dallman, 1998), and is a meeting area of the Mediterranean, Irano-Turanian, Saharo-Sindian and Sudano-Decanian phytogeographical divisions (Ozturk *et al.*, 1996a, b). Taurus Mountains in the EM are mountainous areas deeply dissected by the rivers with a considerable plant distribution and forests. The altitudes above 2000 m contribute three vertical zonation of vegetation. These mountains begin in the southwest of Anatolia and continue in the E-W direction with summits above 3000 m (Atalay *et al.*, 2014).

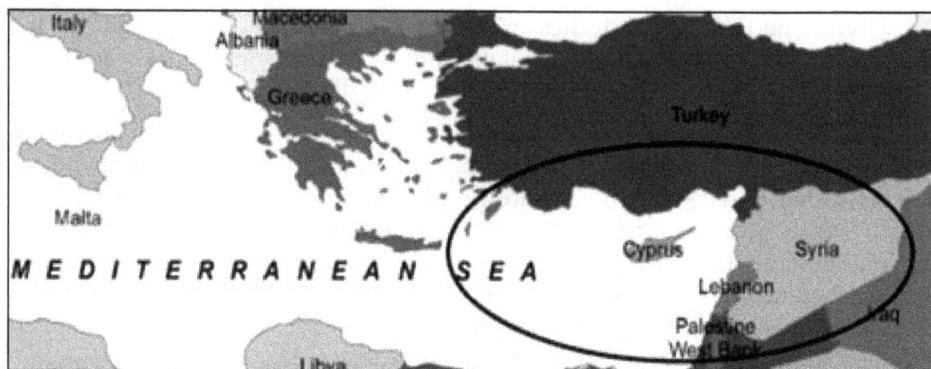

Figure 24.1: Map Showing the East Mediterranean Part of the Mediterranean Basin.

3. RESULTS AND DISCUSSION

Diversity of MAPS

A total of 183 MAPS belonging to 123 genera from 59 families have been identified from this area. The data has been compiled on the distribution area together with the country and altitude, and information on the chorotypes and life forms for each taxon has been included (Table 24.1). The families with the highest number of taxa are Lamiaceae (37), Asteraceae (19) and Rosaceae (18), whereas the genera with the highest number of taxa are *Sideritis* (7), *Origanum, Teucrium* (5 each) and *Juniperus, Quercus, Hypericum, Mentha, Salvia* (4 each). Insofar as the life forms are concerned, hemicryptophytes (31 per cent) and phanerophytes (27 per cent) form the largest groups on the basis of Raunkier's life form spectrum (Figure 24.2). The highest percentage of elements among MAPS are represented by the Mediterranean (31 per cent) followed by Euro-Siberian (20 per cent). Other phytogeographical elements included here are Irano-Turanian (12) and Mediterranean (11 taxa). There are 11 endemic taxa and 3 taxa show a cosmopolitan feature. The phytogeographical identity of 104 taxa is not clear and these are listed as unknown (Figure 24.3). The data compiled on the MAPS reveals that; 177 taxa are distributed between 1000-1500 m, 105 taxa between 1500-2000 m, 58 taxa between 2000-2500 m, 26 taxa between 2500-3000 m and 2 taxa at 3000-3500 m.

A total of 183 plant taxa with medicinal and aromatic values are listed in alphabetical order with their botanical name, part used, ailment treated and preparations used (Table 24.2). A majority of these plants are used for the treatment of urinary system disorders (18.49 per cent), followed by stomach (10.79 per cent) and respiratory disorders (8.32 per cent). Most widely used parts of the plants are leaves (21.85 per cent), followed by flowers (13.31 per cent) and fruits (9.54 per cent) (Figure 24.4). The common preparation of the traditional folk medicine is infusion (32.89 per cent), followed by decoction (25.43 per cent), boiled (11.40 per cent) and poultice (6.14 per cent). Other types of preparations and the percentage values are given in Figure 24.5.

Table 24.1: MAPS from the High Mountains in the EMED and their Ecological Features

Sl.No.	Family/Taxa	Country	Chorotypes	Life Forms	Altitude (m)
	AMARANTHACEAE				
1.	*Amaranthus viridis*	C, P, T	IN	T	1300
	AMARYLLIDACEAE				
2.	*Galanthus elwesii*	T	EM	C	1000-1300
	ANACARDIACEAE				
3.	*Pistacia terebinthus* ssp. *palaestina*	C, L, J, P, S, T	EM	P	1100-1500
4.	*Rhus coriaria*	L, J, P, S, T	IN	P	1100-1900
	APIACEAE				
5.	*Ferula elaeochytris*	L, S, T	EM	H	1000-1900
	ARALIACEAE				
6.	*Hedera helix*	C, L, J, P, S, T	IN	P	1000-1500
	ASPIDIACEAE				
7.	*Dryopteris filix-mas*	C, P, S, T	IN	C	1000-1100
	ASPLENIACEAE				
8.	*Ceterach officinarum*	L, P, S, T	IN	C	1000-1300
	ASTERACEAE				
9.	*Achillea nobilis* ssp. *densissima*	T	EM, EN	C	1300-1600
10.	*Achillea wilhelmsii*	S, T	IR	H	1000-2200
11.	*Anthemis pseudocotula*	C, L, P, S, T	IN	T	1000-1500
12.	*Anthemis tinctoria* var. *tinctoria*	L, S, T	IN	H	1000-1830
13.	*Artemisia absinthium*	T	IN	CH	1000-2600
14.	*Bellis perennis*	C, L, P, S, T	ES	H	1000-2000
15.	*Carduus nutans* ssp. *nutans*	T	IN	H	1000-2130
16.	*Cichorium intybus*	C, L, J, P, S, T	IN	H	1000-3050
17.	*Cirsium arvense*	T	IN	H	1000-2000
18.	*Conyza bonariensis*	C, T	IN	T	1000-1300
19.	*Gundelia tournefortii* var. *tournefortii*	C, L, S, T	IN	H	1000-2300
20.	*Helichrysum plicatum* ssp. *plicatum*	L, S, T	IN	H	1400-2850
21.	*Inula heterolepis*	L, T	EM	C	1000-1500
22.	*Inula viscosa*	L, J, P, S, T	M	CH	1000-1200
23.	*Scorzonera phaeopappa*	L, P, T	IR	H	1000-2050
24.	*Sonchus oleraceus*	C, T	IN	T	1000-1300
25.	*Tragopogon dubius*	T	IN	H	1000-2300
26.	*Tussilago farfara*	L, S, T	ES	C	1000-2400
27.	*Xanthium strumarium* ssp. *cavanillesii*	C, S, T	IN	T	1000-1300

Contd...

Table 24.1–*Contd...*

Sl.No.	Family/Taxa	Country	Chorotypes	Life Forms	Altitude (m)
	BERBERIDACEAE				
28.	*Berberis crataegina*	T	IN	P	1000-2700
29.	*Leontice leontopetalum* ssp. *leontopetalum*	C, L, J, P, S, T	IN	C	1000
	BORAGINACEAE				
30.	*Alkanna orientalis* var. *orientalis*	C, L, P, S, T	IR	H	1000-2000
31.	*Alkanna tinctoria* ssp. *anatolica*	T	EM	H	1000-1400
32.	*Echium italicum*	T	IN	H	1100-1950
33.	*Heliotropium hirsutissimum*	C, P, S, T	EM	T	1000-2200
	BRASSICACEAE				
34.	*Cardaria draba* ssp. *draba*	C, T	IN	H	1000-2700
	BUXUCEAE				
35.	*Buxus sempervirens*	T	ES	P	1080-2000
	CAPRIFOLIACEAE				
36.	*Sambucus ebulus*	C, L, S, T	ES	H	1000-2000
37.	*Sambucus nigra*	C, L, P, S, T	ES	P	1000-1700
	CARYOPHYLLACEAE				
38.	*Silene vulgaris* var. *vulgaris*	C, J, T	IN	H	1000-3000
	CHENOPODIACEAE				
39.	*Chenopodium album* ssp. *album* var. *album*	C, P, T	IN	T	1000-2000
40.	*Chenopodium botrys*	C, L, T	IN	T	1000-2500
41.	*Chenopodium foliosum*	L, T	IN	T	1200-2800
	CISTACEAE				
42.	*Cistus salviifolius*	C, L, J, P, S, T	IN	P	1000-1400
	CONVOLVULACEAE				
43.	*Convolvulus lineatus*	L, P, S, T	IN	CH	1000-2135
	CORNACEAE				
44.	*Cornus mas*	T	IN	P	1070-1500
	CRASSULACEAE				
45.	*Sedum album*	L, T	IN	CH	1000-2700
	CUPRESSACEAE				
46.	*Juniperus drupacea*	L, S, T	IN	P	1000-1500
47.	*Juniperus excelsa*	L, P, S, T	IN	P	1000-2300
48.	*Juniperus foetidissima*	S, T	IN	P	1000-1900
49.	*Juniperus oxycedrus* ssp. *oxycedrus*	L, J, P, S, T	IN	P	1000-2100
	DIOSCOREACEAE				
50.	*Tamus communis* ssp. *cretica*	C, L, J, P, S, T	IN	C	1265

Contd...

Table 24.1–*Contd...*

Sl.No.	Family/Taxa	Country	Chorotypes	Life Forms	Altitude (m)
	EQUISETACEAE				
51.	*Equisetum telmateia*	T	IN	C	1000-1200
	ERICACEAE				
52.	*Erica manipuliflora*	C, L, S, T	IN	P	1000-1500
	EUPHORBIACEAE				
53.	*Euphorbia macroclada*	P, S, T	IR	H	1200-1860
54.	*Euphorbia peplus* var. *peplus*	C, L, P, S, T	IN	T	1300
	FABACEAE				
55.	*Astragalus hamosus*	C, L, P, S, T	IN	T	1000-1400
56.	*Astragalus lydius*	T	EN, IR	CH	1000-1400
57.	*Lotus corniculatus* var. *tenuifolius*	C, L, P, T	IN	H	1000-2750
58.	*Trifolium repens* var. *repens*	C, L, P, T	IN	H	1000-2700
	FAGACEAE				
59.	*Quercus coccifera*	C, T	M	P	1000-1400
60.	*Quercus infectoria* ssp. *boissieri*	L, P, S, T	IN	P	1100-1400
61.	*Quercus ithaburensis* ssp. *macrolepis*	T	EM	P	1000-1400
62.	*Quercus robur* ssp. *robur*	T	ES	P	1160
	GERANIACEAE				
63.	*Erodium cicutarium* ssp. *cicutarium*	C, L, P, S, T	IN	T	1000-1700
64.	*Pelargonium endlicherianum*	S, T	IN	C	1300-1400
	HYPERICACEAE				
65.	*Hypericum confertum* ssp. *confertum*	T	IN	CH	1700
66.	*Hypericum perforatum*	C, L, S, T	IN	H	1000-2500
67.	*Hypericum scabrum*	L, P, S, T	IR	H	1400-2850
68.	*Hypericum tetrapterum*	S, T	IN	H	1100-1200
	JUGLANDACEAE				
69.	*Juglans regia*	L, J, P, S, T	IN	P	1000-1650
	JUNCACEAE				
70.	*Juncus articulatus*	C, T	ES	C	1100-2750
71.	*Juncus inflexus*	T	IN	C	1300-2000
	LAMIACEAE				
72.	*Ajuga chamaepitys* ssp. *chia* var. *chia*	P, T	IN	T	1000-2200
73.	*Ballota nigra* ssp. *uncinata*	S, T	M	H	1000-1300
74.	*Cyclotrichium origanifolium*	L, T	EM	CH	1100-2700
75.	*Lamium album*	T	ES	H	1100-1200
76.	*Melissa officinalis* ssp. *officinalis*	C, L, P, S, T	IN	H	1250

Contd...

Table 24.1–*Contd...*

Sl.No.	Family/Taxa	Country	Chorotypes	Life Forms	Altitude (m)
77.	*Mentha aquatica*	C, L, P, T	IN	C	1000
78.	*Mentha longifolia* ssp. *longifolia*	C, L, P, T	IN	C	1200
79.	*Mentha longifolia* ssp. *typhoides* var. *typhoides*	C, L, J, S, T	IN	C	1000-1500
80.	*Mentha spicata* ssp. *spicata*	C, T	IN	C	1000-2200
81.	*Micromeria fruticosa* ssp. *brachycalyx*	T	EM	CH	1000-1830
82.	*Micromeria myrtifolia*	C, L, P, S, T	EM	CH	1000-1900
83.	*Origanum majorana*	C, T	EM	CH	1000
84.	*Origanum minutiflorum*	T	EM, EN	CH	1000-1800
85.	*Origanum onites*	S, T	EM	CH	1000-1500
86.	*Origanum syriacum*	C, L, J, T	EM	CH	1000-2700
87.	*Origanum vulgare* ssp. *hirtum*	T	EM	H	1000-1500
88.	*Phlomis pungens* var. *hirta*	T	IN	H	1300-1640
89.	*Phlomis pungens* var. *hispida*	S, T	IN	H	1000-1450
90.	*Salvia multicaulis*	S, T	IR	C	1000-2600
91.	*Salvia tomentosa*	L, S, T	M	CH	1000-2300
92.	*Salvia verbenaca*	C, P, S, T	M	C	1600
93.	*Salvia virgata*	T	IR	H	1000-1400
94.	*Satureja cuneifolia*	L, P, T	M	CH	1000-2000
95.	*Sideritis arguta*	T	EM, EN	H	1000-1100
96.	*Sideritis congesta*	T	EM, EN	H	1000
97.	*Sideritis erythrantha* var. *erythrantha*	T	EM, EN	H	1220-1500
98.	*Sideritis libanotica* ssp. *linearis*	T	EN	H	1100-2200
99.	*Sideritis perfoliata*	L, P, S, T	EM	H	1000-2000
100.	*Sideritis pisidica*	T	EM, EN	H	1000-2100
101.	*Sideritis syriaca* ssp. *nusariensis*	S, T	EM	H	1200-2100
102.	*Teucrium chamaedrys* ssp. *lydium*	T	EM	C	1000-1500
103.	*Teucrium chamaedrys* ssp. *syspirense*	T	IR	C	1000-1750
104.	*Teucrium chamaedrys* ssp. *tauricolum*	T	EN, EM	C	1000-2500
105.	*Teucrium montanum*	T	IN	CH	1000-2750
106.	*Teucrium polium*	J, S, T	IN	CH	1000-2000
107.	*Thymbra spicata* var. *intricata*	T	EM, EN	H	1000-1520
108.	*Thymus sibthorpii*	T	ES	CH	1000-1500
	LAURACEAE				
109.	*Laurus nobilis*	C, S, T	M	P	1000-1200
	LILIACEAE				
110.	*Asparagus acutifolius*	C, L, P, T	M	H	1220-1500

Contd...

Table 24.1–*Contd...*

Sl.No.	Family/Taxa	Country	Chorotypes	Life Forms	Altitude (m)
111.	*Lilium candidum*	L, J, S	EM	C	1300
112.	*Ruscus aculeatus* var. *angustifolius*	C, L, P, S, T	IN	C	1000-1400
113.	*Smilax aspera*	C, L, J, P, S, T	IN	P	1000-1200
	LORANTHACEAE				
114.	*Viscum album* ssp. *abietis*	T	IN	P	1400-2000
115.	*Viscum album* ssp. *austriacum*	T	IN	P	1000-1600
	MALVACEAE				
116.	*Alcea pallida*	T	IN	H	1000-2300
117.	*Malva neglecta*	T	IN	T	1000-2700
	OLEACEAE				
118.	*Fraxinus ornus* ssp. *cilicica*	S, T	EM	P	1000-1460
	ORCHIDACEAE				
119.	*Orchis anatolica*	C, L, J, P, S, T	EM	C	1000-1650
	PAEONIACEAE				
120.	*Paeonia mascula* ssp. *mascula*	T	IN	C	1000-2200
	PAPAVERACEAE				
121.	*Fumaria densiflora*	C, L, P, S, T	IN	T	1150-1300
122.	*Fumaria parviflora*	C, L, S, T	IN	T	1200-1300
123.	*Glaucium flavum*	L, P, S, T	IN	H	1500-1700
124.	*Glaucium leiocarpum*	P, S, T	IN	H	1100-2250
	PHYTOLACCACEAE				
125.	*Phytolacca americana*	C, T	IN	H	1300
126.	*Phytolacca pruinosa*	L, S, T	EM	P	1000-1700
	PINACEAE				
127.	*Abies cilicica* ssp. *cilicica*	L, S, T	EM	P	1000-2000
128.	*Cedrus libani*	L, P, T	EM	P	1000-2000
129.	*Pinus nigra* ssp. *pallasiana*	S, T	IN	P	1000-1700
	PLANTAGINACEAE				
130.	*Plantago major* ssp. *intermedia*	C, L, P, S, T	IN	H	1050-1950
	PLUMBAGINACEAE				
131.	*Acantholimon acerosum* var. *acerosum*	S, T	IR	CH	1000-2750
132.	*Plumbago europaea*	C, L, J, P, S, T	ES	CH	1000-1900
	POACEAE				
133.	*Avena barbata* ssp. *barbata*	L, T	M	T	1450
134.	*Cynodon dactylon* var. *dactylon*	C, L, P, T	IN	C	1300
135.	*Hyparrhenia hirta*	C, T	IN	H	1000-1450

Contd...

Table 24.1–*Contd...*

Sl.No.	Family/Taxa	Country	Chorotypes	Life Forms	Altitude (m)
136.	*Lolium perenne*	C, T	ES	H	1000-1700
137.	*Panicum miliaceum*	T	IN	T	1000-1300
138.	*Phragmites australis*	C, S, T	ES	C	1000-1300
	POLYGONACEAE				
139.	*Polygonum bistorta* ssp. *bistorta*	T	ES	C	1200-3500
140.	*Polygonum cognatum*	S, T	IN	CH	1000-3000
141.	*Polygonum equisetiforme*	C, J, L, P, T	IN	H	1100
142.	*Rumex acetosella*	S, T	CO	H	1000-2300
143.	*Rumex angustifolius*	L, S, T	IN	H	2500-2750
144.	*Rumex patientia*	S, T	IN	H	1000-2000
	PRIMULACEAE				
145.	*Cyclamen cilicicum* var. *cilicicum*	T	IN	C	1000-1500
146.	*Cyclamen coum* var. *coum*	L, S, T	IN	C	1000-1300
	RANUNCULACEAE				
147.	*Ranunculus arvensis*	C, T	IN	T	1000-1850
	RHAMNACEAE				
148.	*Paliurus spina-christii*	L, P, S, T	IN	P	1000-1400
149.	*Rhamnus lycioides*	T	IN	P	1000
	ROSACEAE				
150.	*Amygdalus communis*	L, P, T	IN	P	1000-1800
151.	*Amygdalus orientalis*	L, S, T	IR	P	1000-1500
152.	*Cotoneaster nummularia*	L, S, T	IN	P	1000-2700
153.	*Crataegus monogyna* ssp. *azarella*	S, T	IN	P	1700-2100
154.	*Crataegus monogyna* ssp. *monogyna*	C, L, J, P, S, T	IN	P	1000-1500
155.	*Crataegus orientalis* var. *orientalis*	P, S, T	IN	P	1000-2240
156.	*Fragaria vesca*	J, S, T	IN	C	1000-2450
157.	*Geum urbanum*	L, S, T	ES	C	1000-1700
158.	*Potentilla reptans*	L, P, S, T	IN	C	1000-2300
159.	*Prunus divaricata* ssp. *divaricata*	T	IN	P	1000-2450
160.	*Pyrus syriaca* var. *microphylla*	T	EN	P	1000-1400
161.	*Rosa canina*	J, S, T	IN	P	1080-2300
162.	*Rosa pulverulenta*	S, T	IN	P	1000-2750
163.	*Rubus canescens* var. *canescens*	T	IN	P	1000-1500
164.	*Rubus sanctus*	C, L, P, S, T	IN	P	1000-1400
165.	*Sorbus torminalis*	S, T	IN	P	1000-1200
166.	*Sorbus umbellata* var. *cretica*	T	IN	P	1000-2500
167.	*Sorbus umbellata* var. *umbellata*	L, P, S, T	IN	P	1000-2500

Contd...

Table 24.1–*Contd...*

Sl.No.	Family/Taxa	Country	Chorotypes	Life Forms	Altitude (m)
	RUBIACEAE				
168.	*Cruciata taurica*	L, S, T	IR	CH	1000-2750
	SCROPHULARIACEAE				
169.	*Digitalis ferruginea* ssp. *ferruginea*	L, T	ES	H	1000-2700
	SOLANACEAE				
170.	*Atropa bellodonna*	T	ES	H	1000-1800
171.	*Physalis alkekengi*	T	IN	C	1050-2100
172.	*Solanum dulcamara*	L, J, P, S, T	ES	H	1050-2300
173.	*Solanum nigrum* ssp. *nigrum*	C, L, P, T	CO	T	1000-1500
	STYRACACEAE				
174.	*Styrax officinalis*	C, P, S, T	IN	P	1000-1500
	TAXACEAE				
175.	*Taxus baccata*	S, T	IN	P	1800-1900
	THYMELAEACEAE				
176.	*Thymelaea hirsuta*	L, P, T	M	P	1000
	TILIACEAE				
177.	*Tilia argentea*	T	ES	P	1000-1500
	ULMACEAE				
178.	*Celtis australis*	C, L, J, P, S, T	M	P	1000
179.	*Celtis glabrata*	S, T	IN	P	1000-1200
	URTICACEAE				
180.	*Parietaria judaica*	C, L, P, T	IN	H	1000-2000
181.	*Parietaria officinalis*	L, P, S, T	ES	H	1000-1200
182	*Urtica dioica*	L, P, S, T	ES	H	1000-2700
	VERBENACEAE				
183.	*Verbena officinalis*	C, P, T	CO	H	1000-1800

Country: **C**: Cyprus; **J**: Jordan; **L**: Lebanon; **P**: Palaestina; **S**: Syria; **T**: Turkey.

Chorotypes: **CO**: Cosmopolitan; **EM**: East Mediterranean; **EN**: Endemic; **ES**: Euro-Siberian; **IN**: Imperfectly Known; **IR**: Irano-Turanian; **M**: Mediterranean.

Life forms: **C**: Cryptophytes; **CH**: Chamaephytes; **H**: Hemicryptophytes; **P**: Phanerophytes; **T**: Therophytes.

4. CONCLUSIONS

In 2050, the Earth will have a population of at least 10 billion, and many of the plant species will disappear due to different degradative impacts. Ethnobotanical research is key in the development of drugs from natural sources. The information obtained on MAPS will dramatically facilitate the search for new drugs. Traditional knowledge about their uses for treatment purposes is the backbone of such programmes. Any decline in the regional knowledge on MAPS will directly effect

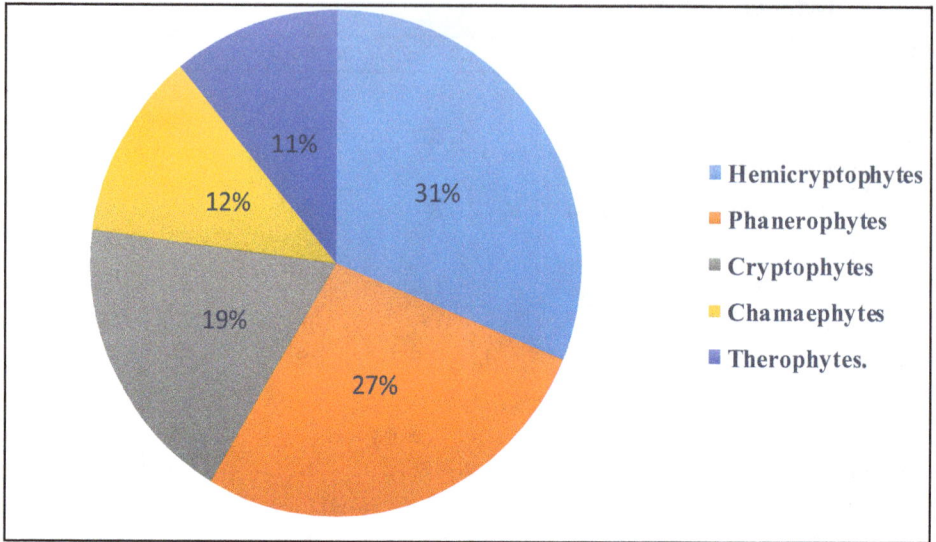

Figure 24.2: Life Forms of MAPS from EMED.

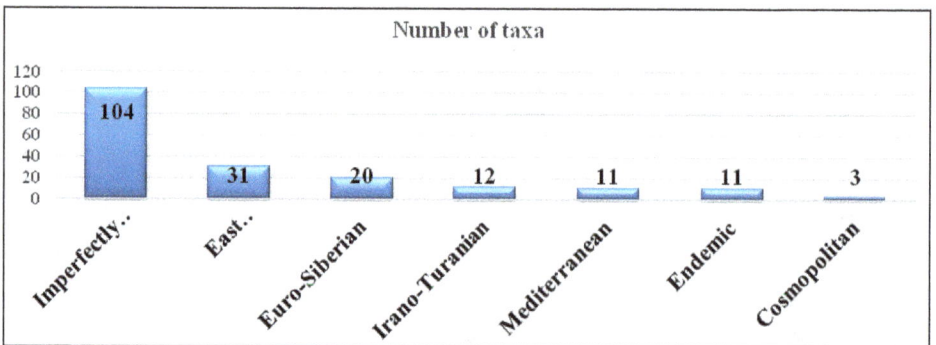

Figure 24.3: Chorotypes of MAPS from EMED.

the drug discovery (Gonzales-Tejero *et al.*, 2008; Öztürk *et al.*, 2011). We are face to face with a dilemma of antibiotic resistant strains (Khalil *et al.*, 2014a). A major problem for herbal therapies is a mixing up of indigenous knowledge with modern medical practices due to lack of scientific data regarding the safety and efficacy of the herbals. It is therefore of paramount urgency to document and authenticate the available indigenous knowledge using modern day scientific principles (Khalil *et al.*, 2014b).

Majority of the MAPS used by the herbal drug industry come from wild collection. The over-exploitation of natural biological resources of MAPs is frequently destructive, which ultimately leads to the extinction of many species. Recent policy interest in the importance of traditional medicine; throughout the industrialising world; has called attention to the significance of this topic for the health of the poor and indigenous groups, as well as in meeting thee pluralistic health

Percent (%)

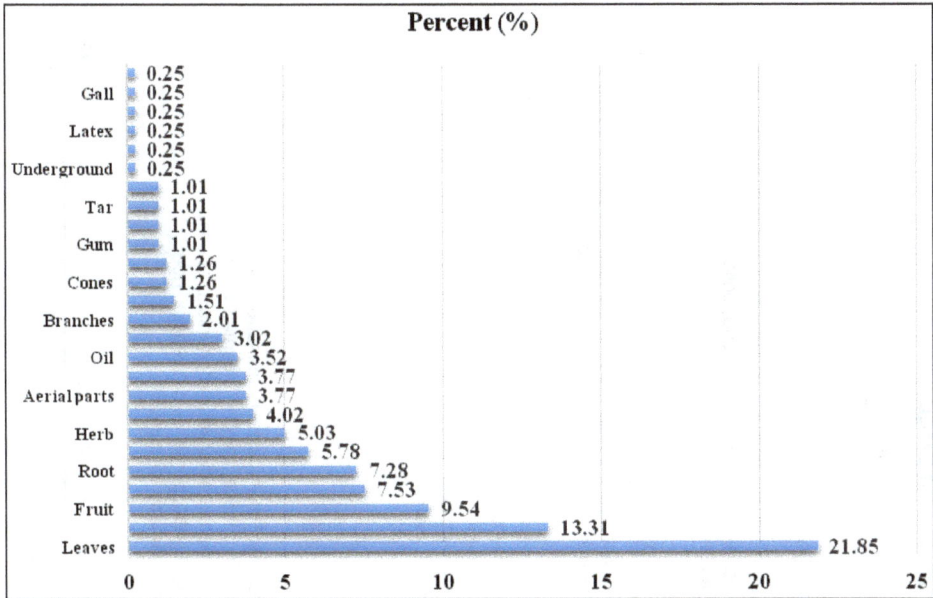

Figure 24.4: The Parts of the MAPS Used in the Treatments on Percentage Basis.

Percent (%)

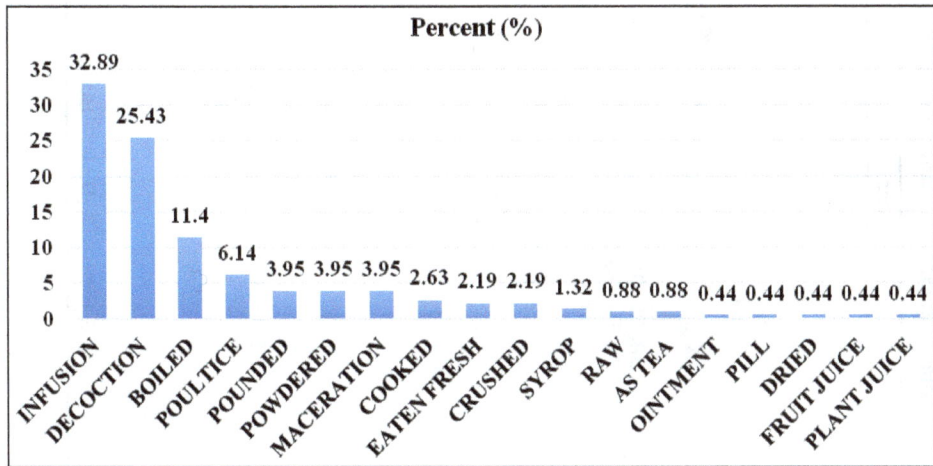

**Figure 24.5: Application and Consumption Means of
EMED MAPS in the Traditional Folk Medicine.**

Some of the EMED MAPS distributed at high altitudes are used as spices by the locals. Most important ones are; *Origanum onites, Mentha aquatica, Mentha longifolia* ssp. *typhoides* var. *typhoides, Origanum syriacum, Satureja cuneifolia, Thymus kotschyanus* var. *glabrescens, Mentha longifolia* ssp. *longifolia* (Lamiaceae); *Cistus salviifolius* (Cistaceae); *Rhus coriaria* (Anacardiaceae) and *Laurus nobilis* (Lauraceae) are used as spice in the study area (Sayar *et al.,* 1995; Ýlçim and Varol, 1996; Duran, 1998; Abay and Kýlýç, 2001; Bulut, 2006; Tuzlacý, 2006; Mart and Türkmen, 2008; Al-Qura'n, 2010; Altay and Karahan, 2012; Akaydýn *et al.,* 2013; Arýcan *et al.,* 2013; Yücel, 2012, 2014).

Table 24.2: MAPS Distributed on the High Mountains of EMED

Sl.No.	Famiy/Taxa	Part Used	Preparation	Treatment	Source
	AMARANTHACEAE				
1	*Amaranthus viridis*			Antipyretic, alexiteric, emollient, expectorant, laxative, stomachic	1-3
	AMARYLLIDACEAE				
2	*Galanthus elwesii*	TU	PU	Fistula, heart strengthener, stomachic, stimulating, boils ripening, gynecological diseases (for women)	4-6
	ANACARDIACEAE				
3	*Pistacia terebinthus ssp. palaestina*	BR, GL, GU, FL, FR, OO, RE	BO	Respiratory system diseases, for softening of joints and muscles, expectorant, diuretic, antiseptic, constipation, chronic coughing, bronchitis, asthma, restorative, for cold and flu, stomachic disorder, allergic	7-12
4	*Rhus coriaria*	BA, FR, LE	BO, DE, IN, PU	Diarrhea, antipyretic, gingiva and throat inflammations, weeping, antiseptic, skin lession, to increase saliva, to treat edema in leg (for animal), kidney stone, dysentery, throat and gum disorders, constipation, mouthwash, styptic, muscle contraction, liver diseases, urinary system, stimulates perspiration, cholesterol, kidney disorders	5, 6, 10, 12-16
	APIACEAE				
5	*Ferula elaeochytris*	RO	PW	Aphrodisiac	16, 17
	ARALIACEAE				
6	*Hedera helix*	FB, FR, LE	DE, IN	Laxative, neural diseases, rheumatic pain, menstrual cycle regulator, anthelmintic, exudative, swollen inflamed wound, vesicant, anticancer, antispasmodic, antimutagenic, bronchial asthma, stimulant, diaphoretic, burns, skin diseases, gout, pain killer, aphrodisiac, narcotic, emetic	5-7, 10, 12, 16, 18-24
	ASPIDIACEAE				
7	*Dryopteris filix-mas*	FS, RH	DE	Human intestinal parasites, rheumatism, gout	10, 23

Contd...

Table 24.2–Contd...

Sl.No.	Famiy/Taxa	Part Used	Preparation	Treatment	Source
	ASPLENIACEAE				
8	Ceterach officinarum	AP, HE, LE, WP	DE, IN, MA	Diuretic, hemorrhoid, kidney stone, joint diseases and wounds, ulcers of duodenum, constipation	5-7, 10, 13, 15, 18, 23
	ASTERACEAE				
9	Achillea nobilis ssp. densissima	LE, SH		Gastric chill, urinary infection, cold	10
10	Achillea wilhelmsii	FL, HE, LE	DE	Bronchitis, abdominal pain (for horses), stomachic, indigestion	13, 25
11	Anthemis pseudocotula	FL	DE, IN	Stimulant, carminative, gynecological disease, anticancer, bactericide, antifungal, anti-diabetic, antitumor, stomachic disorders, gas removed	12, 17, 21, 24
12	Anthemis tinctoria var. tinctoria	FL	IN	Dysmenorrhea, carminative, hair reinforce, hemorrhoid, stimulant, gynecological disease	4, 12
13	Artemisia absinthium	FL, LE, WP	BO, DE, IN	Antipyretic, common cold, stomach disorders, abdominal pain, diuretic, nervine, anthelmintic, rheumatism, arthritis, gastritis, intestinal disorders, dizziness, insomnia, against stings, tonic, appetizing	5-8, 13, 18, 23, 26
14	Bellis perennis	WP	BO	Cold, flu, hair loss	25
15	Carduus nutans ssp. nutans	FL	IN	Bronchitis, cough	13
16	Cichorium intybus	FL, LE, RO, WP	CO, DE, IN	Diuretic, constipation, diaphoretic, stomach pain, laxative, orexigenic, tonic, biligenic, stomachic, sedative, internal hemorrhage, antiseptic, anemia, liver cirrhosis, eczyma, diabetes, diarrhea, vomiting, jaundice and liver enlargement, skin diseases, asthym, ulcer, cholesterol, gout, rheumatism	4-6, 10, 16, 21, 27
17	Cirsium arvense	FL		Peptic ulcer	13
18	Conyza bonariensis	AP, FB, FL	DE	Rheumatism, vermifuge, antispasmodic, antidiuretic, uterine bleeding, diarrhea, entiritis	21, 23
19	Gundelia tournefortii var. tournefortii	GU		Strengthen gums, stomachic, toothache, appetizing	5, 6, 16

Contd...

Table 24.2–Contd...

Sl.No.	Family/Taxa	Part Used	Preparation	Treatment	Source
20	Helichrysum plicatum ssp. plicatum	FL, HE	DE, IN	Kidney stone, jaundice, dysuria	13
21	Inula heterolepis	LE, SH	IN	Appetizing	10
22	Inula viscosa	HE, FB, LE, RO, WP	DE, IN, MA	Eye diseases, stomachic disorders, anthelmintic, wound, ulcer, to stop bleeding, pain, respiratory tract infection, hemorrhoids, fractured bones, diabetes, backaches, gum disorder, skin fungi, appetizer, dysentery, antidermatosic, muscle relaxation and infertility, anticancer, lung disorders, nerve tonic, skin diseases, joint diseases, colds	7, 12, 14-16, 24, 27-37
23	Scorzonera phaeopappa	LE, WP		Indigestion, digestive, excretory system	25
24	Sonchus oleraceus	WP		Tonic, diuretic, febrifuge, cathartic	21
25	Tragopogon dubius	WP		Indigestion, digestive, excretory system	25
26	Tussilago farfara	FL, LE	DE, IN	Antitussive, to promote maturation of abscess, rheumatism, dyspepsia, gastrointestinal infection, expectorant, antiinflammatory, bronchitis, bronchitis asthma, cough, chest softener	5, 6, 10, 13, 16, 18, 23, 26
27	Xanthium strumarium ssp. cavanillesii	FR, LE, RO, SE	DE, IN, PU	Malaria, rheumatism, kidney diseases, tubercularis, antibacterial, antispasmodic, antitussive, stomachic, allergic rhinitis, constipation, diarrhea, leprosy, scrofullosus tumour, wound	38-41
	BERBERIDACEAE				
28	Berberis crataegina	BA, HE, RB, RO	BO, DE, IN	Antipyretic, astringent, hepatitis, urinary and kidney infection, hemorrhoids, dysuria (for animal), to remove itching and reddening of eyes, tension, antihelmintic (for animal)	8-10, 13, 25
29	Leontice leontopetalum ssp. leontopetalum	TU		Emmenagogue, epilepsy	16, 21
	BORAGINACEAE				
30	Alkanna orientalis var. orientalis	RO, WP		Wounds, emmenagogue	12, 16
31	Alkanna tinctoria ssp. anatolica	RO	IN	Constipation, skin lesion	10
32	Echium italicum	HE, LE	PO, PU	To treat maturation of abscess, rheumatic pain, wounds, cholesterol	5, 6, 13
33	Heliotropium hirsutissimum	AP	CR	Scorpion bites	42

Contd...

Table 24.2—Contd...

Sl.No.	Famiy/Taxa	Part Used	Preparation	Treatment	Source
	BRASSICACEAE				
34	*Cardaria draba* ssp. *draba*	LE		Wounds, facilitate digestion, bronchi cleaner, cough, anemia, diabetes, sedative	5, 6, 43
	BUXUCEAE				
35	*Buxus sempervirens*	LE	DE	Diuretic, bile expectorant, antipyretic, diaphoretic, antihelmintic	18
	CAPRIFOLIACEAE				
36	*Sambucus ebulus*	AP, FL, FR, LE	CR, DE, IN, PO, PU	To treat a common cold, high fever, sunstroke, to treat snake bite, rheumatic pain, gout, rheumatism	13, 23, 43
37	*Sambucus nigra*	BA, FL, FR, HE, LE, ST	DE, IN, MA	Diuretic, respiratory, cardiovascular, diaphoretic, laxative, analgesic, rheumatic pain, rheumatism, sciatica, arthritis, hypotension, constipation, spasmolytic, stomach disorders, anaemia, influenza, depression, migraine, eye pains, cough, common cold, anti-inflammatory, detoxification, febrifuge, gas removed, aphrodisiac, antihelmintic, skin disorders	5, 6, 10, 12, 13, 18, 23, 26, 44
	CARYOPHYLLACEAE				
38	*Silene vulgaris* var. *vulgaris*	WP		Excretory or genitourinary system diseases	16
	CHENOPODIACEAE				
39	*Chenopodium album* ssp. *album* var. *album*	AP, FL, LE, SH, WP	BO, DE, PU	Anaemia, an open sore, anthelmintic, antiphlogistic, rheumatism, laxative, bug bites, sunstroke, arthritis, gynecological diseases	9, 10, 21, 23, 45-48
40	*Chenopodium botrys*			Antiinflammations, wound	5, 6
41	*Chenopodium foliosum*	LE	IN	Diuretic, purgative	10
	CISTACEAE				
42	*Cistus salviifolius*	FB, LE	DE, IN	Constipation, expectorant, rheumatism, bronchitis	10, 21, 23
	CONVOLVULACEAE				
43	*Convolvulus lineatus*	FL, RO	DE	Diuretic, laxative, wound, antipyretic	49

Contd...

Table 24.2–Contd...

Sl.No.	Famiy/Taxa	Part Used	Preparation	Treatment	Source
	CORNACEAE				
44	*Cornus mas*	BA, FR, LE	DE, IN, SY	Abdominal pain, antypyretic, diarrhoea, against intestinal wound, styptic, mouth sores, stomachic, constipation, antihelmintic, diabetes, tonic, for boils	5, 6, 10, 16, 18, 25, 50
	CRASSULACEAE				
45	*Sedum album*	HE	PO	Treat warts	13
	CUPRESSACEAE				
46	*Juniperus drupacea*	BR, CO, SE, TR	DE, PW	Tonic, antiseptic, restorative, aphrodiasic, intestinal parasite, abdominal pain, diarrhea	4, 8, 13, 18
47	*Juniperus excelsa*	BR, CO, SE	BO, DE	Diabetes, cough, asthma, rheumatism, lumbago, sciatica, arthritis	13, 23, 25, 50
48	*Juniperus foetidissima*	BR, OO, SE	BO	Diuretic, cough, cold, stomachic disorders, eczema, ringworm, alopecia, psoriasis	7, 11, 25
49	*Juniperus oxycedrus* ssp. *oxycedrus*	BA, BR, CO, SE, TR	BO, DE, IN, MA, PO	Animal diseases, skin disorders, cold, heart failure, treatment of heel spur, skin lesion, manginess treatment, antiseptic, abdominal pain, cough, inhale, calcinosis in joint, to treat catarrh, expectorant, hemorrhoids, urinary inflammations, bronchitis, anal fistula, rheumatism	8, 10, 13, 23, 42, 43
	DIOSCOREACEAE				
50	*Tamus communis* ssp. *cretica*	FR, RH, RO	PO	Rheumatism	8, 13
	EQUISETACEAE				
51	*Equisetum telmateia*	AP, HE	DE	Kidney stone, dysuria, rheumatism, arthritis	13, 23
	ERICACEAE				
52	*Erica manipuliflora*	BR, FL, LE, SH	BO, DE, IN	To lose weight, diuretic, urethritis, constipation, arthritis	4, 10, 23, 51

Contd...

Table 24.2–Contd...

Sl.No.	Famiy/Taxa	Part Used	Preparation	Treatment	Source
	EUPHORBIACEAE				
53	*Euphorbia macroclada*	LA		Hemorrhoids, anthelmintic, laxative, rheumatism, warts	11-13
54	*Euphorbia peplus var. peplus*	FL		Diuretic, expectorant, laxative	21
	FABACEAE				
55	*Astragalus hamosus*	FL, FR		Demulcent, treatment of baldness	14, 21
56	*Astragalus lydius*			Antibacterial, antifungal, antiviral	5, 6
57	*Lotus corniculatus var. tenuifolius*	HE, WP	DE	Sedative	7, 18
58	*Trifolium repens var. repens*	FL	DE	Rheumatism, arthritis	23
	FAGACEAE				
59	*Quercus coccifera*	BA, FR, RO	BO, CO, DE	Skin, hair and eye diseases, burns, constipation, antitussive, astringent, peptic ulcer, digestive system, anti-diabetes, mouth gargle, stomachic diseases	7, 11, 21, 27, 31, 34, 50, 52, 53
60	*Quercus infectoria ssp. boissieri*	BA, FR		Skin, hair and eye diseases, diabetes	7, 53
61	*Quercus ithaburensis ssp. macrolepis*	FR		Constipation, tonic, antiseptic	12
62	*Quercus robur ssp. robur*	FR		Constipation, tonic, antiseptic	12
	GERANIACEAE				
63	*Erodium cicutarium ssp. cicutarium*	WP		Laxative	7
64	*Pelargonium endlicherianum*			Treatment of human intestinal parasites	10
	HYPERICACEAE				
65	*Hypericum confertum ssp. confertum*	FL, LE, OO		Asthým, bronchitis, burn, mental distress	25

Contd...

Table 24.2–Contd...

Sl.No.	Famiy/Taxa	Part Used	Preparation	Treatment	Source
66	*Hypericum perforatum*	AP, FB, FL, HE, LE, OO, SH, WP	BO, DE, DR, IN, MA, PW	Hemorrhoids, prostatitis, diabetes, hypertension, urinary infections, diaper rash, rheumatism, osteoporosis, skin lesion, sunburn, knife cut, antiseptic, antispasmodic, constipation, taenia fudge, stomach-ache, laxative, for burn wounds, ulcer, stomach disorders, antiseptic, antidiarrheal, sedative, antihelminthic, arthritis, dyspepsia, depression, insomnia, expectorant, abdominal pain, hepatitis, hemostatic, herpesgout, jaundice, tuberculosis and lung therapy, asthym	4-7, 10, 12, 13, 16-18, 23, 26, 50
67	*Hypericum scabrum*	FL, LE, OO, SH		Skin lesion, sunburn, knife cut, antiseptic	10
68	*Hypericum tetrapterum*	FB		Antispasmodic, constipation, sedative, antihelmintic, antiseptic, wounds	18
	JUGLANDACEAE				
69	*Juglans regia*	BA, FL, FR, LE, OO, RO, SE, SH	CR, DE, IN, PU	Antiseptic, eye disorders, diabetes, to increase breast milk, whopping cough (pertussis), appetizing, astringent, tonic, anthemintic, purgative, diuretic, glycemia, skin diseases, hair reinforce, dyspepsia, arthritis, orexigenic, sunstroke, to treat eruptions caused by dermatophytes between fingers and toes, ease cough, laxative, stomachic, asthma, sexual weakness, rheumatism, nutritional, sciatica, eczema, wounds, diarrhoea, inflammation, facial skin improvement, constipation	4, 7, 8, 10, 12-16, 18, 21, 23, 31, 33
	JUNCACEAE				
70	*Juncus articulatus*	ST	DE	Diuretic, sedative, febrifuge, antiphlogistic, sore throat, jaundice, oedema, acute urinary tract infection	47, 54
71	*Juncus inflexus*	ST	DE	Diuretic, sedative, febrifuge, antiphlogistic, sore throat, jaundice, oedema, acute urinary tract infection	47, 54
	LAMIACEAE				
72	*Ajuga chamaepitys ssp. chia var. chia*	FL, HE, LE, SH	BO, DE, IN	Gynecology (regularize menstrual disorders), painkiller, astringent, antitussive, hemorrhoids, gout, rheumatism, abdominal pain, diabetes	5, 6, 10, 11, 13
73	*Ballota nigra ssp. uncinata*	LE	PU	Flatulence, stomach upset	13

Contd...

Table 24.2—Contd...

Sl.No.	Family/Taxa	Part Used	Preparation	Treatment	Source
74	Cyclotrichium origanifolium	FL, LE		Abdominal pain	8
75	Lamium album	FL	IN	Constipation, tonic, the cleaner urinary tract, blood regulators, wound	4-6, 18
76	Melissa officinalis ssp. officinalis	AP, FL, LE	IN	Antiseptic, sedative, relaxing, stomach disorders, intestinal disorders, cardiovascular disease, digestive, diarrhea, migraine, analgesic, carminative, ear ache, brain stimulant, common cold, mental-nervous, intestine pain, inflammation, cardiotonic, hypertension, tachycardia, antiemetic, colic, dyspepsia, spasmolytic, stomach tonic, antiaging, depression, headache, nervous tonic, appetizer, diaphoretic	4, 7, 15, 18, 25, 26, 42, 44
77	Mentha aquatica	LE		Antiseptic	7
78	Mentha longifolia ssp. longifolia	LE, SH		Stomach pain, nausea and vomiting, fever, grib, hair loss	25
79	Mentha longifolia ssp. typhoides var. typhoides	LE	AT	Antiseptic, nausea, abdominal pain	7, 37
80	Mentha spicata ssp. spicata	AP, BR, FL, LE, OO, SE, SH	CO, CR, DE, IN, RW	Cold and flu, diarrhoea, indigestion, diarrhea, nausea, digestive, respiratory, arthritis, cancer, flatulence, neutralizes acidity, sugar in blood	10, 12, 14, 23, 44, 55
81	Micromeria fruticosa ssp. brachycalyx	LE, SH	IN	Cold and flu, diarrhoea, indigestion	10
82	Micromeria myrtifolia	FL, LE, WP	BO, IN	Respiratory system, skin diseases, heart diseases, digestive system and asthma, antispasmodic, female sterility, sedative	9, 15, 31, 34
83	Origanum majorana	FB	IN, PW	Sedative, stomachic, diuretic, gas removed, diaphoretic, constipation	18
84	Origanum minutiflorum	FL, LE, SH	IN	Cold and flu, throat infection	10
85	Origanum onites	AP, FL, HE, LE, OO, SH	IN	Antiseptic, abdominal ailments, cold, high cholesterol, sedative, cardio-vascular disease, antiparasitic, throat infection, as a moth-repellent	7, 10, 13, 25, 42
86	Origanum syriacum	FL, LE, OO, SE	MA	Rheumatism, eye ailment, burns, stomach troubles, sedative	21, 23
87	Origanum vulgare ssp. hirtum	FL, LE, SH	IN	To reduce cholesterol and blood sugar, digestive and respiratory system disease, ulcer and intestines disease, arthrosis and waist pain	10

Contd...

Table 24.2–Contd...

Sl.No.	Family/Taxa	Part Used	Preparation	Treatment	Source
88	Phlomis pungens var. hirta	LE	IN	Constipation, stomach ache, appetizing, cold and flu	10
89	Phlomis pungens var. hispida	LE	IN	Constipation, stomach ache, appetizing, cold and flu	10
90	Salvia multicaulis	LE	IN	Wound healing	12
91	Salvia tomentosa	LE	IN	Gas removed, antiseptic, tonic, stimulant	18
92	Salvia verbenaca	LE		Respiratory	7
93	Salvia virgata	LE		Respiratory	7
94	Satureja cuneifolia	LE	BO	Upper respiratory tract disorders, mental fatigue, tonic, increase sexual desire, breast disease, stomach booster, facilitation pregnancy	5, 6, 8
95	Sideritis arguta	FL, LE, SH	IN	Gas removed, sedative, antidiarrhoeal, constipation, stomachache, appetizing, painkiller, throat inflammation, neural appeaser, cold	8, 10
96	Sideritis congesta	FL, LE, SH	IN	Constipation, stomachache, appetizing, painkiller, throat inflammation, neural appeaser, cold	10
97	Sideritis erythrantha var. erythrantha	FL, LE, SH	IN	Constipation, stomachache, appetizing, painkiller, throat inflammation, neural appeaser, cold	10
98	Sideritis libanotica ssp. linearis	FL, LE, SH	IN	Gas removed, sedative, antidiarrhoeal, constipation, stomachache, appetizing, painkiller, throat inflammation, neural appeaser, cold	8, 10
99	Sideritis perfoliata	AP, FL, LE, SH	IN	Constipation, stomachache, appetizing, painkiller, throat inflammation, neural appeaser, cold, digestive, mental-nervous, dyspepsia, stomach disorders, anaemia, influenza, diaphoretic, calmative, bronchitis, cough, diuretic, aphrodisiac	10, 26, 44
100	Sideritis pisidica	FL, HE, LE, SH	IN, PU	Stomachic disorders, stimulant, constipation, stomachache, appetizing, painkiller, throat inflammation, neural appeaser, cold, abdominal pain	7, 10, 13
101	Sideritis syriaca ssp. nusariensis	FL, LE, SH	IN	Constipation, stomachache, appetizing, painkiller, throat inflammation, neural appeaser, cold	10
102	Teucrium chamaedrys ssp. lydium	FB	IN	Appetizing, stomach pain, stimulant, tonic, diabetes	18
103	Teucrium chamaedrys ssp. syspirense	FB	IN	Appetizing, stomach pain, stimulant, tonic, diabetes	18

Contd...

Table 24.2–Contd...

Sl.No.	Famiy/Taxa	Part Used	Preparation	Treatment	Source
104	*Teucrium chamaedrys* ssp. *tauricolum*	FB	IN	Appetizing, stomach pain, stimulant, tonic, diabetes	18
105	*Teucrium montanum*	FB	IN	Appetizing, stomach pain, stimulant, tonic	18
106	*Teucrium polium*	AP, FL, HE, LE, SH	BO, DE, IN	Diabetes, diarrhea, stomachache, hypertension, appetizing, pain, tonic, cold, sore throat, stomach, toothache, foot pain, abdominal pain, high fever, rheumatic pain, indigestion, kidney stones, liver diseases, stomach and intestine inflammation, anti-inflammatory, astringent, kidney, spasm, antiflatulence, digestive system, abdominal colics, headache, vermifuge, depurative, constipation, urinary tract inflammations, obesity	5-7, 10, 13, 15, 17, 18, 21, 27, 28, 31, 33-35, 43, 50, 56
107	*Thymbra spicata* var. *intricata*	FB, LE, OO		Stomach upset, appetizing, chest softener, asthma, cough, tonsillitis and pain	7, 8
108	*Thymus sibthorpii*	FL, LE, SH		Diabetes, ulcer, atherosclerosis, hypertension, shortness of breath, fever, grib	25
	LAURACEAE				
109	*Laurus nobilis*	BA, FR, LE, OO, SE	BO, CR, DE, IN, MA	Upper respiratory tract disorders, stomach disorders, analgesic, sedative, muscle to match, cough, herniated disk, skin care and cleaning, diabetes, rheumatoid pain, appetizing, indigestion, digestive system diseases, diaphoretic, antiseptic, sweaty, skin disease, enteritis, calmative, muscular pain, cancer, hair loss, burns, igmoritis, throat ache, bloating, colic, dyspepsia, mouth infections, anti scabies, urinary system, stones, hypertension, diarrhea, general weakness, arthritis, dandruff, indigestion, grib, chronic bronjit, headache, diuretic	4, 7, 8, 10, 14-18, 23-27, 31, 42, 44, 50, 51, 56, 57
	LILIACEAE				
110	*Asparagus acutifolius*	FR, LE, RO, SH, UG	CO, DE, IN	Diabetes, diuretic, painkiller, to dry inflammation of kidney, antipyretic, lithontriptic, gout, rheumatism, tonic	7, 10, 16, 23, 25, 43, 44

Contd...

Table 24.2–Contd...

Sl.No.	Famiy/Taxa	Part Used	Preparation	Treatment	Source
111	Lilium candidum	BU	IN	Diuretic, expectorant, boils ripening	18
112	Ruscus aculeatus var. angustifolius	FR	BO	Diuretic, sand removal	9, 11
113	Smilax aspera	FR	EF	Purifies the blood	9
	LORANTHACEAE				
114	Viscum album ssp. abietis	BR, HE	DE	Tension	8, 13
115	Viscum album ssp. austriacum			Cancer, diabetes, asthým	25
	MALVACEAE				
116	Alcea pallida	FL, SE	BO	Chest emollient, expectorant, throat, mouth and gum diseases, colds and bronchitis, cough	5, 6, 9, 11
117	Malva neglecta	AP, HE, LE	BO, CO, DE, IN, PU	Constipation, sore throat, abdominal pain, hemorrhoids, gout, rheumatism, constipation, antitussive, softener, to relieve the pain of wounds	5, 6, 13, 23, 27, 50
	OLEACEAE				
118	Fraxinus ornus ssp. cilicica	LE	IN	Constipation	10
	ORCHIDACEAE				
119	Orchis anatolica	TU	BO	Spleen disorders, tremors in the hands and feet, internal hemorrhage, stomach, intestine, lung, uterus, urinary system, diarrhea, dysentery, nervine tonic, aphrodisiac, chronic fever, abdominal catarrhs	21, 51
	PAEONIACEAE				
120	Paeonia mascula ssp. mascula	RO	IN	Appeaser, antitussive, epilepsy, sedative, antidiarrheal	10, 16
	PAPAVERACEAE				
121	Fumaria densiflora	WP		Eczema, liver diseases, hemorrhoid	21
122	Fumaria parviflora	FB, LE, ST		Eczema, hemorrhoids, liver diseases, diaphoric, diuretic	21
123	Glaucium flavum	FL, WP	PW	Wound, sedative, antitussive	16, 17
124	Glaucium leiocarpum	FL	PO	To treat goitre, cardiac slowing	5, 6, 13

Contd...

Table 24.2–*Contd...*

Sl.No.	Famiy/Taxa	Part Used	Preparation	Treatment	Source
	PHYTOLACCACEAE				
125	*Phytolacca americana*	FR, RO		Stimulant, exudative, emetic, irritant, diarrhea, laxative, constipation	5, 6, 12, 16
126	*Phytolacca pruinosa*	FR, RO		Emetic, diarrhea, laxative	18
	PINACEAE				
127	*Abies cilicica* ssp. *cilicica*	CO, RE, SH	DE, OI, PO	Wound healing, vascular diseases, gastric ulcer, bronchitis, common cold, tuberculosis	13
128	*Cedrus libani*	LE, RE, RO, TR	DE, PO	Antiseptic, diabetes mellitus, to treat fistulas on hand/foot, abdominal pain, diarrhea, rheumatism, snake, scorpion bite, bronchitis, common cold, diaphoretic, pain, muscle relaxant, constipation and softening, swelling, inflammation, diuretic, prostate diseases, wound healing, antimicrobial	11, 13, 23, 25
129	*Pinus nigra* ssp. *pallasiana*	CO, RE, SH, TR	BO, PI, PW	Internal diseases, common cold, cough, gastric ulcers, burns, fractrured bones, abdominal pain, scorpion bite, wound healing, antihelmintic, remove spine from skin, antiseptic, antimicrobials, urinary tract, and respiratory diseases, rheumatism, psoriasis, skin disorders, expectorant, lung cancer	5, 6, 13, 25
	PLANTAGINACEAE				
130	*Plantago major* ssp. *intermedia*	LE, OO, SE	DE, IN, PW	Hypertension, common cold, kidney stone, blood purification, wound, reducing blood cholesterol, ague, sore eyes, emollient, constipation, expectorant, diuretic, laxative, diarrhea, hemorrhoids, promote maturation of abscess	4, 13, 18, 38, 39, 43, 58
	PLUMBAGINACEAE				
131	*Acantholimon acerosum* var. *acerosum*			Strengthens the immune system, stress, colds	5, 6
132	*Plumbago europaea*	RO		Skin diseases, mouth, gum, teeth care, dermal disese, wounds	21, 31
	POACEAE				
133	*Avena barbata* ssp. *barbata*	LE, SE		Vaginal infection, diabetes, stomach, intestinal catarh	21

Contd...

Table 24.2–Contd...

Sl.No.	Famiy/Taxa	Part Used	Preparation	Treatment	Source
134	*Cynodon dactylon var. dactylon*	RO	DE, IN, PJ	Diuretic, dropsy, secondary syphilis, to stop bleeding, skin diseases, hysteria, epilepsy, chronic diarrhea, dysentery, tumor, cough, head-ache, hypertension, urinary system, cramps, cystitis, hemorrhage, warts hand wounds, kidney, bleeding of piles, astringent, catarrah	1, 21, 38, 44, 59-61
135	*Hyparrhenia hirta*			Bran, sciatica, rheumatism	23
136	*Lolium perenne*	WP	IN	Rheumatism	23
137	*Panicum miliaceum*	FL	DE	Kidney stone, diuretic	50
138	*Phragmites australis*	RH	DE	Rheumatism, diuretic, antipyretic, blood purifier, urinary tract diseases, arthritis, antipyretic, antiemetic, stomach ailments, jaundice, sweaty, gout, dental pain, bronchitis	5-7, 18, 21, 23
	POLYGONACEAE				
139	*Polygonum bistorta ssp. bistorta*	RO		Antidiarrheal, antiseptic, diuretic, mouth sore, antifibrinolytic	16
140	*Polygonum cognatum*	HE, RO	DE, PU	Anti-diabetes, oxyuriasis, worms, internal diseases, diuretic	12, 13, 16
141	*Polygonum equisetiforme*	AP, RO	DE	Kidney stone, urinary system, against intestinal, gastric, nephretic infections	1, 15, 21, 31
142	*Rumex acetosella*	HE, LE, RO	IN, PU, PW	Constipation, intestine gaseous, treatment of human intestinal parasites (worm), to increase ballast, headache, anemia, loss of appetite, fistula, diuretic, anti-inflammatory, cholagogue, vesicant	4, 10, 13, 16
143	*Rumex angustifolius*	LE, RO	PU	To promote maturation of abscess	13
144	*Rumex patientia*	RO, SE		Laxative, swollen inflamed wounds, eczema, diarrhea	12
	PRIMULACEAE				
145	*Cyclamen cilicicum var. cilicicum*	TU	IN	Inflammatory gynecological diseases, purgative	8, 10
146	*Cyclamen coum var. coum*	TU	IN	Emetic, purgative, stimulant	18
	RANUNCULACEAE				
147	*Ranunculus arvensis*	HE, RO	DE, PO	Rheumatic pain, sciatica, lumbago	13, 23

Contd...

Table 24.2–Contd...

Sl.No.	Famiy/Taxa	Part Used	Preparation	Treatment	Source
	RHAMNACEAE				
148	*Paliurus spina-christii*	BA, FR, SE	BO, DE, IN	For warts, constipation, kidney stone, hepatitis, lung inflammation, stomach pain, dysentery, antidiarrheal, diuretic, lithontriptic, kidney inflammation	5-11, 16, 43
149	*Rhamnus lycioides*	GU, LE, ST	PU	Toothache, boils ripening	9
	ROSACEAE				
150	*Amygdalus communis*	FR, LE, OO, SE	CO, DE, EF, RW	Skin lession, burn, sore throat, coughing, anthelmintic, diuretic, laxative, local paralisis and hair loss, stomach and intestine, lung inflammation, cancer, wound healing, softening	10, 12, 15, 16, 18, 55
151	*Amygdalus orientalis*	FR, OO		Skin lession, burn	10
152	*Cotoneaster nummularia*	FR		Appetizing, stomachic, expectorant	18
153	*Crataegus monogyna* ssp. *azarella*	FR	EF	Sedative	50
154	*Crataegus monogyna* ssp. *monogyna*	FL, FR, LE	BO, IN	Pain relief, stone and sand, sedative, hypertension, prostate disease, hypotensive, diuretic, antidiarrheal, heart disease, constipation, asthma, palpitations, insomnia, against atherosclerosis	6, 7, 10, 11, 16
155	*Crataegus orientalis* var. *orientalis*	FL, LE	IN	Hypertension, prostate disease, sedative, against palpitations and insomnia, atherosclerosis, heart disease	5, 6, 10
156	*Fragaria vesca*	FR, LE, RH, RO, WP	DE, FJ, IN	Appetizing, pimple, arthritis, gout, rheumatism, laxative, diuretic, astringent, antidiarrheic, kidney stone, strengthening the uterus, indigestion, antipyretic, aphrodisiac, edema	5, 6, 10, 21, 23
157	*Geum urbanum*	RO	IN	Peptic pain, antipyretic, antiseptic, antidiarrheal, nervine, stomachic, appetizer, intestinal disorders, pain, constipation, tonic	5, 6, 10, 16, 18
158	*Potentilla reptans*	WP		Tonic	7
159	*Prunus divaricata* ssp. *divaricata*	FL, FR, LE		Laxative, constipation, diuretic, anthelmintic	12
160	*Pyrus syriaca* var. *microphylla*	FR	EF	Diarrhea	25

Contd...

Table 24.2–Contd...

Sl.No.	Famiy/Taxa	Part Used	Preparation	Treatment	Source
161	*Rosa canina*	BA, FL, FR, LE, SH	DE, IN, PW, SY	Tonic, antitussive, diabetes, cardiotonic, kidney stone, rheumatism, bronchitis, osteoclasis, source of vitamins, stomach disorders, cold, constipation, glycemi, kidney troubles, ease cough, to treat burns, abdominal pain, diarrhea, nerve, heart disorders, shortness of breath, bone loss, sore throat, stomach pain, constipation, bladders disorders, astringent, diuretic, anti-scorbutic, pain, antihelmintic	4-11, 13, 16, 21, 25, 42
162	*Rosa pulverulenta*	FR	IN	Cold and flu	10
163	*Rubus canescens var. canescens*	FR	SY	Intestinal	51
164	*Rubus sanctus*	LE, RO, SH	BO, DE, IN	Diuretic, kidney stone, urinary infection, diabetes, stomachache, constipation, birthmark, cardiovascular, nutritional, tonic, tonsillitis, wound healing	5-8, 10, 43, 44, 50
165	*Sorbus torminalis*	FR, LE	IN	Chest softener	5, 6
166	*Sorbus umbellata var. cretica*	FR, LE	IN	Diabetes, against blood coagulation	10
167	*Sorbus umbellata var. umbellata*	FR, LE	IN	Arteriosclerosis, inhaler, diabetes, blood coagulation against	8, 10
	RUBIACEAE				
168	*Cruciata taurica*	SH	AT	Sand bladder, jaundice, boils and pimples, burns	5, 6
	SCROPHULARIACEAE				
169	*Digitalis ferruginea* ssp. *ferruginea*	SH	AT	Laxative	25
	SOLANACEAE				
170	*Atropa bellodonna*	LE		Analgesic, antispasmodic, pain	5, 6, 16
171	*Physalis alkekengi*	FR	DE	Diuretic, antipyretic, sedative	18
172	*Solanum dulcamara*	FR, LE, WP		Diuretic, stimulant, central nervous system, blood disorder, aphrodisiac, antirrheumatic, laxative, kidney, gallbladder disorders, bronchitis, chest pain, cough, intestinal disorder	5, 6, 21
173	*Solanum nigrum* ssp. *nigrum*	FB, FL, LE	IN	Rheumatism, cough, dysentery, chronic bronchitis, tumors, cancer, laxative, antipyretic, aphrodisiac, sedative, pain, bloating	5, 6, 11, 18

Contd...

Table 24.2–Contd...

Sl.No.	Famiy/Taxa	Part Used	Preparation	Treatment	Source
	STYRACACEAE				
174	*Styrax officinalis*	BA, FL, GU, LE, SE	IN	Mental-nervous, expectorant, diptheria, leucorrhoea, sedative, anti-septic, dermal troubles, skin rash, leprosy, antimicrobial, wound healing	5, 6, 14, 21, 27, 44
	TAXACEAE				
175	*Taxus baccata*	LE	MA	Stomachic, carminative, sedative, gynecological disorder, arthritis, sedative, gas removed	5, 6, 12, 16, 18, 23
	THYMELAEACEAE				
176	*Thymelaea hirsuta*	LE		Skin diseases, antihelmintic, hydragogue, cathartic, expectorant	1, 15, 21
	TILIACEAE				
177	*Tilia argentea*	FL		Diuretic, sedative, soporific, expectorant, sore throat, exudative	12
	ULMACEAE				
178	*Celtis australis*	FR, LE	IN	Against amenorrhoea, colic pain, stomach pain, wound healing, diuretic, foot sweating, kidney stones, cough, constipation	5, 6, 18, 21
179	*Celtis glabrata*	FR	DE	Astringent, constipation, indigestion, diarrhoea	10
	URTICACEAE				
180	*Parietaria judaica*	LE	DE	Nerve system and respiratory system, sedative, diuretic, laxative, kidney and bladder stones	5, 6, 15
181	*Parietaria officinalis*	FB		Diuretic, softener	18
182	*Urtica dioica*	AP, LE, RO, SE, SH, WP	BO, DE, EF, IN, MA	Diabetes, cancer, hemorrhoids, urinary infection, alopecia, antypyretic, painkiller, high tension, diuretic, dyspnea, antirheumatism, orexigenic, blood purification, appetizer, hair health, stomach ache, kidney stones, inflammatory wounds, gynecological inflammations, stomachic, arthritis, colds, relieving edema, liver failure, asthma, gas removed	4-6, 9-12, 16, 18, 23, 43, 50, 57

Contd...

Table 24.2–Contd...

Sl.No.	Famiy/Taxa	Part Used	Preparation	Treatment	Source
	VERBENACEAE				
183	*Verbena officinalis*	LE, ST, WP	IN	Antipyretic, stimulant, menstrual flow and milk secretion, female sterility, liver diseases, stomach pain, fever and menstrual cramps, sedative, diuretic, diaphoretic, pain, fatigue, insomnia	5-7, 15, 27

PART USED: **AP**: Aerial parts; **BA**: Bark; **BR**: Branches; **BU**: Bulb; **CO**: Cones; **FB**: Flowering branches; **FL**: Flower; **FR**: Fruit; **FS**: Fronds; **GL**: Gall; **GU**: Gum; **HE**: Herb; **LA**: Latex; **LE**: Leaves; **OO**: Oil; **RB**: Rootbarks; **RH**: Rhizome; **RE**: Resin; **RO**: Root; **SE**: Seed; **SH**: Shoots; **ST**: Stem; **TR**: Tar; **TU**: Tuber; **UG**: Underground; **WP**: Whole plants.

PREPARATION: **AT**: As tea; **BO**: Boiled; **CO**: Cooked; **CR**: Crushed; **DE**: Decoction; **DR**: Dried; **EF**: Eaten fresh; **FJ**: Fruit juice; **IN**: Infusion; **MA**: Maceration; **OI**: Oinment; **PI**: Pill; **PJ**: Plant juice; **PO**: Poultice; **PW**: Powdered; **RW**: Raw; **SY**: Syrop.

SOURCE: **1**: Auda, 2012; **2**: Chopra *et al.*, 1958; **3**: Yusuf *et al.*, 1994; **4**: Akbulut and Bayramoğlu, 2013; **5**: Yücel, 2012; **6**: Yücel, 2014; **7**: Sayar *et al.*, 1995; **8**: Duran, 1998; **9**: Eºen, 2008; **10**: Fakir *et al.*, 2009; **11**: Saday, 2009; **12**: Öztürk *et al.*, 2013; **13**: Yeºilada *et al.*, 1995; **14**: Lev and Amar, 2002; **15**: Said *et al.*, 2002; **16**: Öztürk *et al.*, 2012; **17**: Tuzlacý, 2006; **18**: Baytop, 1984; **19**: Elias *et al.*, 1990; **20**: Trute *et al.*, 1997; **21**: Oran and Al-Eisawi, 1998; **22**: Hofmann *et al.*, 2003; **23**: Marc *et al.*, 2008; **24**: Afifi-Yazar *et al.*, 2011; **25**: Bulut, 2006; **26**: Karousou and Deirtmentzoglou, 2011; **27**: Al-Qura'n, 2009; **28**: Dafni *et al.*, 1984; **29**: Krispil, 1987; **30**: Suspluges *et al.*, 1995; **31**: Ali-Shtayeh *et al.*, 2000; **32**: Wang *et al.*, 2004; **33**: Azaizeh *et al.*, 2006; **34**: Aburjai *et al.*, 2007; **35**: Hudaib *et al.*, 2008; **36**: Talib and Mahasneh, 2010; **37**: Altay and Karahan, 2012; **38**: Chopra *et al.*, 1986; **39**: Bown, 1995; **40**: Moerman, 1998; **41**: Foster *et al.*, 2002; **42**: Arýcan *et al.*, 2013; **43**: Mart and Türkmen, 2008; **44**: González-Tejero *et al.*, 2008; **45**: Duke and Ayensu, 1985; **46**: Foster and Duke, 1990; **47**: Stuart, 1995; **48**: Manandhar, 2002; **49**: Al-Qudat and Qadir, 2011; **50**: Akaydýn *et al.*, 2013; **51**: Abay and Kýlýç, 2001; **52**: Al-Khalil, 1995; **53**: Keskin and Alpýnar, 2002; **54**: Yeung, 1985; **55**: Ali-Shtayeh *et al.*, 2011; **56**: Abu-Irmaileh and Afifi, 2003; **57**: Ýlçim and Varol, 1996; **58**: Grieve, 1984; **59**: Suwal, 1993; **60**: Arshad and Rao, 1998; **61**: Ali-Shtayeh and Jamous, 2008.

Table 24.3: Therapeutic Uses of the Medicinal and Aromatic Plant Taxa

Name of the Disease	Per cent (Per cent)
Urinary system disorders	18.49
Stomach disorders	10.79
Respiratory diseases	8.32
Skin disorders	7.40
Nervous system	3.93
Internal diseases	3.78
Pains	3.39
Cardiac diseases	3.16
Rheumatism	3.00
Cold and flu	2.93
Mouth, Teeth and throat diseases	2.70
Bone, joint and muscles diseases	2.62
Diabetes	2.39
Sedative	2.39
Antipyretic	2.00
Anthelmintic	1.85
Tonic	1.77
Antiseptic	1.77
Gynecological diseases (For women)	1.69
Stimulant	1.16
Hypertension, Hypotension, Tension	1.08
Diaphoretic	1.08
Hair health	1.08
Aphrodisiac	0.85
Scorpition bites	0.85
Cancer	0.77
Antispasmodic	0.77
Antibacterial, antifungal, antitumor and antiviral	0.77
Eye diseases	0.61
Gas removed	0.61
Jaundice	0.54
Cholesterol-lowering	0.46
For veterinary purposes	0.46
Others	4.54

requirements of more affluent consumers internationally. The possible solutions are; sustainable use of the natural resources of MAPS, domestication and introduction of different species into cultivation, which will provide reliable botanical identification and at the same time guarantee a steady source of raw material. Ex situ and in-situ conservational measures are needed to prevent our MAPS (Shinwari and Gilani, 2003; Shinwari and Qaiser, 2011). The global market of MAPS has increased and is expected to reach 5 trillion dollars by 2050 (Shinwari, 2010).

Native plant species occurring in alpine and subalpine areas are collected for medicinal, food and spices purposes as well as ornamentals and for other purposes by the local population especially shepherds. Some traditional knowledge on medicinal or other useful plants is still present and people with traditional medical skills are still available who frequently visit these areas. Today natural herbs still maintain their popularity (Grabherr, 2009). But, tourism, unsustainable agriculture and forestry, developments like hydroelectric dam constructions and several other anthropogenic activities like mountain and winter sports, mining as well as wrong land use practices such as animal grazing lead to increased erosion in the alpine and subalpine areas and degradation of habitats both at local and regional scales (Sarý, 2010).

Medicinal plants in this context need a priority within the scope of habitat protection.The solution is conservation efforts, informative activities of the planned cooperation with the local people in the context of sustainable development (Svetlana *et al.*, 2012). There is need for urgent protection of genetic resources at high altitudes in the EMED as well like other high altitude habitats. Determination of the ecological and economic values in such habitats in terms of sustainability should be the cornetstone of our future investigations as economic indicators (Costanza and Farber, 2002; Farber *et al.*, 2006).

Unless alternatives are developed to collect plants from nature including the endemic and rare plants in particular, and "the nature and species protection" is not understood well, the laws cannot be applied effectively. This alternative application must not be an application valid for the that are in the wild. Many native plants are facing the danger of extinction with a decline in their populations in natural habitats due to gathering and overconsumption. There is an urgent need for developing alternative applications for these resources. A cultivation of these at alarge scale in future will be very much helpful to protect this genetic heritage (Bayram *et al.*, 2010; Öztürk *et al.*, 2011). Different uses of MAPS need to be well organized on long-term basis among the industry and related stakeholders, based on their production requirements. We must also think for future in the light of climate change scenarios. like impacts from drought, flooding, erosion and other natural disasters, ecosystem viability and sustainable land management, market preferences and demand trends, genetic resources and biological diversity, variety development work, organic products, planning related to the production and trade, establishment of researchers inventory and collaboration platform, developing a close cooperation among different research groups (Bayram *et al.*, 2010; Ozturk *et al.*, 2015). The loss of soil and water resources due to climate change, future demographic outburst will result in the reduction of economic plants as well (Svetlana *et al.*, 2012). Climate change will affect the world's mountainous areas equally, as documented in the Scandes, Ural and Balcanic mountains (Meshinev *et al.*, 2000; Kullman, 2003; Moiseev and Shiyatov, 2003; Öztürk *et al.*, 2011, 2012).

In fact the studies related to the exploration of the potential of natural MAPS will offer a great opportunity for human welfare (Kala and Ratajc, 2012). Alpine and subalpine areas that are valuable and sensitive ecosystems in terms of biodiversity are recognized as one of the topics of awareness in many countries. Even there are organizations engaged in research related to the alpine and subalpine biodiversity (Sarý, 2010). However, these organizations should get linked with each other for better evaluations for our future generations.

5. REFERENCES

1. Abay, G., Kılıç, A., 2001. Pürenbeleni ve Yanıktepe (Mersin) yörelerindeki bazı bitkilerin yöresel adları ve etnobotanik özellikleri. *Ot Sistematik Botanik Dergisi*, 8 (2): 97-104.

2. Abu-Irmaileh, B.E., Afifi, F.U., 2003. Herbal medicine in Jordan with special emphasis on commonly used herbs. *Journal of Ethnopharmacology*, **89**: 193-197.

3. Aburjai, T., Hudaib, M., Tayyem, R., Yousef, M., Qishawi, M., 2007. Ethnopharmacological survey of medicinal herbs in Jordan, the Ajloun Heights region. *Journal of Ethnopharmacology*, **110**: 294-304.

4. Afifi-Yazar, F.U., Kasabri, V., Abu-Dahab, R., 2011. Medicinal plants from Jordan in the Treatment of Cancer: Traditional uses vs. In vitro and In Vivo evaluations-Part 1. *Planta Med.*, **77**: 1203-1209.

5. Akaydın, G., Şimşek, I., Arıtuluk, Z.C., Yeşilada, E., 2013. An ethnobotanical survey in selected towns of the Mediterranean subregion (Turkey). *Turkish Journal of Biology*, 37: 230-247.

6. Akbulut, S., Bayramoðlu, M.M., 2013. The trade and use of some medicinal and aromatic herbs in Turkey. *Ethno Med.*, **7** (2): 67-77.

7. Ali-Shtayeh, M.S., Yaniv, Z., Mahajna, J., 2000. Ethnobotanical survey in the Palestinian area: a classification of the healing potential of medicinal plants. *Journal of Ethnopharmacology*, **73**: 221-232.

8. Ali-Shtayeh, M.S., Jamous, R.M., 2008. Traditional Arabic Palestinian Herbal Medicine. TAPHM. Biodiversity and Environmental Research center (BERC). Till. Nablus. Palestine. p. 221.

9. Ali-Shtayeh, M., Jamous, R.M., Jamous, R.M., 2011. Herbal preparation use by patients suffering from cancer in Palaestine. *Complementary Therapies in Clinical Practice*, **17**: 235-240.

10. Al-Khalil, S.A. 1995. Survey of plants used in Jordanian traditional medicine. *Int J Pharmacognosy*, **33**: 317-323.

11. Al-Qudat, M., Qadir, M., 2011. The halophytic flora of Syria. International Center for Agricultural Research in the Dry Areas, Aleppo, Syria. pp. 1-186.

12. Al-Quran, S., 2009. Ethnopharmacological survey of wild medicinal plants in Showbak, Jordan. *Journal of Ethnopharmacology*, **123**: 45-50.

13. Al-Qura'n, S.A., 2010. Ethnobotanical and ecological studies of wild edible plants in Jordan. *Libyan Agriculture Research Center Journal International*, 1 (4): 231-243.

14. Altay, V., Karahan., F., 2012. Tayfur Sökmen Kampüsü (Antakya-Hatay) ve çevresinde bulunan bitkiler üzerine etnobotanik bir araþtýrma. *The Black Sea Journal of Sciences*, 2 (7): 13-28.

15. Anderson, D., Salick, J., Moseley, R.K., Xiaokun, O.,. 2005. Conserving the sacred medicine mountains: a vegetation analysis of Tibetan sacred sites in Northwest Yunnan. *Biodiversity and Conservation*, 14: 3065-3091.

16. Arýcan, Y.E., Yeþil, Y., Genç, G.E., 2013. A preliminary ethnobotanical survey of Kumluca (Antalya). *J. Fac. Pharm. Istanbul*, 43 (2): 95-102.

17. Arshad, M., Rao, A., 1998. Medicinal plants of Cholistan desert. Cholistan Institute of Desert Studies, Islamic University of Bahwalpur, Bahwalpur.

18. Arya, P.Y. 2010. The Tibetan medicinal herbs growing in the European Alps. (http://www.tibetanmedicineedu.org. (accessed 10 Fep. 2015).

19. Atalay, I., 2004. Mountain Ecosystems of Turkey. Proc. of the 7th International Symposium on High Mountain Remote Sensing Cartography, Bishkek, Kyrgyzstan July, 2002. Institute for Cartography Dresden University of Technology, Germany, 29-38.

20. Atalay, I., 2006. The Effects of Mountainous Areas on Biodiversity: A Case Study from the Northern Anatolian Mountains and the Taurus Mountains. Grazer Schriften der Geographie und Raumforschung. Band, 41: 17-26.

21. Atalay, I., 2008. Ekosistem Ekolojisi ve Coðrafyasý. Çevre ve Orman Bakanlýðý Yay. No: 327, Cilt: I-II, Meta Basým, Ýzmir, Türkiye.

22. Atalay, I., Efe, R., M. Öztürk, M., 2014. Effects of topography and climate of the ecology of Taurus Mountains in the Mediterranean Region of Turkey. *Procedia - Social and Behavioral Sciences*, 120: 142-156.

23. Atay, S., Güleryüz, G., Orhun, C., Seçmen, Ö., Vural, C., 2009. Daðlarýmýzdaki Zenginlik Türkiye'nin 120 Alpin Bitkisi. Dönence Basým ve Yayýn Hizmetleri, Ýstanbul, Türkiye.

24. Auda, M.A, 2012. Medicinal plant diversity in the flora Gazza Valley, Gaza Strip, Palestine. *An-Najah Univ J. Res*. (N. Sc.), 26: 61-84.

25. Azaizeh, H., Saad, B., Khalil, K., Said, O., 2006. The state of the art of traditional Arab Herbal Medicine in the Eastern region of the Mediterranean: A Review. eCAM, 3(2): 229-235.

26. Bahn, M., Körner, C., 2003. Recent increases in summit flora caused by warming in the Alps. *In: Nagy, L., Grabherr, G., Körner, C., Thompson, D.B.A. (eds.)*, Alpine Biodiversity in Europe-A Europe-wide Assessment of Biological Richness and Change Ecological Studies, Springer, Berlin, Vol. 167, pp. 437-441.

27. Bayram, E., Kýrýcý, S., Tansý, S., Yýlmaz, G., Arabacý, O., Kýzýl, S., Telci, Ý., 2010. Týbbi ve Aromatik Bitkiler Üretimi-nin Arttýrýlmasý Olanaklarý. Türkiye

Ziraat Mühendisliði VII. Teknik Kongresi Bildiriler Kitabý - I., pp. 437-456, Ankara.

28. Baytop, T., 1984. Therapy with medicinal plants in Turkey (past and present), Publication of the Istanbul University, No: 3255. Ýstanbul Üniversitesi, Istanbul, Türkiye.

29. Bown, D. 1995. Encyclopaedia of herbs and their uses. Dorling Kindersley, London.

30. Bulut, Y., 2006. Manavgat (Antalya) yöresinin faydalý bitkileri. Süleyman Demirel Üniversitesi Fen Bilimleri Enstitüsü, Biyoloji Anabilim Dalý, Yüksek Lisans Tezi. Isparta-Türkiye.

31. Casazza, G., Barberis, G., Minutol., L., 2005. Ecological Characteristics and rarity of endemic plants of the Italian maritime Alps. *Biological Conservation*, **123** (3): 361-371.

32. Chopra, R.N., Chopra, I.C., Handa, K.L., Kapur, L.D., 1958. Indigenous drugs of India. Calcutta: Academic Publishers.

33. Chopra, R.N., Nayar, S.L., Chopra, I.C., 1986. Glossary of Indian medicinal plants. Council of Scientific and Industrial Research, New Delhi.

34. Costanza, R., Farber, S., 2002. Introduction to the special issue on the dynamics and value of ecosystem services: integrating economic and ecological perspectives. *Ecological Economics*, **41**: 367-373.

35. Dafni, A., Yaniv, Z., Palevitch, D., 1984. Etnobotanical survey of medicinal plants in Northern Israel. *J Ethnopharmacol*, **10**: 295-310.

36. Dallman, P.R., 1998. Plant life in the World's Mediterranean climates. University of California Press, California, p. 258.

37. Duke, J.A., Ayensu, F.S., 1985. Medicinal plants of China. Publications Inc., Algonac. ISBN 0-917256-20-4.

38. Duran, A., 1998. Akseki (Antalya) ilçesindeki bazý bitkilerin yerel adlarý ve etnobotanik özellikleri. Ot Sistematik Botanik Dergisi. **5**(1): 77-92.

39. Duran, C., 2013. Türkiye'nin Bitki Çeþitliliðinde Daðlýk Alanlarýn Rolü. Biyoloji Bilimleri Araþtýrma Dergisi, **6**(1): 72-77.

40. Elias, R., De Meo, M., Vidal-Ollivier, E., Laget, M., Balansard, G., Dumenil, G., 1990. Antimutagenic activity of some saponins isolated from *Calendula officinalis* L., *C. arvensis* L. and *Hedera helix* L. Mutagenesis, **5**: 327-331.

41. Erschbamer, B., Mallaun, M., Unterluggauer, P., 2006. Plant diversity along altitudinal gradients in the Southern and Central Alps of South Tyrol and Trentino Italy. *Gredleriana*, **6**: 47-68.

42. E°en, B., 2008. Aydýnlar Köyü ve Çevresinin (Erdemli/Mersin) Etnobotanik Özellikleri. Selçuk Üniversitesi Fen Bilimleri Enstitüsü Yüksek Lisans Tezi Biyoloji Anabilim Dalý. Konya-Türkiye.

43. Fakir, H., Korkmaz, M., Güller, B., 2009. Medicinal plant diversity of Western Mediterranean Region in Turkey. *Journal of Applied Biological Sciences*, 3(2): 33-43.

44. Farber, S., Costanza, R., Childers, D.L., 2006. Linking ecology and economics for ecosystem management. *BioScience*, 56: 121-133.

45. Foster, S., Hobbs, C., Peterson, R.T., 2002. A field guide to Western medicinal plants and herbs. Houghton Miffl in Harcourt. Perterson Field Guide, Mifflin.

46. Foster, S., Duke, J.A.A., 1990. Field guide to medicinal plants of Eastern and Central North America. Houghton Miffl in Co, Boston.

47. Gairola, S., Shariff, N.M., Bhatt, A., Kala, C.P., 2010. Influence of climate change on production of secondary chemicals in high altitude medicinal plants: Issues needs immediate attention. *Journal of Medicinal Plants Research*, 4 (18): 1825-1829.

48. Gemici, Y., G. Görk, G., Acar, I., 1994. Batý ve Güney Anadolu Yüksek Dağ Vejetasyonu ve Florasý, I. Vejetasyon. TUBITAK projesi, TBAG-993, Ankara.

49. González-Tejero, M.R., Casares-Porcel, M., Sánchez-Rojas, C.P., Ramiro-Gutiérrez, J.M., Molero-Mesa, J., Pieroni, A., Giusti, M.E., *et al.*, 2008. Medicinal plants in the Mediterranean area: Synthesis of the results of the Project Rubia. *Journal of Ethnopharmacology*, 116: 341-357.

50. Grabherr, G., 2009. Biodiversity in the high ranges of the Alps: Ethnobotanical and climate change perspectives. *Global Environmental Change*, 19: 167-172.

51. Grabherr, G., Nagy, L., Thompson, D.B.A., 2003. An outline of Europe's Alpine areas. *In: Nagy, L., Grabherr, G., Körner, Ch., Thompson, D.B.A. (eds.),* Alpine Biodiversity in Europe. Springer, Berlin, pp. 3-12.

52. Grieve, A., 1984. Modern herbal. Penguin, London.

53. Guleryuz, G., Gucel, S., Ozturk, M., 2010. Nitrogen mineralization in a high altitude ecosystem in the Mediterranean phytogeographical region of Turkey. *The Journal of Environmental Biology*, 31: 503-514.

54. Heywood, V.H. (ed). 1995a. Global diversity assessment. Cambridge University Press, Cambridge.

55. Heywood, V.H., 1995b. The Mediterranean flora in the context of world diversity. Ecol Mediterr, 21: 11-18.

56. Heywood, V.H., 2003a. The future of floristics in the Mediterranean region. Isr J Plant Sci, 50: 5-13.

57. Heywood, V.H., 2003b. Mediterraneans plant collections: need and options. Setting the scene: what we have inherited. Bocconea, 16: 283-287.

58. Hofmann, D., Hecker, M., Volp, A., 2003. Efficacy of dry extract of ivy leaves in children with bronchial asthma-a review of randomised controlled trials. *Phytomedicine*, 10: 213-220.

59. Hudaib, M., Mohammad, M., Bustanji, Y., Tayyem, R., Yousef, M., Abuirjeie, M., Aburjai, T., 2008. Ethnopharmacological survey of medicinal plants in Jordan,

Mujib Nature Reserve and surrounding area. *Journal of Ethnopharmacology*, **120**: 63-71.

60. Ýlçim, A., Varol, Ö., 1996. Hatay ve K. Maraþ (Türkiye) illerindeki bazý bitkilerin etnobotanik özellikleri. *Ot Sistematik Botanik Dergisi*. **3**(1): 69-74.

61. Kadereit, J.W., Licht, W., Uhink, C.H., 2008. Asian relationships of the flora of the European Alps. *Plant Ecol Divers.*, **1**(2): 171-179.

62. Kala, C.P., 2002. Medicinal plants of Indian trans-Himalaya: focus on Tibetan use of medicinal resources. Bishen Singh Mahendra Pal Singh, Dehradun, p. 200.

63. Kala, C.P., 2005. Indigenous uses, population density, and conservation of threatened medicinal plants in protected areas of the Indian Himalayas. *Conserv Biol.*, **19** (2): 368-378.

64. Kala, C.P., 2008. High altitude medicinal plants: A promising resource for developing herbal sector. *Hima-Paryavaran*, **20**(2): 7-9.

65. Kala, C.P., Mathur, V.B., 2002. Patterns of plant species distribution in the trans-Himalayan region of Ladakh. *India J Veg Sci*, **13**(6): 751-754.

66. Kala, C.P., Ratajc, P., 2012. High altitude biodiversity of the Alps and the Himalayas: ethnobotany, plant distribution and conservation perspective. *Biodivers Conserv.*, **21**: 1115-1126.

67. Karousou, R., Deirmentzoglou, S., 2011. The herbal market of Cyprus: Traditional links and cultural exchanges. *Journal of Ethnopharmacology*, **133**: 191-203.

68. Keskin, M., Alpýnar, K., 2002. Kýþlak (Yayladaðý-Hatay) hakkýnda etnobotanik bir araþtýrma. Ot Sistematik Botanik Dergisi. **9**(2): 91-100.

69. Khalil, A.T., Khan, I., Ahmad, K., Khan, Y.A., Khan, J., Shinwari, Z.K., 2014a. Antibacterial activity of honey in northwest Pakistan against select human pathogens. *J. Tradit. Chin. Med.*, **34**: 86-89.

70. Khalil, A.T., Shinwari, Z.K., Qaiser, M., Marwat, K.B., 2014b. Phyto-therapeutic claims about Euphorbeaceous plants belonging to Pakistan; An ethnomedicinal review. *Pak. J. Bot.*, **46**(3): 1137-1144.

71. Klanderud, K., Birks, H.J.B., 2003. Recent increases in species richness and shifts in altitudinal distributions of Norwegian mountain plants. *The Holocene*, **13**: 1-6.

72. Körner, C., 1995. Alpine Plant Diversity: A Global Survey and Functional Interpretations. *In: Chapin, F.S., Körner, C. (eds.)*, Arctic and Alpine Biodiversity: Patterns, Causes and Ecosystem Consequences, *Ecological Studies* 113, Springer, Berlin, pp. 45-62.

73. Körner, Ch., 2003. Alpine Plant Life: Functional Plant Ecology of High Mountain Ecosystems, 2nd ed. Springer, Berlin.

74. Körner, C., Spehn, E.M., 2002. Mountain Biodiversity: A Global Assessment. Parthenon Publishing, London, New York.

75. Krispil, N., 1987. The medicinal and useful plants of Palaestina. Yara Publishing House, Jerusalem.

76. Kullman, L., 2003. Recent reversal of neoglacial climate cooling trend in the Swedish Scandes as evidenced by birch tree-limit rise. Global and Planetary Change, **36**: 77-88.

77. Kürschner, H., Parolly, G., Raab-Straube, E.V., 1998. Phytosociological studies on high mountain plant communities of the Taurus Mountains (Turkey). 3. Snow-patch and meltwater communities. *Feddes Repertorium*, **109** (7-8): 581-616.

78. Lev, E., Amar, Z., 2002. Ethnopharmacological survey of traditional drugs sold in the Kingdom of Jordan. *Journal of Ethnopharmacology*, **82**: 131-145.

79. Manandhar, N.P., 2002. Plants and people of Nepal. Timber Press, Oregon.

80. Marc, E.B., Nelly, A., Annick, D.D., Frederic, D., 2008. Plants used as remedies antirheumatic and antineuralgic in the traditional medicine of Lebanon. *Journal of Ethnopharmacology*, **120**: 315-334.

81. Mart, S., Türkmen, N., 2008. Bahçe ve Hasanbeyli (Osmaniye) bölgesinin etnobotanik kültürü. *Ot Sistematik Botanik Dergisi*. **15**(2): 137-150.

82. Meshinev, T., Apostolova, I., Koleva, E., 2000. Influence of warming on timberline rising: a case study on *Pinus peuce* Griseb. in Bulgaria. *Phytocoenologia*, **30**: 431-438.

83. Moerman, D., 1998. Native American ethnobotany. Timber Press, Oregon.

84. Moiseev, P.A., Shiyatov, S.G., 2003. Vegetation dynamics at the treeline ecotone in the Ural highlands, Russia. *In: Nagy, L., Grabherr, G., Körner, Ch., Thompson, D.B.A. (eds.)*, Alpine Biodiversity in Europe. Springer, Berlin, pp. 3-12.

85. Oran, S.A., Al-Eisawi, D.M., 1998. Check-List of Medicinal Plants in Jordan. *Medicinal and Biological Sciences*, **25**(2): 84-112.

86. Ozenda, P., Borel, J-L., 2003. The Alpine vegetation of the Alps. *In: Nagy, L., Grabherr, G., Körner, Ch., Thompson, D.B.A. (eds.)*, Alpine Biodiversity in Europe. Springer, Berlin, pp. 53-64.

87. Öztürk, M., Gemici, Y., Seçmen, Ö., Görk,G., 1990. The mountain flora and vegetation of Mediterranean part of Turkey. Third Plant Life of Southwest Asia Symposium, 3-8 September, Berlin-Germany.

88. Ozturk, M., Gemici, Y., Gork, G., Seçmen, O., 1991. A general account of high mountain flora and vegetation of mediterranean part of Turkey. *Ege Univ. Sci. Fac. Jour.*, **13**: 51-59.

89. Ozturk, M., Seçmen, Ö., Gork, G. (eds). 1996a. Plant life in Southwest and Central Asia. Ege University Press, Izmir, Vol 1, pp. 1-499.

90. Ozturk, M., Seçmen, Ö., Gork, G. (eds). 1996b. Plant life in Southwest and Central Asia. Ege University Press, Izmir, Vol 2, pp. 500-1093.

91. Ozturk, M., Gucel, S., Sakcali, S., Gork, Ç., Yarci, C., Gork, G., 2008. An overview of plant diversity and land degradation interactions in the eastern

Mediterranean. *In: Efe, R., et al. (eds)*, Natural environment and culture in the Mediterranean region. Cambridge Scholars Publisher, Cambridge, pp. 215-239.

92. Öztürk, M., Gücel, S., Altundað, E., Çelik, S., 2011. Turkish Mediterranean Medicinal Plants in the Face of Climate Change. *In: Ahmad, A. et al. (eds.)*, Medicinal Plants in Changing Environment, ISBN: 81-85589-14-3. New Delhi.

93. Ozturk, M., Gucel, S., Altundag, E., Mert, T., Gork, C., Gork, G., Akcicek, E. 2012. An Overview of The Medicinal Plants of Turkey. *In.: Singh, R.H. (ed)*, Genetic resources, chromosome engineering, and crop improvement, CRC Press, Boca Raton, Vol.: 6, pp. 181-206.

94. Öztürk, M., Uysal, I., Gücel, S., Altundað, E., Doðan, Y., Baþlar, S., 2013. Medicinal uses of natural dye-Yielding plants in Turkey. *RJTA*, **17** (2): 69-80.

95. Ozturk,M.,Hakeem,K.R.,Faridah-Hanum,I.,Efe,R. (Eds.) 2015. Climate Change Impacts on High-Altitude Ecosystems. Springer Science+Business Media, NY,736 pp.

96. Parolly,G. 2015. The High-Mountain Flora and Vegetation cf the Western and Central Taurus Mts. (Turkey) in the Times cf Climate Change. In: Climate Change Impacts on High-Altitude Ecosystems (Eds.Ozturk *et al.*), Springer Science+Business Media, NY, pp: 99-133.

97. Pauli, H., Gottfried, M., Reiter, K., Ch. Klettner, Ch., Grabherr, G., 2007. Signals of range expansions and contractions of vascular plants in the high Alps: Observations 1994-2004 at the GLORIA* master site Schrankogel, Tyrol, Austria. *Global Change Biology*, **13**: 147-156.

98. Pickering, C.M., Armstrong, T., 2003. Potential impact of climate change on plant communities in the Kosciuszko alpine zone. *Victorian Naturalist*, **120**: 263-272.

99. Price, M.F., G.R. Neville, G.R., 2003. Designing strategies to increase the resiliance of alpine/montane systems to climate change. *In: Hansen, L., Biringer, J., Hoffmann, J. (Eds.)*, Buying Time: A User's Manual for Building Resistance and Resilience to Climate Change in Natural Systems. WWF International, Gland, pp. 73-94.

100. Saday, H., 2009. Güzeloluk Köyü ve Çevresinin (Erdemli/Mersin) Etnobotanik Özellikleri. Selçuk Üniversitesi Fen Bilimleri Enstitüsü Yüksek Lisans Tezi, Biyoloji Anabilim Dalý. Konya-Türkiye.

101. Said, O., Khalil, K., Fulder, S., Azaizeh, H., 2002. Ethnopharmacological survey of medicinal herbs in Israel, the Golan Heights and the West Bank region. *Journal of Ethnopharmacology*, **83**: 251-265.

102. Salick, J., Byg, A., Amend, A., Gunn, B., Law, W., Schmidt, H., 2006. Tibetan medicine plurality. *Economic Botany*, **60**: 227-253.

103. Salick, J., Zhendong, F., Byg, A., 2009. Eastern Himalayan Alpine Plant Ecology, Tibetan Ethnobotany, and Climate Change. *Global Environmental Change*, **19**: 147-155.

104. Sarý, D., 2010. Biyoçeþitlilik ve Floristik Çeþitlilik Açýsýndan Alpin Alanlarýn Önemi. III. Ulusal Karadeniz Ormancýlýk Kongresi, 20-22 Mayýs 2010, Vol: IV, pp. 1447-1455.

105. Sayar, A., Güvensen, A., Özdemir, F., Öztürk, M., 1995. Muðla (Türkiye) ilinde bazý türlerin etnobotanik özellikleri. *Ot Sistematik Botanik Dergisi.* **2** (1): 151-160.

106. Shrestha, P.M., Dhillion, S.S., 2003. Medicinal plant diversity and use in the highlands of Dolakha district, Nepal. *J. Ethnopharmacol.*, **86**: 81-96.

107. Shinwari, Z.K., 2010. Medicinal plants research in Pakistan. *J. Medi. Plants. Res.*, **4**: 161-176.

108. Shinwari, Z.K., Gilani, S.S, 2003. Sustainable harvest of medicinal plants at Bulashbar Nullah, Astore (Northern Pakistan). *J. Ethnopharmacol.*, **84**: 289-298.

109. Shinwari, Z.K., Qaiser, M., 2011. Efforts on Conservation and Sustainable Use of Medicinal Plants of Pakistan. *Pak. J. Bot.*, **43** (SI): 5-10.

110. Stuart, G.A., 1995. Chinese materia medica. Encyclopedia of herbs and their uses. Southern Materials Centre, Taipei.

111. Suspluges, C., Balansard, G., Rossi J.S., *et al.*, 1995. Evidence of anathematic action of aerial parts from *Inula viscosa. ISHS, Acta Hortic.*, **96**.

112. Suwal, P.N., 1993. Medicinal plants of Nepal. Department of Medicinal plants, Ministry of Forest and Soil Conservation, Thapathali, Nepal.

113. Svetlana, A.P., Atayevna, M.G., Ozturk, M., Gucel, S., Ashyraliyeva, M., 2012. An overview of the ethnobotany of Turkmenistan and use of *Juniperus turcomanica* in phytotherapy. In.: Singh, R.H. (ed.), *Genetic resources, chromosome engineering, and crop improvement*, CRC Press, Boca Raton, Vol.: 6, pp. 207-220.

114. Talib, W.H., Mahasneh, A.M., 2010. Antiproliferative activity of plant extracts used against cancer in traditional medicine. *Sci Pharm.*, **78**: 33-45.

115. Tecimen, H.B., Sevgi, O., Altundað, E., 2012. Kaz Daðlarýnda Yükseltiye Baðlý Azot Minerallеþmesinin Deðiþimi. *Journal of the Faculty of Forestry Istanbul University*, **62** (1): 19-29.

116. Theurillat, J.-P., Guisan,A., 2001. Potential impacts of climate change on vegetation in the European Alps: a review. *Climatic Change*, **50**: 77-109.

117. Trute, A., Gross, J., Mutschler, E., Nahrstedt, A., 1997. In vitro antispasmodic compounds of the dry extract obtained from *Hedera helix. Planta Med.*, **63**: 125-129.

118. Tuzlacý, E., 2006. Þifa Niyetine-Türkiye'nin Bitkisel Halk ilaçlarý. Alfa Yayýnlarý, Ýstanbul, Türkiye.

119. Vare, H., Lampinen, R., C. Humphries,C., P. Williams, P., 2003. Taxonomic diversity of vascular plants in the European Alpine areas. *In: Nagy, L., Grabherr, G., Körner, Ch., Thompson, D.B.A. (eds.)*, Alpine Biodiversity in Europe. Springer, Berlin, pp. 133-148.

120. Walther, G.-R., Beißner, S., Burga, C.A., 2005. Trends in upward shift of alpine plants. *Journal of Vegetation Science*, **16**: 541-548.

121. Wang, W., Ben-Daniel, B.H., Cohen, Y., 2004. Control of plant disease by extracts of *Inula viscosa. Phytopathological*, **94**: 1042-1047.

122. Yeşilada, E., Honda, G., Sezik, E., Tabata, M., Fujita, T., Tanaka, T., Takeda, Y., Takaishi, Y., 1995. Traditional medicine in Turkey. V. Folk medicine in the inner Taurus Mountains. *Journal of Ethnopharmaecology*, **46**: 133-152.

123. Yeung, H.C., 1985. Handbook of Chinese herbs and formulas. Institute of Chinese Medicine, Los Angeles.

124. Yusuf, M., Chowdhury, J.U., Wahab, M.A., Begum, J., 1994. Medicinal plants of Bangladesh. Chittagong Bangladesh Council for Science and Research (BCSIR).

125. Yücel, E., 2012. Tıbbi Bitkiler 1 (A-L). Cetemenler Dijital, Eskişehir. ISBN: 978-975-93746-3-1.

126. Yücel, E., 2014. Growing in Turkey guide to medicinal plants. Türmatsan Organize Matbaacılık, Eskişehir. ISBN: 978-975-93746-8-6.

Chapter 25

Traditional Herbs and Allergy: An Investigation from Vietnam

Nguyen Van Hung[1] and Nguyen Van Doan[2]

[1]MD, PhD. Associate Professor and Dean,
Faculty of Pharmacy,
Haiphong University of Medicine and Pharmacy
E-mail: nvhung@hpmu.edu.vn
[2]MD, PhD, Associate Professor and Director
The Center of Allergology – Clinical Immunology, Bach Mai Hospital

ABSTRACT

With market- economy oriented policies, dramatic changes in Vietnam's health care system during the last two decades have led to the exploitation of the drug market which include western medicines and traditional medicines or herbs. Drug allergies are symptoms manifested by allergic reactions to a drug; and the usage of traditional medicine/herbs is not an exception. Recently, many life-threatening acute allergic reactions (anaphylaxis) after taking herbs were reported. However, there has been a limited effort to report/analyze the situation of traditional medicines-related complications in general and allergy in particular. Community awareness on the possibility of herb- induced adverse reactions should be based on evidence as many of the cases were wrongly interpreted.

After reviewing 119 cases admitted to Bachmai Hospital in the past 10 years, we found that it was unclear whether traditional medicine caused the allergic reactions, or whether these reactions were caused by other drugs/substances. Patients visiting traditional healers were often given "Packets", for diseases with unclear reasons (tired, insomnia…) which were often linked to drug abuse. In 41 serious cases patients presented late after using traditional medicines (11,6 days), with complicated syndromes, including Stevens-Johnson and Lyell syndromes. Skin lesions, systemic rash, hepatitis and nephritis were common symptoms. It was not clear in many cases which traditional medicines could be the cause as often combined/mixed "packet" of herbal products were given to these patients. Self-medication was another problem as this resulted in the usage of different medicines. Syndrome/symptoms were not specific manifestations related to herbs only, as contamination with chemical preservatives (sulfur, phosphorus, desiccants, mold…) also occurred.

Traditional herbs caused serious ADRs in some cases only. However, traditional herbs should be used with greater caution. More research is needed to prove which herbs and/or substances are really causing allergies in order to avoid serious complications.

Keywords: Traditional medicine, Herbs, Allergy, Vietnam.

1. INTRODUCTION

Normally, the perception about "natural therapy" that it is safe. However allergic and toxic reactions to complementary and alternative medicines including herbs are described. "Vietnamese Herbs Cure Vietnamese people" was stated by Tue Tinh, the most well-known Vietnamese traditional physician more than 700 years ago. Allergy caused by traditional medicines was not recognized in Vietnam before the introduction of Western Medicine after the 19th century. Allergy due to traditional herbs was rarely mentioned in Vietnam before "Doimoi" in 1986.

With the market- economy oriented policies, dramatic changes in Vietnam's health care system during the last two decades have led to the exploitation of a drug market which includes western medicines and traditional medicines or herbs. Drug allergies are symptoms manifested by allergic reactions to a drug, and the usage of traditional medicine/herbs is not an exception. Recently, many life-threatening acute allergic reactions (anaphylaxis) after taking herbs were reported. However, there has been a limited effort to report/analyze the situation of traditional medicines-related complications in general and allergy in particular. Community awareness on the possibility of herb- induced adverse reactions should be based on evidence as many of the cases were wrongly articulated. There is a great need to analyze typical herb-related cases of ADRs in order to provide evidence and recommendations for the safer use of traditional medicine in the country.

This research aims at answering the question whether herbs were really the cause of herb-reported ADRs in the past10 years of the most rapidly changing times in the history of Vietnam.

2. MATERIALS AND METHODS

A retrospective study design was applied to analyze patients who were admitted to the hospital diagnosed as traditional medicine-related ADRs.

119 cases admitted to Bachmai[1] Hospital diagnosed as herb-related ADRs over a 10 year period (1990-1999) were analyzed. Herb-related ADRs, especially serious allergic cases, were also reviewed from different sources. The main areas for investigation are:

☆ General features and key symptoms and signs

---

1 Bach Mai hospital is the largest general hospital in the North of Vietnam located in the center of Hanoi which receives most of serious cases from different provinces.

☆ Reasons for admission, what kind of herbs were used and for which diseases

☆ Progress of their treatment at the hospital

☆ Possible causes of the complications (ADRs, especially allergies)

☆ Typical cases admitted to the hospital in the period 2000-2010

3. RESULTS AND DISCUSSION

General Features

Reviewing 119 cases admitted to Bachmai Hospital during 10 years when the country had the fastest change after the renovation (1990-1999), we found out that

☆ There are many cases which were unclear whether traditional medicine caused allergies or other drugs/substances abused by traditional healers for their patients. Patients visiting traditional healers were often given **"Packets"**, for diseases with unclear reasons (tired, insomnia, asthma, arthritis...) which were often linked to use of specific drugs/substances (prednisolone or dexamethasone). Different kinds of traditional medicines which included some types of remedies were also found:

☐ Traditional medications,

☐ Herbal medicine,

☐ Chinese herbs or unknown packets,

☐ Packet of Powder

Traditional herbs' "packets" were a major concern for our research as there was no evidence that these packets contained only herbs, or were mixed with additional Western Medications

What are the Possible Causes of these Allergies/Complications?

It was surprising that, there was no clear answer in many cases as a consequence of the complexity of traditional medicines. The packets given to the patients often combined/mixed "packets" rather than pure herbal medicines.

In case of self-medication by patients, the practice was often linked with the use of different medicines (including western medication/drugs).

Syndrome/symptoms were not specific manifestations related to herbs only.

Contamination with chemical preservatives (sulfur, phosphorus, desiccants, mold...) was repeatedly reported as raising awareness for the people. However, there was no evidence and yet it was reported as a contaminant.

Discussion

Traditional herbs cause serious ADRs. Yes, may be. However, it could be proved in some cases only. In other cases, ADRs linked to abuse of "Western" medicines were found. In particular, we were not sure in many cases because of the various combinations between different drugs and/or contaminated substances that were being used. Other herb-related complications may not be reported. However, traditional herbs should be used with greater caution. One should be careful with a "packet of powder" as these are not always traditional herbs. More research is needed to prove which herbs and/or substances are really causing the allergic reaction to avoid serious complications.

Mr. H. said: "*My son and I suffered from neuralgia; we visited a traditional healer and was provided traditional medicines. After taking 10 packets, my son found vesicles and blisters on his hands that spread quickly throughout the body. I also took these packets but I was not affected in the same way as my son, except that I had a rash in some areas. When my son's situation became more severe, I asked the healer and he told me that my son would recover due to the severity of his condition! I did not believe him and took my son to the district hospital where doctors advised me to send him to Bach Mai hospital as soon as possible!*"

These symptoms were linked to the use of "packets" given by traditional healers as adverse reactions/side-effects.

However, there are many possible causes which may or may not relate to the use of traditional herbal medicines:

☆ These herbal remedies caused adverse reactions due to their toxic reactions or allergens

☆ These herbs in "packets" were contaminated by toxic substances

☆ "Packets" given by healers may have added conventional drugs, like corticosteroids, pain killers or other medicines

It's difficult to find evidence about the toxic substances in the "packets" as the cause of the situation of Mr. H's son. The major problem in this case was that the contents of the "packet" were not disclosed and we could not find which herbs were used for this patient.

Table 25.1: Number of Herb-Related ADRs in the Fastest Changing Period of Vietnam

Period	Reported Cases at the Clinic	Admitted to the Hospital (in patients)
1990-1994	32	18
1995-1999	87	41
Total	110	59

Among 110 cases, 41 serious cases were analyzed and they often presented to the hospitals late after using traditional medicines (11,6 days), with complicated syndromes including Stevens-Johnson and Lyell syndromes. Skin lesions, systemic rash, hepatitis and nephritis were common symptoms. Patients with specific allergic diseases such as asthma are more likely than others to become sensitive to different allergens in herbal medicines.

Herbal Medicines for which Diseases?

Typically, a patient with an unknown disease, such as being tired or with insomnia..., would come to visit a traditional healer asking for an examination and treatment. They would often be given several "packets" of traditional medicines, normally herbal medicine or sometimes in a powder form.

There are some common diseases that are often seen by traditional healers, including asthma, arthritis and in these situation, most of the cases are linked with possible abuse of dexamethasone.

One of the common features of herbal medicine-related ADRs was that the patients would suffer from conventional "Western Medicines", especially corticosteroid abuse:

Chu Trong D, 46 years old who during the last 3 years was provided with packets of white powder by a traditional healer for asthma.

He was admitted to the hospital in 28/09/2006 with clinical symptoms as shown in his picture.

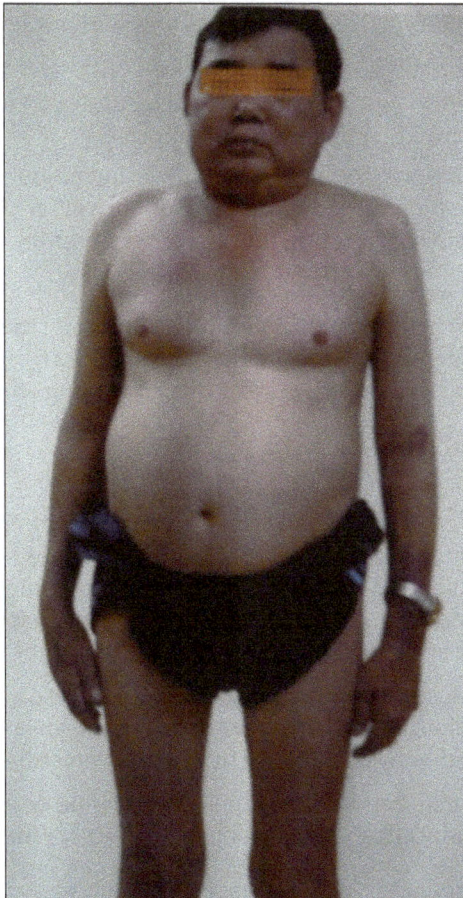

Figure 25.1: Pseudo Cushing's Syndrome.

Pseudo Cushing Syndrome appeared and dexamethasone was detected from packets given to him proving that traditional medicines were not found in these "packets", but, rather that there was an abuse of conventional medicine usage in this case.

At the national level, according to the National Center of Drug Information and Adverse Drug Reactions Monitoring of Vietnam, among 6016 reported ADRs in 2013, traditional herbs account for 47 cases (0,8 per cent). However, this data was not verified to see if herbal medicines were really causing the ADRs that were reported.

In general, we could not find clear answers for ADRs being casually related to herbal medicine usage, recognizing the fact that:

☆ The complexity of traditional medicines: often combined/mixed "packet" given to the patients and it is not clear which herbs could be the cause, or other substances?

☆ Self-medication by the patients often linked with the use of different medicines (including western ones decided by patients themselves or by his/her doctors). In addition, it is a fact that many patients go to pharmacies to buy drugs for themselves without prescriptions

☆ Syndrome/symptoms are not specific manifestations related to herbs only

☆ Contamination with chemical preservatives (sulfur, phosphorus, desiccants , mold...) is repeatedly reported

☆ Drug interactions could be one of the major risks of adverse events occurring as patients may also use not only other OTC drugs, but other substances such as alcohol, this being very popular now in the country.

There is a great need to improve the monitoring and supervision as well as different management activities to ensure that traditional medicines will be used appropriately. Misunderstanding about the ADRs caused by herbs may negatively affect the use of these herbal remedies including consulting with traditional healers.

4. CONCLUSIONS

Adverse drug reactions caused after being treated by herbal remedies were recognized in many cases and need to be given more attention by physicians, patients and government agencies. Traditional medicines should be used with greater caution by patients. Monitoring from the government agencies should be strengthened. However, there is limited evidence showing that traditional medicines were the causes of reported ADRs.

There is a great need to conduct more systematic research and investigation to see which forms of herbal medicines are safe and which are harmful, especially which are often abused and/or wrongly mixed with potential toxic substances.

New drugs discovered from herbal medicines would be a very important way to maximize the potential effect of traditional medicines for the health of the people.

5. ACKNOWLEDGEMENTS

Special thanks are to doctors and staff of The Center of Allergology - Clinical Immunology, Bach Mai Hospital, Hanoi who had contributed greatly in providing treatment and support for these patients and participated in analyzing herbal medicine-related ADRs.

6. REFERENCES

1. Adverse Drug Reactions Advisory Committee. (1999). 'An adverse reaction to the herbal medication milk thistle (*Silybum marianum*)'. *Med J Aust*, 170: 218-9.

2. Angiola Crivellaro, M., Senna, G., Riva, G., Cislaghi, C., Falagiani, P., Walter Canonica, G. and Passalacqua, G. (2000). 'Pollen mixtures used as health food may be a harmful source of allergens'. *J Investig Allergol Clin Immunol*, 10: 310-1.

3. Nguyen Van Doan, Nguyen Nang An (2000). Considerations in using herbal medicines. *Journal of Medical Information*. Vol 5. pp. 15-17.

4. Nguyen Van Doan (2007). Illustrative pictures of skin lesions due to drug-induced complications. Medical Publishing House. Ministry of Health Vietnam.

5. Nguyen Van Doan (2011). Drug allergy. Medical Publishing House, Ministry of Health Vietnam.

Mysore - Ooty Resolution on Herbal and Medicine Knowledge for Health and Wellness

WHILE EXPRESSING GRATITUDEtothe **National Science Centre, Ministry of Science,WHILE EXPRESSING GRATITUDE to** the Centre for Science and Technology of the Non-Aligned and Other Developing Countries (NAM S&T Centre) for organising the on 'Science Centres in PromotingA Knowledge and Innovative Society for SustaInternational Training Workshop on 'Herbal Medicine: Drug Discovery from Herbs - Approaches, Innovations and Applications' at Mysore/ Ooty, India during 30th March - 3rd April 2015;

EXPRESSING APPRECIATION to the JSS University, Mysore and JSS College of Pharmacy, Ooty for co-organising and hosting the International Training Workshop on Herbal Medicine;

RECOGNISING that large sections of the population in the developing countries depend on the traditional medicines and the use of traditional medicines is not limited to these developing countries alone, as the public interest in natural therapies has increased greatly in recent years all over the world with expanding use of ethno botanicals;

FURTHER RECOGNISING that extensive research is required to meet the present day challenges by identifying the plants to develop advanced techniques for future discovery of potent pharmaceutical agents and to document these techniques and already existing and widely used herbal remedies in English and other languages;

HAVING CONSIDERED that there is a need for establishing standards for preparing the herbal formulations, characterisation of the herbal drugs, appropriate storage and archiving methods, safety and toxicity profiles, safe use of herbal drugs, determining the mechanistic basis of action and conducting and implementing the clinical trials and establishing research laboratories to help modern drug testing processes in NAM and other developing countries;

HAVING DELIBERATED on the need to establish the appropriate standards and methods mentioned above to promote research and development activities as well as to archive the representative herbal materials and conduct clinical trials, and extensively debating various aspects of herbal medicines in promoting better health and wellness, improving the quality of life by global networking including the power of herbals, innovations, formulations, characterisation, compositions and safety and toxicity.

WE, THE PARTICIPANTS OF THE WORKSHOP, representing the institutions and agencies from Afghanistan, Botswana, Cambodia, Cameroon, Egypt, India, Indonesia, Iran, Iraq, Malaysia, Mongolia, Myanmar, Nigeria, Oman, Pakistan, Qatar, Rwanda, South Africa, Slovenia, Tanzania, Turkey, Uganda, Vietnam and Zimbabwe;

UNANIMOUSLY RESOLVE AND RECOMMEND

☆ A Global team effort by interfacing with the knowledge of each country, each region, the continents and subcontinents to usher in a new world with less disease and a healthier society.

☆ Achievable, doable and affordable interfacing through academia, R and D and industry interactions with the right policies and dynamic framing regulations and the individual government policies in harmony with global policies involving the NGOs, Self Help Groups and the consumers for the safety of this herbal approach.

☆ In conjunction with using the most modern scientific tools to understand the mechanism of action of the drugs and development of newer drugs, molecules and cluster of molecules from herbs and a wide variety of plants covering the large biodiversity of species to ensure a sustainable livelihood not only for the low socio-economic countries, but also for the benefit of all countries in respect of advancing knowledge.

☆ An integrated approach that will take the system through a long and sustainable mechanism right from the sourced integration of herbals to the trading of cost effective herbal products for tomorrow's market (*Farm to Folk*).

☆ Knowledge transfer of traditional and modern medicine, pharmaceuticals, nutraceuticals and phytochemistry with support of advanced science and technology for validating traditional use of herbal medicine.

☆ Developing best practices and procedures in cultivation, harvesting and processing of medicinal herbs with due consideration for conserving the environment and further protecting threatened and endangered species.

☆ Contribution of the scientists, technologists, pharmacists and herbal pharmacists, practitioners, social groups, medical doctors, farmers, policy makers and other stakeholders together with information technologists with modern scientific knowledge to a great level of achievable goals in herbals through scientific adaptation of commitment for the claims and principles of herbal products to reach out to the society through joint projects between NAM and other developing countries.

The participants urged the Member Countries of the NAM S&T Centre and other developing countries to individually and collectively support the above recommendations.

It was proposed by JSS University to host a joint NAM S&T Center – JSS University Herbal Medicine Fellowship scheme for the training of the developing country scientists, researchers, professionals and practitioners for a period of two weeks with International fare being paid by the selected fellows themselves and local hospitality accorded by JSS University, Mysuru, India with its constituent colleges free of any cost including accommodation and food etc., subject to approvals from concerned agencies. The proposal was highly welcomed and applauded by the participants of the training workshop.

THUS, RESOLVED IN OOTY (TAMIL NADU), INDIA ON THIS DAY, THE 3rd of APRIL 2015.

www.ingramcontent.com/pod-product-compliance
Lightning Source LLC
Chambersburg PA
CBHW050507190326
41458CB00005B/1463